T0310067

Complexity Challenges in Cyber Physical Systems

Complexity Challenges in Cyber Physical Systems

Using Modeling and Simulation (M&S) to Support
Intelligence, Adaptation and Autonomy

Saurabh Mittal
The MITRE Corporation
Fairborn
OH, USA

Andreas Tolk
The MITRE Corporation
Hampton
VA, USA

This edition first published 2020
© 2020 John Wiley & Sons, Inc.

The right of Saurabh Mittal and Andreas Tolk to be identified as the authors has been asserted in accordance with law.

Registered Office
John Wiley & Sons, Inc., 111 River Street, Hoboken, NJ 07030, USA

Editorial Office
111 River Street, Hoboken, NJ 07030, USA

For details of our global editorial offices, customer services, and more information about Wiley products visit us at www.wiley.com.

Wiley also publishes its books in a variety of electronic formats and by print-on-demand. Some content that appears in standard print versions of this book may not be available in other formats.

Library of Congress Cataloging-in-Publication Data

Names: Mittal, Saurabh, author. | Tolk, Andreas, author.
Title: Complexity challenges in cyber physical systems : using modeling and
    simulation (M&S) to support intelligence, adaptation and autonomy /
    Dr. Saurabh Mittal, The MITRE Corporation, McLean, OH, US, Dr. Andreas Tolk,
    The MITRE Corporation, Hampton, VA, US.
Description: First edition. | Hoboken, NJ : Wiley, 2020. | Series: Stevens
    institute series on complex systems and enterprises | Includes
    bibliographical references and index.
Identifiers: LCCN 2019027671 (print) | LCCN 2019027672 (ebook) |
    ISBN 9781119552390 (cloth) | ISBN 9781119552468 (adobe pdf) |
    ISBN 9781119552499 (epub)
Subjects: LCSH: Cooperating objects (Computer systems) | Automatic
    control–Simulation methods.
Classification: LCC TJ213 .M5335 2020 (print) | LCC TJ213 (ebook) |
    DDC 006.2/2–dc23
LC record available at https://lccn.loc.gov/2019027671
LC ebook record available at https://lccn.loc.gov/2019027672

Cover Design: Wiley
Cover Images: © v_alex/Getty Images, © wayra/Getty Images

Set in 9.5/12.5pt STIXTwoText by SPi Global, Pondicherry, India

Printed in United States of America

V10015728_112019

*To the Infinite Intelligence that created things simple, just,
and accommodating enough, which manifests itself in
complex universes, both within and without, that we all
share, enjoy, and strive to understand.*

Saurabh Mittal

*To all scientists and researchers who dare to leave the
comfort of their home discipline and seek collaboration with
like-minded partners to create transdisciplinary teams
inspiring progress in our complex work.*

Andreas Tolk

Contents

Preface

The various definitions for Cyber Physical Systems (CPSes) all focus on their computational and physical components, integrating sensors, networks, motors, and more. But we often overlook that CPS will significantly change the way we access systems and our environment. They are ubiquitous: cars self-park, recognize street signs and react accordingly, know the distance to other cars and keep the correct distance, and more. CPS allows a new family of medical devices, from surgical assisting tools to smart prostheses. Smart houses observe the comfort level of people and control the air conditioning accordingly. They are learning when people are home, can prepare their meals and keep the meals warm in case of a traffic jam. If the house were part of the smart city, sensors would have learned about the jam and diverted the traffic, automatically reconfiguring the traffic lights and communicating the news to the smart cars. First responders as well as soldiers are getting accustomed to their colleague CPS in human cyber teams, where the CPS can assess regions too dangerous or otherwise not reachable for the human team partners. However, all this support comes with a price: growing complexity! How can we either manage or govern such intelligent, adaptive, and autonomous systems? How can we take advantage of positive emergence, and avoid the major consequences of negative ones?

We went through a similar dramatic change before, namely when the Internet changed our view on searching for and gaining access to information. Many CPSes are using the Internet to gather and change information as well; and again, it comes with a price. Before the Internet era, many complex systems had both the software and hardware components, but they were shielded from cyberattacks due to a lack of network access. Due to the additional capability of connectedness of these components across varied networks (both within and without the organization that own the CPS), new challenges have emerged. Some of the challenges include cyber security, control, test, degree of connectivity, constant vigilance and operation, degree of autonomy, intelligence-based behavior, resilience, and impact on the socioeconomic fabric.

As is the case in many current publications, CPS and Internet of Things (IoT) are used interchangeably but there are some subtle differences between the two. We understand CPS as domain-specific versions of IoT, so the difference lies in terms of scale, societal impact and the propagation of effects. CPS are more focused towards a specific domain such as aviation, health, military, defense, manufacturing, etc. Due to the domain-specific nature, CPS can be studied in more detail at both the operational technology and information technology levels. However, the resulting danger is that CPS within their domains share neither their insights nor benefit from insights of other CPS from other domains. A domain-agnostic common theory providing common methods that lead to domain specific solutions would be advantageous, and some candidates exist and will be discussed, but no common formalism in support of this idea has been widely accepted yet.

Modeling and Simulation (M&S) has emerged as a mechanism by which various CPS challenges can be studied in a virtual environment. Model-based engineering (MBE) and simulation-based engineering are two distinct activities, even though the simulation activity subsumes the modeling activity. A model is an abstract representation of the system and is evaluated in an environment that may be a live (people using real systems), a virtual (people using simulated systems), or a constructive (simulated people and systems) environment. The simulation infrastructure ensures the model system is provided the right environment for evaluating the capabilities, which are essentially the system's capabilities that need to be tested and evaluated.

We started our journey in Fall 2017, when we gratefully received some internal MITRE research funding to research the challenges of hybrid simulation in support of CPS. We allocated part of the funding to bring experts to a panel discussion. The experts belonged to disparate domains who employ M&S to address the CPS challenges and conduct CPS engineering together. Interestingly, this spawned some collaboration, as we discovered similarities in our challenges and solutions, from which this book ultimately emerged. This book tries to organize the obtained insights and report the latest in the use of M&S for CPS engineering. We address the subject in five parts: Introduction, Modeling Support to CPS Engineering, Simulation-Based CPS Engineering, The Cyber Element, and The Way Forward.

The Part I begins with a chapter from us and provides an overview of complexities associated with the application of M&S to CPS Engineering. Castro et al. in the second chapter, provide a more detailed description of the challenges in the operation and design of intelligent CPS. The third chapter by Mazal et al. discusses M&S in the context of autonomous systems involvement within the North Atlantic Treaty Organization (NATO). Part II begins with a chapter from Traoré on multi-perspective modeling and holistic simulation for very complex systems analysis. The next chapter by Barros describes a unifying framework for hierarchical co-simulation of CPS. This is followed by model-based system of systems engineering tradeoff analytics by Markina-Khusid et al. The next chapter by Mittal et al. considers a larger version of CPS, i.e. IoT, and the complexities associated with developing a risk assessment framework. Part III begins with a chapter from Castrol et al. on simulation model continuity for efficient development of embedded controllers in CPS. This is followed by another practical application by Henares et al. on CPS design methodology for prediction of symptomatic events in chronic diseases. In this chapter they present the entire lifecycle methodology for CPS engineering from concept to cloud deployment and execution. The next chapter by Bhadani et al. applies model-based engineering to the subject of autonomy in CPS. Part IV begins with a chapter by Furness on providing various perspectives on securing CPS. This is supported by the next chapter by Haque et al. on CPS resilience and discusses frameworks, complexities, and future directions on resilient systems engineering. The next chapter by Suarez and Demareth discusses the creation of social structures employing CPS. The Part V incorporates another chapter by the editors on the way forward and provides a research agenda for addressing complexity in application of M&S for CPS engineering.

Editing this book was a rewarding journey that offered plenty of opportunities to learn and discover. We invite you to share the exciting journey of CPS engineering that offers a wealth of opportunities for advancement at various levels. CPS are going to shape our lives: observing the well-being of the elderly, observing our health, observing and optimizing our production systems, and many more opportunities. Just as our children can hardly imagine finding information on certain topics of interest – mainly due to home work or college projects – before Internet and Google, the new generation may no longer imagine how often we had to practice parallel parking, or how parcels were delivered only once a day. We hope to contribute to the efficient development of future CPS solutions with this compendium, and hopefully generate some ideas for scholars and researchers as well.

Saurabh Mittal[1], PhD
The MITRE Corporation,
Fairborn, OH, USA

Andreas Tolk[2], PhD
The MITRE Corporation,
Hampton, VA, USA

1, 2 The author's affiliation with the MITRE Corporation is provided for identification purposes only, and is not intended to convey or imply MITRE's concurrence with, or support for, the positions, opinions, or viewpoints expressed by the author. Approved for Public Release, Distribution Unlimited. Case: PR_18-2996-3.

Foreword

Several important global trends are occurring with regards to the advancement of cyber physical systems. Worldwide, significant technology-driven advances are being pursued that address increasing cyber physical system performance, safety, and security while achieving design, development, and operational efficiencies that reduce cost. These trends include:

- Significant investment in higher levels of automation for physical systems, including autonomous systems.
- Increasing research and early applications of Artificial Intelligence to physical systems (AI), including addressing "Dependable AI" for high assurance AI-software design and development.
- Development of advanced static and dynamic analysis tools by the Modeling and Simulation community. The resulting Model-based Systems Engineering (MBSE) analysis tools and methods address considerations related to the growing complexity of highly integrated System-of-System architectures.
- Development of cyberattack resilient system architectures that can restore acceptable system operation in response to a real-time detection of a functionally disabling cyberattack.

These initiatives bring with them increased complexity of system designs, with a corresponding set of risks that need to be addressed when designing new or significantly upgraded systems. These risks include:

- Cyberattacks that include supply chain and insider attacks that can directly impact the application layer of physical systems and, in the worst case, can potentially result in operator or user injuries or loss of life.
- Safety-related incidents due to undetected deficiencies in system design.
- Operator errors due to uncertainties related to human–machine roles under anomalous circumstances.

But, perhaps the most concerning risk is the recognized shortage of engineers and scientists who can contribute to the development of these new technologies and tools, as well as the shortage of the workforce that can productively employ the analysis tools that are designed to enable high productivity in the development and evaluation of new cyber physical system designs. This book helps to address this risk by providing a well-constructed, selective set of articles that together offer the reader an integrated view of the state-of-the-art in addressing complex cyber physical system design and development. By integrating the diverse set of articles, the book serves to compliment the education curriculums at Universities, which tends to separate the subjects discussed above into the curriculums of different departments (e.g. Mechanical Engineering for physical systems, Computer Science for AI and cybersecurity, Systems Engineering for complex system design analysis, etc.). As a result, I believe that books of this kind can play a significant role in enabling engineers to build on their formal education and prior experience in a manner that supports the greatly needed enhanced design and evaluation skills that the trends in cyber physical systems are calling for.

Reading this book is something that I highly recommend for engineers and scientists who are interested in becoming important participants in the global trends related to advancing the automation levels of cyber physical systems!

Barry Martin Horowitz

Member of the National Academy of Engineering

Munster Professor Systems and Information Engineering

University of Virginia

Previously CEO of The MITRE Corporation

Previously Virginia Cybersecurity Commissioner

March 2019

About the Editors

SAURABH MITTAL is Chief Scientist for Simulation, Experimentation, and Gaming Department at The MITRE Corporation in Fairborn, OH, Vice President-Memberships and member of Board of Directors for Society of Modeling and Simulation (SCS) International in San Diego, CA. He holds a PhD and MS in Electrical and Computer Engineering with dual minors in Systems and Industrial Engineering, and Management and Information Systems from the University of Arizona, Tucson. He has co-authored over 100 publications as book chapters, journal articles, and conference proceedings including 3 books, covering topics in the areas of complex systems, system of systems, complex adaptive systems, emergent behavior, modeling and simulation (M&S), and M&S-based systems engineering across many disciplines. He serves on many international conference program/technical committees, as a referee for prestigious scholastic journals and on the editorial boards of Transactions of SCS, Journal of Defense M&S and Enterprise Architecture Body of Knowledge. He is a recipient of Herculean Effort Leadership award from the University of Arizona, US DoD's highest civilian contractor recognition: Golden Eagle award, and Outstanding Service and Professional Contribution awards from SCS.

ANDREAS TOLK is a Senior Divisional Staff Member at The MITRE Corporation in Hampton, VA, and adjunct Full Professor at Old Dominion University in Norfolk, VA. He holds a PhD and MSc in Computer Science from the University of the Federal Armed Forces of Germany. His research interests include computational and epistemological foundations and constraints of modeling and simulation as well as mathematical foundations for the composition of model-based solutions in computational sciences. He published more than 250 peer reviewed journal articles, book chapters, and conference papers, and edited 10 textbooks and compendia on Modeling and Simulation and Systems Engineering topics. He is a Fellow of the Society for Modeling and Simulation and Senior Member of IEEE and the Association for Computing Machinery.

List of Contributors

Jose L. Ayala
Complutense University of Madrid
Madrid
Spain

Fernando J. Barros
Department of Informatics
Engineering
University of Coimbra
Coimbra
Portugal

Rahul Bhadani
Department of Electrical and
Computer Engineering
University of Arizona
Tucson
AZ
USA

Marco Biagini
NATO Modelling & Simulation Center
of Excellence (M&S COE)
Italy

Agostino Bruzzone
Genoa University
Genoa
Italy

Matt Bunting
Department of Electrical and
Computer Engineering
University of Arizona
Tucson
AZ
USA

Sheila A. Cane
Quinnipiac University
Hamden
CT
USA

Sebastian Castro
MathWorks
Natick
MA
USA

Rodrigo Castro
Departamento de Computación,
FCEyN
Universidad de Buenos Aires
and Instituto de Ciencias de la
Computación, CONICET
Buenos Aires
Argentina

Fabio Corona
NATO Modelling & Simulation Center
of Excellence (M&S COE)
Italy

Judith Dahmann
The MITRE Corporation
McLean
VA
USA

Loren Demerath
Department of Sociology
Centenary College of Louisiana
Shreveport
LA
USA

Zach Furness
INOVA Health Systems
Sterling
VA
USA

Juan I. Giribet
Departamento de Ingeniería
Electrónica y Matemática, FIUBA
Universidad de Buenos Aires, and
Instituto Argentino de Matemática
Alberto Calderón, CONICET
Buenos Aires
Argentina

Md Ariful Haque
Computational Modeling and
Simulation Engineering
Old Dominion University
Norfolk
VA
USA

Richard B. Harris
The MITRE Corporation
McLean
VA
USA

Kevin Henares
Complutense University of Madrid
Madrid
Spain

Ryan Jacobs
The MITRE Corporation
McLean
VA
USA

Jason Jones
NATO Modelling & Simulation Center
of Excellence (M&S COE)
Italy

Bheshaj Krishnappa
Risk Analysis and Mitigation
ReliabilityFirst Corporation
Cleveland
OH
USA

Ezequiel Pecker Marcosig
Departamento de Ingeniería
Electrónica, FIUBA
Universidad de Buenos Aires
and Instituto de Ciencias de la
Computación, CONICET
Buenos Aires
Argentina

Aleksandra Markina-Khusid
The MITRE Corporation
McLean
VA
USA

Jan Mazal
NATO Modelling & Simulation Center
of Excellence (M&S COE)
Italy

Saurabh Mittal
The MITRE Corporation
Fairborn
OH
USA

Pieter J. Mosterman
MathWorks
Natick
MA
USA

Josué Pagán
Technical University of Madrid
Madrid
Spain

Akshay H. Rajhans
MathWorks
Natick
MA
USA

José L. Risco-Martín
Complutense University of Madrid
Madrid
Spain

Charles Schmidt
The MITRE Corporation
McLean
VA
USA

Sachin Shetty
Computational Modeling and
Simulation Engineering
Old Dominion University
Norfolk
VA
USA

Jonathan Sprinkle
Department of Electrical and
Computer Engineering
The University of Arizona
Tucson
AZ
USA

E. Dante Suarez
School of Business, Department of
Finance and Decision Sciences
Trinity University
San Antonio
TX
USA

Andreas Tolk
The MITRE Corporation
Hampton
VA
USA

Mamadou K. Traoré
IMS UMR CNRS
University of Bordeaux
Bordeaux
France

John Tufarolo
Research Innovations, Inc.
Alexandria
VA
USA

Michele Turi
NATO Modelling & Simulation Center
of Excellence (M&S COE)
Italy

Marina Zapater
Swiss Federal Institute of Technology
Lausanne
Lausanne
Switzerland

Author Biography

Aleksandra Markina-Khusid is a Principal Systems Engineer in The MITRE Corporation Systems Engineering Technical Center, supporting several SoS modeling efforts for DoD and DHS. She is the MITRE Model Based Engineering Capability Area Team leader. Dr. Markina-Khusid holds a BS degree in Physics, MS and PhD degrees in Electrical Engineering, and an MS in Engineering & Management, all from the Massachusetts Institute of Technology.

Agostino G. Bruzzone is Full Professor at DIME University of Genoa, Director of M&S Net (International Network involving 34 Centers) Director of the MISS McLeod Institute of Simulation Science – Genoa Center (over 28 Centers distributed worldwide) founder member and president of the Liophant Simulation, he served as Vice President and Member of the Board of MIMOS (Movimento Italiano di Simulazione), member of the NATO MSG, Executive VP of the Society for Modeling and Simulation International. He works on innovative modeling, AI techniques, application of Neural Networks, GAs and Fuzzy Logic to industrial plant problems using Simulation and Chaos Theory. He is member of several International Technical and Organization Committees (i.e. AI Application of IASTED, AI Conference, ESS, AMS) and General Coordinator of Scientific Initiatives (i.e. General Chair of SCSC and I3M). He teaches "M&S" for the DIMS PhD Program (Doctorship in Integrated Mathematical M&S). He is Director of the Master Program in Industrial Plants & Technologies for the University of Genoa and founder and chair of STRATEGOS International MSc in Engineering Technologies for Strategy and Security (http://www.itim.unige.it/strategos). He served as Project Leader for the NATO Science & Technology Organization at the Centre for Maritime Research and Experimentation (CMRE) founding the new research track on Modeling and Simulation. He has been the 10th scientist worldwide to enter into the Modeling and Simulation Hall of Fame as top Lifetime Achievement Awards of the Society for Modeling and Simulation International.

Akshay H. Rajhans is Principal Cyber-Physical Systems Research Scientist at MathWorks in the Advanced Research & Technology Office, where his research focuses on technical computing for and model-based design and analysis of cyber-physical systems (CPS). Previously, he worked on research and development and application engineering of electronic control systems for diesel-engine applications at Cummins, and invented a model-based approach to non-intrusive load monitoring at Bosch Research and Technology Center. Dr. Rajhans has been involved in leadership capacities in top research conferences in CPS and modeling and simulation communities, including as the inaugural CPS Track Chair at both the Winter Simulation Conference (2017) and the Spring Simulation Conference (2019), and as a Co-Chair of the International Workshop on Monitoring and Testing of CPS (2019). He is a recipient of the 2011 IEEE/ACM William J. McCalla Best Paper Award and his work has been recognized as a Research Highlight in "Communications of the ACM," ACM's flagship magazine. Dr. Rajhans has a PhD in Electrical and Computer Engineering from Carnegie Mellon University and an MS in Electrical Engineering from the University of Pennsylvania. He is a member of IEEE and ACM.

Bheshaj Krishnappa is currently working as a Principal at ReliabilityFirst Corporation. Mr. Krishnappa is responsible for risk analysis and mitigation of threats to bulk power system reliability and security across a large geographic area in the United States. He has over 22 years of professional experience working for large- and mid-sized companies in senior roles implementing and managing information technology, security, and business solutions to achieve organizational objectives. He is a business graduate knowledgeable in sustainable business practices that contribute to the triple bottom line of social, environmental, and economic performance. He is motivated to apply his vast knowledge to achieve individual and organizational goals in creating a sustainable positive impact.

Charles Schmidt is a Group Lead at The MITRE Corporation. He has over 17 years of experience in cybersecurity, security automation, and standards development. He holds a BS in Mathematics and Computer Science from Carleton College and an MS in Computer Science from the University of Utah.

Ezequiel Pecker Marcosig received a degree in Electronic Engineering from the Faculty of Engineering (FIUBA) of the University of Buenos Aires, Argentina. He is currently a PhD student at FIUBA and ICC-CONICET working on modeling and simulation based design of hybrid controllers for cyber-physical systems. His work is supported with a PhD Fellowship from the Peruilh Foundation. Since 2013, he is a Teaching Assistant in the Department of Electronic Engineering at FIUBA in the area of Automatic Control. His academic interests include automatic control, cyber-physical systems, modeling and simulation, and hybrid systems.

E. Dante Suarez is Associate Professor for Finance and Decision Sciences at Trinity University, USA. He holds PhD and MS in Economics from Arizona State University, USA. Suarez's main research field is international finance, where he studies the integration of international financial markets, such as the relationship between American depositary receipts and their corresponding underlying stocks. This research is aimed at understanding how markets around the world interact with each other in this age of globalization. Other research areas include econometrics, European studies and Latin American business practices.

Fabio Corona is employed in the Concept Development and Experimentation Branch (CD&E) at NATO Modelling & Simulation Centre of Excellence in Rome. His primary interests at the Centre are emerging technologies and concepts regarding Autonomous System and M&S as a Service. He earned a PhD degree in Electrical Engineering from "Politecnico di Torino" and a MSc degree in Electronics Engineering from "Roma Tre" University. After joining the Italian Army, his employment ranged from the internetworking field under the Italian Army Signal Headquarter to the maintenance and procurement of optoelectronic and communication systems under the Italian Army Logistics Headquarter. During his PhD study, the main research field was in efficiency and power quality of photovoltaic systems under mismatching conditions.

Fernando Barros is a Professor in the Department of Informatics Engineering at the University of Coimbra. He holds a PhD in Electrical Engineering from the University of Coimbra. His research interests include theory of modeling and simulation, hybrid systems and dynamic topology models. He has published more than 80 contributions to journals, book chapters, and conference proceedings. Fernando Barros is a member of IEEE.

Jan Mazal is graduate of the Faculty of Military Systems Management of the Military College of Ground Forces in Vyskov. In 2003 he graduated the Academic Course of Military Intelligence in Fort Huachuca, Arizona, USA. Since 2005 he is a doctor in the field of the theory of the defence management and since 2013 he is Associate professor in the problematic of military management and C4ISR systems. He is former deputy chief of the Department of Military Management and Tactics at the University of Defence in Brno, currently he works as Doctrine Education and Training Branch Chief at NATO Modelling & Simulation Centre of Excellence in ROME. He is focused on the issue of military intelligence and reconnaissance, C4ISR systems, and Operational Decision Support. He is the author and co-author of more than 70 professional publications, he solved more than 10 scientific projects, and he is the author of a number of functional samples and application software. In his previous military practice, he held command and staff functions at the tactical level and also he took part in the foreign missions as EUFOR (2006) and ISAF (2010).

Jason M. Jones is the Deputy Director for the NATO Modelling & Simulation Centre of Excellence. He has been a U.S. Army Functional Area 57, Simulations

Operations Officer, since 2003 and has worked in all aspects of simulations: training, planning, world-wide simulation distribution, testing, and experimentation. Areas of expertise include: live and constructive training; missile defense and logistics simulation; and knowledge management. He has a Master's Degree in Modeling, Virtual Environments, and Simulations from the Naval Postgraduate School in Monterey, CA, where his thesis examined the use of commercial gaming software for training infantry squads.

John Tufarolo is the Technical Director for Systems Engineering at Research Innovations. He holds a BS in Electrical Engineering from Drexel University, and an MS in Systems Engineering from George Mason University, and has more than 32 years of experience providing systems engineering project work, planning, and leadership in complex distributed systems.

Jonathan Sprinkle is the Litton Industries John M. Leonis Distinguished Associate Professor of Electrical and Computer Engineering at the University of Arizona. In 2013 he received the NSF CAREER award, and in 2009, he received the UA's Ed and Joan Biggers Faculty Support Grant for work in autonomous systems. His work has an emphasis for industry impact, and he was recognized with the UA "Catapult Award" by Tech Launch Arizona in 2014, and in 2012 his team won the NSF I-Corps Best Team award. His research interests and experience are in systems control and engineering, and he teaches courses ranging from systems modeling and control to mobile application development and software engineering.

José L. Ayala got his PhD in Electronic Engineering from Technical University of Madrid and is currently an Associate Professor in the Department of Computer Architecture and Automation at Complutense University of Madrid. During his career, he has collaborated and performed research stays in University of California in Irvine, University of California in Los Angeles, EPFL, and University of Bologna. He is currently the VP New Initiatives of the IEEE Council of Electronic Design Automation; CEDA representative in IEEE IoT initiative and IEEE Smart Cities initiative; and steering committee of several international conferences (IEEE Smart Cities Conference, IEEE GLSVLSI, VLSI-SoC, PATMOS, IEEE ASAP, etc). His research interests focus on IoT and edge solutions for personalized medicine approaches, including health monitoring, wireless sensor networks, and disease modeling.

José L. Risco-Martín is Associate Professor at Complutense University of Madrid. He is head of the Department of Computer Architecture and Automation. Previously, he was Assistant Professor at Colegio Universitario de Segovia and Assistant Professor at C.E.S. Felipe II de Aranjuez. Dr. Risco-Martín served as General Chair for SummerSim'17, Program Chair for SummerSim'15, General

Chair for Summer Computer Simulation Conference 2016 and Vice General Chair for SummerSim'16. He has co-authored more than 100 articles in various international conferences and journals. His research interests focus on design methodologies for integrated systems and high-performance embedded systems, including new modeling frameworks to explore thermal management techniques for Multi-Processor System-on-Chip, dynamic memory management and memory hierarchy optimizations for embedded systems, Networks-on-Chip interconnection design, low-power design of embedded systems and more generally Computer-Aided Design in M&S of Complex System, with emphasis on DEVS-based methodologies and tools.

Josué Pagán is a Teaching Assistant Professor in the Universidad Politécnica de Madrid. He got his PhD with honors in Computer Science at Complutense University of Madrid in 2018. His work focuses on develop robust methodologies for information acquisition in biophysical and critical scenarios. He has worked developing models for prompt prediction and classification of neurological diseases. On summer 2016 he did a 12-week research stay at the Embedded Pervasive Systems Lab at Washington State University under the supervision of Prof. Hassan Ghasemzadeh. Previously, on fall 2015 he did a 16-week research stay at the Pattern Recognition Lab. at Friedrich Alexander University under the supervision of Prof. Bjoern Eskofier. He achieved his MSc at Universidad Politécnica de Madrid in September 2013 with Honor Mention. He also achieved a Bachelor in Telecommunication Engineering by the Universidad Publica de Navarra in 2010.

Juan I. Giribet is a Researcher in the Instituto Argentino de Matemática (IAM-CONICET), Argentina. He is also Associate Professor in the Faculty of Engineering of the University of Buenos Aires, Argentina. He holds an MS and PhD in Electronic Engineering from the University of Buenos Aires. He published more than 70 contributions to journals and conference proceedings in topics covering electronic engineering and applied mathematics. He is Director of the master's program in Engineering mathematics at University of Buenos Aires. He is a senior member of IEEE.

Judith Dahmann is a Principal Senior Scientist in the MITRE Corporation Center for The MITRE Systems Engineering Technical Center and the Capability Action Team leader for Systems of Systems (SoS). Dr. Dahmann holds a Bachelor's Degree from Chatham College in Pittsburgh, PA (1972), spent a year as a special student at Dartmouth College (1971–1972), a Master's Degree from The University of Chicago (1973), and a Doctorate from Johns Hopkins University (1984). Dr. Dahmann is an INCOSE Fellow and the co-chair of the INCOSE Systems of Systems Working Group and the DoD liaison and co-chair of the National Defense Industry Association SE Division SoS SE Committee.

Kevin Henares is a PhD student at the Complutense University of Madrid (UCM). He received an MSc in Computer Engineering in the same university (Spain, 2018) and a University Degree in Computer Engineering at the University of Vigo (Spain, 2016). His work focuses on the development of robust modeling and simulation methodologies to study the behavior of complex systems, and the generation of models to classify and predict critical events in neurological diseases. His email address is khenares@ucm.es.

Loren Demarath is Professor and Chair of the Sociology Department at Centenary College of Louisiana. He is currently working on information processing theory and model of emergence and complexity. He is the author of numerous publications examining how the evolution of meaning is guided by aesthetic responses to order. His book, *Explaining Culture: The Social Pursuit of Subjective Order*, describes the emergent nature of culture.

Marina Zapater is a Post-Doctoral researcher in the Embedded Systems Laboratory (ESL) at Ecole Polytechinique Federale de Lausanne (EPFL), Switzerland, since 2016. She was non-tenure-track Assistant Professor in the Computer Architecture Department of Universidad Complutense de Madrid (UCM), Spain, in the academic year 2015–2016. She received her PhD degree in Electronic Engineering from Universidad Politécnica de Madrid, Spain, in 2015, and an MSc in Telecommunication Engineering and a MSc in Electronic Engineering, both from Universitat Politècnica de Catalunya (UPC), Spain, in 2010. Her research interests include thermal and power optimization of heterogeneous architectures, and energy efficiency in data centers. In this area, she has co-authored over 50 publications in top-notch international conferences and journals, and she has participated in several international research projects, including five European H2020 projects. She is IEEE member, and the current Young Professionals representative of IEEE CEDA. She has served as TPC of several conferences, including DATE, ISLPED, and VLSI-SoC.

Mamadou K. Traoré is Full Professor at University of Bordeaux in France. He holds an MS and PhD in Computer Science from the Blaise Pascal University in Clermont-Ferrand, France. His contributions are in formal specifications, symbolic manipulation and automated code synthesis of simulation models. He received the International DEVS M&S Award in 2011. He is a member of ACM and SCS.

Marco Biagini is the Concept Development and Experimentation (CD&E) Branch Chief at NATO Modelling & Simulation Centre of Excellence. He has a PhD in Mathematics, Engineering, and Simulation and master degrees in strategic studies, peace keeping and security studies, and new media and communication. He has more than 15 years of experience in the M&S field. He was Battalion

Commander at the Italian Army Unit for Digitization Experimentation (USD) and Section Chief at the Italian Army Simulation and Validation Centre. He is chairing the NATO M&S Group (NMSG) 150, and member of NMSG 145, NMSG 136, and NMSG 147.

Matt Bunting is a PhD student in the Department of Electrical and Computer Engineering at the University of Arizona. He earned his BS (2010) in Electrical Engineering from the University of Arizona. His research interests include the study of modeling techniques for domain-specific modeling languages for development of Cyber-Physical Systems. Mr. Bunting is also a co-founder of the Safkan Health medical device company.

Md Ariful Haque is a PhD student in the Department of Modeling Simulation and Visualization Engineering at Old Dominion University. He is currently working as a graduate research assistant in the Virginia Modeling Analysis and Simulation Center. He has earned Master of Science (MS) in Modeling Simulation and Visualization Engineering from Old Dominion University (ODU) in 2018. He has received Master of Business Administration (MBA) degree from the Institute of Business Administration (IBA), University of Dhaka in 2016. He holds a BS degree in Electrical and Electronic Engineering from Bangladesh University of Engineering and Technology in 2006. Before joining Old Dominion University as a graduate student, he has worked in the Telecommunication industry for around seven years. His research interests include but not limited to cyber-physical system security, cloud computing, machine learning, and Big data analytics.

Michele Turi is a Colonel of Italian Army and he is currently in charge as Director of the NATO M&S Center of Excellence. He has a PhD from Genoa University on Engineering, Mathematics, and Simulation PhD program and long experience in using M&S in operational environment. He started his career as mechanized Artillery Officer, covering roles in the ranks of Captain and Major also as Chief of Brigade Intelligence Section, Security Officer, Training Officer and C3I-Computer & ICT Section. He gained further experience in the Command & Control and Military Decision Making Process when he was on duty at the IT – Army Staff College as C4 Military Teacher. He participated to the Project Management Working Group dedicated to the training simulation both for Live and Constructive Simulation systems to develop the Italian Army Constructive Simulation Centers and Live MOUTs & CTCs. He was involved in the working group for studies, projects, training programs, and simulation systems evaluation related to the Training Simulation, M&S, C2 systems interoperability and the VV&A process for concepts and systems in use and also for the future development. During his permanence at the IT Army M&S Section, he has been responsible to manage, perform, and apply the Army M&S policy; managing, assembling, and testing the first experimental Integrated Test Bed cell; coordinate the resources allocated for the International Working Group responsible to define deployable functional solutions, designing, adopting, and implementing system interoperability test plans' for the Italian Army side's of Afghan Mission Network; use project management methodologies, procedures and techniques to develop projects, plans, and define resources and management R&D projects' in the M&S branch. His areas of expertise include among the others Information Technology, C3I, M&S, VV&A, CD&E, Training Simulation, and Intelligence.

Pieter J. Mosterman is Chief Research Scientist and Director of the Advanced Research & Technology Office at MathWorks in Natick, Massachusetts, where he works on computational methodologies and technologies for technical computing and model-based design tools. He also held an Adjunct Professor position at the School of Computer Science of McGill University. Prior to this, he was a research associate at the German Aerospace Center (DLR) in Oberpfaffenhofen. He earned his PhD in Electrical and Computer Engineering from Vanderbilt University in Nashville, Tennessee, and his MSc in Electrical Engineering from the University of Twente, the Netherlands. Dr. Mosterman developed the Electronics Laboratory Simulator that was nominated for The Computerworld Smithsonian Award by Microsoft Corporation in 1994. In 2003, he was awarded the IMechE Donald Julius Groen Prize for his paper on the hybrid bond graph modeling and simulation environment {\sc HyBrSim}. In 2009, he received the Distinguished Service Award of The Society for Modeling and Simulation International (SCS) for his services as editor in chief of SIMULATION: Transactions of SCS. Dr. Mosterman was guest editor for special issues on computer automated multiparadigm modeling of SIMULATION, IEEE Transactions on Control Systems Technology, and ACM Transactions on Modeling and Computer Simulation.

Rahul Bhadani is a PhD student in the Department of Electrical and Computer Engineering at the University of Arizona. He earned his BS (2012) from Bengal Engineering and Science University and MS (2017) in Computer Engineering from the University of Arizona. His research interests include modeling, simulation, and control of autonomous vehicles, developing novel statistical models for traffic simulation and software engineering. Prior to joining the University of Arizona, Mr. Bhadani worked as a software engineer for Oracle.

Richard B. Harris is a principal cybersecurity policy engineer for the Homeland Security Center, MITRE. He has over 14 years of experience in cybersecurity with the Department of Homeland Security and MITRE, and a perspective on complex risk environments that was seasoned by a 26 year career in the US Marine Corps.

Roberto G. Valenti is a Senior Robotics Research Scientist at MathWorks in the Advanced Research & Technology Office. His research interests include robotics,

robotics sensing for navigation, sensor fusion, mobile autonomous robots (self-driving cars, unnamed aerial vehicles), inertial navigation and orientation estimation, control, computer vision, and deep learning. Previously, he worked as a research and development engineer within the autonomous driving team at Nvidia. He obtained a PhD in Electrical Engineering at the City University of New York, The City College, NY, USA where he focused his research on state estimation and control for autonomous navigation of micro aerial vehicles. Dr. Valenti received his MSc in Electronics Engineering from the University of Catania, Italy. He is a member of IEEE and the Robotics and Automation Society (RAS).

Rodrigo D. Castro is a Researcher with the Instituto de Ciencias de la Computación (ICC-CONICET) and Associate Professor in the Computer Science Department, School of Exact and Natural Sciences of the University of Buenos Aires (UBA), Argentina. Rodrigo holds a degree in Electronic Engineering and a PhD in Engineering from the National University of Rosario (Rosario, Argentina). He was Postdoc Associate at the Swiss Federal Institute of Technology in Zurich (ETH Zurich), Switzerland (Department of Environmental Systems Science and Computer Science). He heads the Laboratory of Discrete Events Simulation and is a member of the Society for Modeling and Simulation international (SCS) and the IEEE.

Ryan B. Jacobs is a Group Leader at The MITRE Corporation Systems Engineering Technical Center. He has over 10 years of experience in systems analysis, modeling and simulation, and model-based engineering. He holds a BS in Aerospace Engineering from Embry-Riddle Aeronautical University, an MS in Aerospace Engineering from the Georgia Institute of Technology, and a PhD in Aerospace Engineering from the Georgia Institute of Technology.

Sachin Shetty is an Associate Professor in the Virginia Modeling, Analysis and Simulation Center at Old Dominion University. He holds a joint appointment with the Department of Modeling, Simulation and Visualization Engineering and the Center for Cybersecurity Education and Research. Sachin Shetty received his PhD in Modeling and Simulation from the Old Dominion University in 2007. Prior to joining Old Dominion University, he was an Associate Professor with the Electrical and Computer Engineering Department at Tennessee State University. He was also the associate director of the Tennessee Interdisciplinary Graduate Engineering Research Institute and directed the Cyber Security laboratory at Tennessee State University. He also holds a dual appointment as an Engineer at the Naval Surface Warfare Center, Crane Indiana. His research interests lie at the intersection of computer networking, network security, and machine learning. His laboratory conducts cloud and mobile security research and has received over $10 million in funding from National Science Foundation, Air Office of Scientific Research, Air Force Research Lab, Office of Naval Research, Department of

Homeland Security, and Boeing. He is the site lead on the DoD Cyber Security Center of Excellence, the Department of Homeland Security National Center of Excellence, the Critical Infrastructure Resilience Institute (CIRI), and Department of Energy, Cyber Resilient Energy Delivery Consortium (CREDC). He has authored and co-authored over 140 research articles in journals and conference proceedings and two books. He has served on the technical program committee for ACM CCS, IEEE INFOCOM, IEEE ICDCN, and IEEE ICCCN. He is an Associate Editor for International Journal of Computer Networks.

Sheila A. Cane is currently an adjunct professor with Quinnipiac University. She practiced industrial and systems engineering for over 30 years, most recently as a project leader and strategic technical advisor for The MITRE Corporation, working on defense and homeland security projects across the US federal government. She has experience in command and control, war game and cyber security modeling and simulation, data analytics, and enterprise architecture of large systems. She holds a BS in Applied Mathematics from Buffalo State College, a MS in Industrial Engineering from SUNY Buffalo, and a DBA in Information Systems Management from Nova Southeastern University.

Sebastian Castro is a Senior Robotics Engineer at MathWorks, where he manages the MathWorks portfolio of robotics student programs with a focus on student competitions at the undergraduate level and higher. He holds a BS and MS in Mechanical Engineering from Cornell University, with a concentration in dynamics, controls, and systems. His research was in high-level control of reconfigurable modular robots using linear temporal logic. His professional experience consists primarily of modeling and simulation of physical systems.

Zach Furness is currently the Director of IT Security at Inova Health System. Prior to joining Inova, he was Technical Director of the National Cybersecurity Federally Funded Research and Development Center (FFRDC), operated by the MITRE Corporation, and sponsored by the National Cybersecurity Center of Excellence. In that role, Mr. Furness oversaw development of standards-based cybersecurity guidance for industry and government. Mr. Furness has also served as the Chief Engineer of MITRE's Global Operations and Intelligence Division where he supported the development and application of technical solutions involving cybersecurity, autonomous systems, and modeling and simulation. Prior to that, Mr. Furness established and led MITRE's modeling and simulation department within its DoD FFRDC. He is the author of numerous papers on the application of cybersecurity technologies, modeling and simulation, and systems engineering.

Part I

Introduction

1

The Complexity in Application of Modeling and Simulation for Cyber Physical Systems Engineering

Saurabh Mittal[1] and Andreas Tolk[2]

[1] *The MITRE Corporation, Fairborn, OH, USA*
[2] *The MITRE Corporation, Hampton, VA, USA*

1.1   Introduction

Cyber Physical Systems (CPS), according to a definition provided by National Science Foundation (NSF) are hybrid networked cyber and engineered physical elements co-designed to create adaptive and predictive systems for enhanced performance. These systems are built from, and depend upon, the seamless integration of computation and physical components. Advances in CPS are expected to enable capability, adaptability, scalability, resiliency, safety, security, and usability that will expand the horizons of these critical systems.

CPS engineering is an activity that brings these elements together in an operational scenario. Sometimes, an operational scenario may span multiple domains, for example, Smart Grid incorporating Power critical infrastructure and Water infrastructure. Intelligent home devices such as a smart washing machine utilize both the infrastructures. Another example would be Smart Transportation system wherein intelligent transportation devices interact with numerous smart vehicles to coordinate large scale traffic behaviors. Numerous such examples exist within the Internet of Things (IoT) perspective. These complex systems involve components at varying level of specifications. The constituent elements are supplied by multiple vendors and composing a solution without a formal test and evaluation infrastructure is a real challenge. Integration of functionality is not happening before the deployment, but after CPS are already deployed. CPS engineering requires a consistent model of operations that need to be supported by the compositions of various CPS contributors. CPS engineering lacks tools to design and experiment within a lab setting. How does one develop a repeatable engineering methodology to evaluate ensemble behaviors and emergent behaviors when larger systems involving critical infrastructure cannot be brought in a lab setting?

This increase in overlapping CPS capability in multitude of domains also introduces a level of complexity unprecedented in other engineered systems. The cross-sector deployment and usage introduces risk that may have cascaded impacts in a highly networked environment. One possibility to reduce the technical risk is to remotely control the systems in the cyber environments, but the sheer number of variables and possible situations introduce complexity at multiple scales. This complexity results in test plans with limited coverage. Additional cyber physical system related issues of intelligence, adaptation, autonomy, and security make the problem even worse. The proposed solution is the enhanced use of Modeling and simulation (M&S). The M&S discipline has supported the development of complex systems since its inception. During the Spring Simulation Multi-Conference 2017, a group of invited experts discussed general challenges in M&S of CPS. In 2018, as follow-on panel was launched dealing with how the combination of various simulation paradigms, methods – so-called hybrid simulation – can be utilized regarding complexity, intelligence, and adaptability of CPS.

While the focus of CPS is both on computation and physical devices, it belongs to the class of super complex systems in a man-made world, where labels such as System of Systems (SoS), Complex Adaptive Systems (CAS), and Cyber CAS (CyCAS) are used interchangeably (Mittal 2014; Mittal and Risco-Martín 2017a). All of them are multi-agent systems. The constituting agents are goal-oriented with incomplete information at any given moment and interact among themselves and with the environment. SoS is characterized by the constituent systems under independent operational and managerial control, geographical separation between the constituent systems and independent evolutionary roadmap. CAS is an SoS where constituent systems can be construed as agents that interact and adapt to the dynamic environment. Cyber CAS is a CAS that exist in a netcentric environment (for example, Internet) that incorporates human elements where distributed communication between the systems and various elements is facilitated by agreed upon standards and protocols. CPS is an SoS wherein the constituent physical and embedded systems are remotely controlled through the constituent cyber components.

Complex systems engineering identified a set of methods needed by systems engineers to govern such complex systems and cope with new challenges, like emergent properties or behavior not known in traditional systems. Many of these methods are rooted in the M&S discipline (Mittal et al. 2018). This chapter will provide an overview on the M&S methods and technologies that aid CPS engineering in the development and testing phase, and CPS governance when they are deployed in complex cyber environments. How to apply such means to enable the full potential of CPS is one of the grand challenges of our days. With this volume, we contribute to the discussion of developing a computational infrastructure for modeling, simulation, experimentation, and analytics in a transdisciplinary CPS context.

The chapter is organized as follows. Section 1.2 provides an overview on multiple modalities of CPS. Section 1.3 describes the fundamental issues with CPS engineering. Section 1.4 describes the current M&S technology, especially the co-simulation methodology, available for CPS engineering for developing a virtual CPS environment. Section 1.5 describes the intelligence, adaptation, and autonomy aspect of CPS and how the computational element in CPS provides opportunities for advanced control and access mechanisms. Section 1.6 concludes the chapter.

1.2 Multimodal Nature of CPS

CPS are also considered as systems with integrated physical and computational capabilities that can interact with humans through variety of modalities (Baheti and Gill 2011). This ability to interact with the physical world through computational means, and by doing so expanding the capabilities of the user of the CPS, allows the CPS to interact within a team, such as enabling human-machine-collaborations, as well as with the environment, such as providing alternative means of locomotion – moving of the CPS –, actuation – positioning of sub-components, such as sensors –, or manipulation – interacting with the environment.

This multimodality, the ability to interact with humans, others CPS, and the environment via a multitude of computational and physical means, is one of main sources for the complexity challenges we are coping with. It allows CPS to work in different domains and sectors, and provide their services to many different users. The same functionality can be accessed via several different interfaces to be applied in a multitude of contexts in various domains, making the validation of the CPS challenging, if not impossible. As observed in (Rajkumar et al. 2010), "... the gap between formal methods and testing needs to be bridged. Compositional verification and testing methods that explore the heterogeneous nature of CPS models are essential. V&V must also be incorporated into certification regimes" (page 735).

But validation is not the only concern. The multimodality leads to a multitude of interconnections between potentially many CPS, users, and components of the environment, creating a system of interlinked and interdependent objects. Combined with capabilities that now can be applied by CPS in the same domain, the overall complexity increases significantly.

The other side is, however, that the amount of options for an appropriate reaction in an unforeseen turn of events increases also. If many CPSs can provide a wide variety of services to the same domain, the likelihood that even under catastrophic circumstances we still have options for the appropriate reactions available increases as well. Multimodality is therefore not only the source for more complexity, it also provides the means to cope with it, as it enables higher agility and flexibility. If one modality fails or is not available, it can be quickly replaced by an alternative invocation structure. If one service usually applied within a domain does not succeed, an alternative service may lead to the desired result as well. The multimodel nature of CPS creates the challenge, but it also provides the means to cope with it.

1.3 Why CPS Engineering Is Complex?

In todays world, engineering has taken on a new meaning when different branches of Science are brought together, for example, biology and physics in Biomedical engineering, cognitive psychology and systems in Intelligent Systems engineering, urban policy and management, and automobile in Transportation engineering, and the most complex of all, Smart City engineering that includes transportation, smart automobiles, intelligent infrastructure, human factors, cybersecurity, and many more. Likewise, the multi-domain warfare involves air, sea, and land falls into this category as well, so do IoT, the Industry 4.0 initiatives, and CPS. This is the world of complex systems engineering. One characteristic property of such complex systems is that there are always some unknowns. There is high variability. One can never have a complete set of variables to apply brute force engineering based on a single branch of science. Consequently, per the Law of Requisite Variety, it is impossible to develop a controller for such a complex system (Mittal and Rainey 2015). This incomplete information and uncertainty in developing control mechanisms lead to emergent behaviors which is the hallmark of any complex system (Mittal 2013; Mittal et al. 2018). Consequently, methodologies are needed that embrace emergent behaviors as features of such a complex system.

Complex systems today are multi-disciplinary systems that require multiple branches of Science to interact with competing, contrasting, or orthogonal theories. A model may be valid in one scientific theory and simultaneously, may be completely invalid in another branch of Science, for example, consider Quantum mechanics and Newtonian mechanics operating in the same computational model. While they work in reality, it is a difficult problem altogether in a computational realm. Yet, M&S is the best we have to test any upcoming scientific theory or engineer a system-at-scale in a lab setting in any multi-disciplinary endeavor (Mittal et al. 2018).

Due to the proliferation of various modalities, both hard sciences (bound by physics and mathematics) and soft sciences (e.g. cognitive science, sociology) need to be brought in a computational environment for verification and validation (V&V), test and evaluation (T&E), and experimentation purposes for such CPSs. Bringing these fundamental sciences together require the incorporation of emergent behavior that may arise as different scientific theories are brought to interact within a CPS use-case. Development of CPS through the fundamentals of Systems Theory supported by an equally robust M&S theory is the preferred way forward (Mittal and Zeigler 2017).

A typical CPS comprises of the following components:

- Sensors
- Actuators
- Hardware platforms (that hosts sensors and actuators)
- Software interfaces (that accesses hardware directly or remotely through a cyber environment)
- Computational software environments (that may act both as a controller or service provider)
- Networked environments (that allows communication across geographical distances)
- End user autonomy (that allows CPS to be used as a passive system or an active interactive system)
- Critical infrastructures (water, power, etc., that provides the domain of operation and operational use-case)
- Ensemble behaviors
- Emergent behaviors

Figure 1.1 shows various aspects of CPS divided into Left-Hand Side (LHS) and Right-Hand Side (RHS). LHS consists of collection of users, systems (both

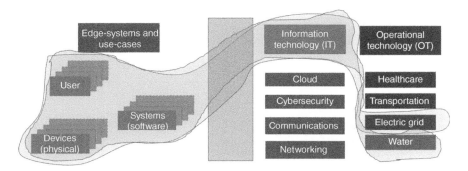

Figure 1.1 CPS landscape.

hardware and software), and devices (physical platforms). Traditional systems engineering practices and end-user use-cases can be developed in LHS. The RHS shows aspects related to infrastructures. Fundamentally, they can be characterized into Information Technology (IT) and Operational Technology (OT). Between the LHS and RHS is the network/cyber environment that allows information exchange between the two. With the network spanning large geographical distances, the presence of large number of entities/ agents and their concurrent interactions in the CPS result in ensemble and emergent behaviors. The infrastructure-in-a-box is largely unavailable but can be brought to bear with various existing domain simulators in an integrated simulation environment.

In M&S, we are not only limited to the computational implementations of models. We distinguish between live simulations in which the model involves humans interacting with one another (role playing, play acting, etc.), virtual simulations where the model is simulated by a fusion of humans and computer-generated experiences, and constructive simulations where the model is entirely implemented in a digital computer and may have increased levels of abstraction. Increasingly, we are mixing the three forms of simulation in what is commonly known as live-virtual-constructive (LVC) simulation (Hodson and Hill 2014). LVC simulations are used mainly for training but they can be adapted for the type of experimentation/exploration needed to investigate emergent behavior (Mittal et al. 2015), as shown by the cyclical process in Figure 1.2, elaborated in (Mittal et al. 2018).

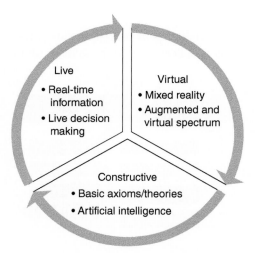

Figure 1.2 Experimental LVC approach for generating emergence.

Table 1.1 CPS contributor and the associated M&S paradigm.

CPS contributor	M&S paradigm	LVC element
Sensors	Continuous, physics-based	L, V, C
Actuators	Continuous, physics-based	L, V, C
Hardware platform	Both continuous and discrete	L, V
Software platform	Discrete	V, C
Network	Discrete	L, V
End user	Discrete, agent-based	L, V, C
Critical infrastructure	Both continuous and discrete (hybrid)	V, C
Ensemble, emergent behaviors	Discrete agent-based	V, C

Table 1.1 associates each of the CPS constituent element with the corresponding M&S paradigm and how it can be incorporated in the LVC environment.

From a systems theoretic perspective, a CPS model is a hybrid system made up of both continuous and discrete systems. A continuous system (CS) is one that operates in continuous time and in which input, state, and output variables are all real values. A discrete (dynamic) system (DDS) is one that changes its state in piece-wise constant event-based manner (which also included discrete-time systems as they are a special case of discrete event systems) (Lee et al. 2015). A typical example of a hybrid system is a CPS in which the computation subsystem is discrete and a physical system is CS. A CyCAS (Mittal 2014) in LVC environment also qualifies as a CPS, with live systems as CS, constructive systems as discrete and virtual systems as hybrid, containing both continuous and discrete. At the fundamental level, there are various ways to model both timed and untimed discrete event systems, all of which can be transformed to, and studied within, the formal Discrete Event Systems (DEVS) theory (Mittal 2013; Mittal and Risco-Martin 2013; Mittal and Martin 2016; Traor et al. 2018; Vangheluwe 2000; Zeigler et al. 2000).

1.4 M&S Technology Available for CPS Engineering

M&S is being considered as a vehicle by which complex systems engineering, including CPS engineering could be done. However, using M&S for CPS engineering is not straight forward due to the inherent complexities residing in both the modeling and the simulation activities. Simulation subsumes modeling. Performing a CPS simulation requires that a CPS model be first built. While CPS modeling is not the focus of this chapter, a recent panel explored the state of the art of CPS modeling and the complexity associated in engineering intelligence, adaptation, and autonomy through M&S. The literature survey conducted in Tolk et al., (2018) enumerate the following active research areas and the associated technologies for CPS modeling and concluded that the need for a common formalism that can be applied by practitioners in the field is not yet fulfilled:

- DEVS formalism: Strong mathematical foundation that support multi-paradigm modeling, multi-perspective modeling, and complex adaptive systems modeling to handle emergent behaviors.
- Process algebra: Provides hybrid processes using multi-paradigm modeling. Models combine behavior on a continuous time scale with discrete state transition behavior at given points in time.
- Hybrid automata: Combines finite state machines with Ordinary Differential Equations (ODE) to account for non-deterministic finite states. Bond graphs are used to govern changes.
- Simulation languages: Combines discrete event and continuous system simulation languages. Involves modular design of hybrid languages, multiple abstraction levels combining different formalisms.
- Business processes: Use of standardized notation languages like Business Process Modeling Notation (BPMN) provides value in securing buy-in from the stakeholders in an efficient manner.
- Interface design for co-modeling: Functional Mock-up Interface (FMI) as a means of integration of various CPS components. DEVS can also be used as a common denominator in a vendor neutral manner.
- Model-driven approaches: Model transformation chains to arrive at a single formal model. Governance is required to develop such automation.
- Agent-based modeling: Paradigm to employ component models at scale with individual behaviors, to study ensemble effects.

The above-mentioned approaches and technologies allow the development of CPS models, albeit in a piece-wise manner. These model pieces and their definitions and specifications are dictated by the cross-domain CPS operational use-case. Assuming we now have a validated model (i.e. a model that has been deemed valid by the stakeholders), next comes the task of executing it on a computational platform, i.e. simulation. The piece-wise model composition sometime does not directly translate into a monolithic simulation environment due to the confluence of both the continuous and discrete system in the hybrid system. In the literature survey (Tolk et al. 2018) as well as in many discussions with the experts, the use of co-simulation was identified as the preferred course of action in support of CPS for development, testing, and eventually training.

Co-simulation is the co-existence of independent simulators to support a common model (Mittal and Zeigler 2017). To understand co-simulation, consider a complex system model comprising of Electric Grid, thousands of smart homes, and data-communication network. This would require modeling to be done for:

1) Continuous system for the Power system (using GridLab-D power flow simulator)
2) Continuous system for Building simulation (using LabView simulator)
3) Discrete system for the smart home behavior (using model-predictive software controllers in Generic Algebraic Modeling Systems (GAMS) language)
4) Discrete system model for data communication network (using OmNet++ simulator)

The simulators used to implements both the discrete and continuous system models in the common Electric Grid hybrid system model are shown in parenthesis. Such hybrid systems demonstrate how a large-scale hybrid modeling could be attempted and can eventually lead to a robust simulation environment, bringing together user(s) in smart homes as agents with big infrastructure such as Electric Grid through the power flow simulator. The first example at Oak Ridge National Lab (ORNL) developed a complex system with item #1 and #4, described in Nutaro et al. (2008). The second example at National Renewable Energy Lab (NREL) developed a system comprising of item #1, #2, and #3, described in Pratt et al. (2015), and with a stronger flavor on co-simulation application in Mittal et al. (2015). Both examples integrated different modeling paradigms and ran simulations on High Performance Computing (HPC) environment in virtual (as-fast-as-possible) and real (wall-clock) time. The NREL effort also integrated air-conditioner hardware with the simulation exercise for a real-time 7-day scenario, described in detail in Pratt et al. (2017). Both the efforts at ORNL and NREL employed the DEVS formalism for integrating the constituent simulators. The survey published in Thule et al. (2008) provides a state-of-the-art overview of co-simulation practices and applications in multiple domains, including CPS.

A scalable M&S architecture has distinct modeling and simulation layers. In order to deploy in cloud environment, sufficient automation is needed at both the simulation layer and the modeling layer. This can now be achieved by current practices in DevOps implemented using Docker technology. DevOps, a recent buzzword, provides methodologies to automate developer operations, such as compiling, building, releasing, testing through executable scripts. Mittal and Risco-Martin (2017b) integrated Docker with the granular service oriented architecture (SOA) Microservices paradigm and advanced the state-of-the-art in model and simulation interoperability. This automated deployment of various "DEVS nodes" under a single administrative control is defined as a DEVS Farm. They described the architecture incorporating DevOps methodologies using containerization technologies to develop cloud-based distributed simulation farm for DDS systems specified using DEVS formalism. The research will extend towards containerization of various other simulators either through DEVS wrappers or Functional Mockup Interface (FMI) standards as functional mockup units (FMUs).

Knowledge engineering is an activity that has not been adequately dealt within cognitive architectures when it comes to bringing high-level knowledge structures into existing architectures. They have largely focused on symbolic representation, memory structure, and symbol manipulation. As the corpus is small, likewise, the environments these cognitive architectures can be put to use, is also limited to simple to moderate operational environments. Further, issues like abductive reasoning, dynamic memory that acquire new conceptual structures, creative aspects of problem solving, emotional processing, along with plausibly related concepts of metacognition and goal reasoning have received little attention (Langley 2017) and may require revision to established cognitive theories. A middle ground using a cognitive framework like Belief-Desire-Intention (BDI) in conjunction with (i) algorithms and heuristics, and (ii) high level knowledge representation, will provide adequate strength for a developing a rational agent, capable of handling complex situations. Prior work at Air Force Research Lab (AFRL) demonstrates the development of integrated cognitive systems for building artificial systems (Douglass and Mittal 2013). Further, work (Mittal and Zeigler 2014a, 2014b) on attention-switching for resource-constrained complex intelligent dynamical systems (RCIDS) provide further evidence of bringing together cybernetics, Systems Theory, Cognitive Science, and Software engineering to develop attention-focusing activity-based systems.

These recent developments bring together cloud technologies, co-simulation methodologies, and hybrid modeling approaches to deliver an M&S substrate that is applicable across the entire CPS landscape (Figure 1.3). The LHS in Figure 1.1 employs traditional Systems Engineering practices, and provides the context

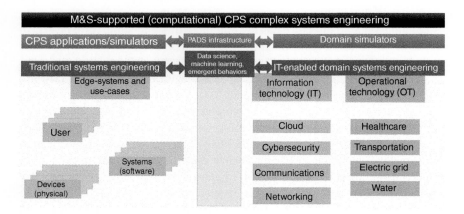

Figure 1.3 M&S supported (computational) CPS engineering test-bed perspective.

use-case for CPS applications. The RHS provides various domain simulators and employ IT and OT to provide "infrastructure-in-a-box" through LVC architectures. To bridge, LHS and RHS, emerging disciplines like Machine Learning and Data Science will need to be employed to get a handle on data-driven approaches that tackle emergent behaviors when LHS and RHS interact in a parallel distributed discrete event co-simulation environment.

1.5   Intelligence, Adaptation, and Autonomy Aspects

Intelligent, adaptive, and autonomous are characteristics often used to describe CPS. In this section, we will provide some of the references to important work supporting to gain a better understanding of these tightly connected topics.

1.5.1   Intelligence

Having to define intelligence, adaption, and autonomy is not an easy task, in particular not for the first term. Scientists and philosophers try to understand for centuries what intelligence is. In the context of computational intelligence, the famous "Imitation Game," as described in Turing (1950), avoids providing definitions of intelligence itself but instead focuses on the question: "Can computers pass a behavioral intelligence test?" For many applications, such a test is suitable, but not all forms of computational intelligence can be tested in form of such Turing tests, in particular not when the modalities give the nature of the system under test away. A smart device that behaves intelligently will nonetheless be immediately be recognized as a device, which likely will bias the evaluating person, who now knows that he is testing a machine, not a human.

However, systems engineering and computer science are collaborating for several years now to develop smart systems. In Tolk et al. (2011), we summarized several abilities an intelligent system needs to expose. This list reflects the collective view of various artificial intelligence viewpoints. It is neither complete nor exclusive, but it is still valid and builds a good foundation for smart CPS understanding as well. This list contains adaptability and autonomy as characteristics for intelligence based systems as well, again showing the close proximity of these concepts pivotal to CPS.

- Intelligent system can *explain* their decision. It is important that smart decision come to an applicable decision, but they should also be able to explain the reasoning behind such decisions. This is particularly important in human-machine-teams, where the human must be able to understand the decision process and interact with it, or eventually even overrule the decision.

Furthermore, the system behavior of smart devices may change due to learning and adaption, so explaining new reasoning is important.

An example is the explanation component of expert systems used for diagnosis tracing the line of reasoning used by the underlying inference engine to answer the questions: Why is the answer to the question the one you recommend? For systems that are able to modify themselves being able to explain their reason is mandatory to ensure credibility.

Furthermore, the decision space of computational systems is always closed by definition, as it is the union of the range of all included computable functions. If a situation comprises a characteristic attribute outside of the dimension of the space, the attribute does not contribute to the decision which hence may be sub-optimal. This is very important if the intelligence is based on data-driven machine learning using neural networks, as such system will always generate a solution based on the selected training data set and its domain.

- Intelligent systems must be *robust*. This characteristic property of a system means that the system behaves well and adequate not only under ordinary conditions, but also under unusual conditions that stress the original requirements and derived assumptions. In other words, robust systems do not break easily, but are able to continue to behave well even under variant circumstances that could lead to failure of system.

  In the more recent systems engineering literature, the term *antifragile* has been introduced to describe systems that are not only robust, but that actually are getting better under stress (Nicholas Taleb 2012). Again, the often circular definitions are showing up here as well, as many of the characteristics captured in this list are used for explaining antifragile systems as well (Jones 2014).

- Another related characteristic is *fault tolerance*. The intelligent systems will continue to behave well and continue to adequately perform even if one or more of its internal system components fail or break. This tolerance is applicable to external causes, such as malevolent behavior of other systems, as well as to internal causes, such as simple wear and attrition.

  Conducting maintenance procedures to avoid wear and having the ability to repair externally caused damage immediately can help, but these are just custodian abilities to provide fault tolerance. The ability to adapt to a new capability set and plan accordingly is also contributing to fault tolerance, and they will be dealt with in more detail in bullets of their own.

- Another characteristic often used to describe intelligent systems is the ability to *self organize*. They can organize their internal components and capabilities in new structures without a central or an external authority in place. Often, this characteristic also applies to the population of several such systems when conducting a mission to reach a common objective. These new structures can be temporal and spatial. In some cases, instead of self-organizing the term self-optimizing is used synonymously. When these self-organizing systems interact within a persistent environment, such as in stigmergic systems, new macro behavioral patterns emerge because the environment becomes an inter-acting agent itself due to a persistent and evolving structure (Mittal 2013).

In particular when complex tasks have to be conducted by many simple systems, the use of cooperative swarm systems applying specialization and self-organization as a special form of distributed artificial intelligence that may lead to emergent structures on the swarm level. Even in defense opera-tions, the use of self-organizing teams has been identified as a good practice in complex environments (Alberts et al. 2010).

- This aspect of *cooperation* generally exposes social capabilities, which is also a characteristic of intelligent behavior.

Cooperative systems interact with other systems and potentially humans as well via some kind of communication language or any other modalities described earlier in this chapter. This interaction is not limited to pure observa-tion, but these systems can exchange plans, distribute tasks, etc. Whiteboard technologies are as often used as direct communication.

- Intelligent systems are also able to *learn* from observing the achieved results and compare them with the desired outcome. Using methods such as reinforce-ment learning, decisions that led to positive results are enforced while those with negative results are avoided.

Learning can also occur by observing other systems and the results of their activities. It is also possible to observe human partners and mimic their behavior.

Learning can imply deductive as well as inductive methods. Systems can learn general principles from the observations of detailed example, and they can apply general behavioral schemes to guide their decision in new environ-ments. When new knowledge is captured in from of applicable models, abduc-tive learning is possible as well (Håkansson and Hartung 2014).

- The final characteristic to be addressed in this list is *agility*. In general, agile systems are able to manage and apply knowledge and their capability effectively so that they behave well and adequate in continuously and unpredictably changing environments.

In particular in complex environments, agility is pivotal to react quickly and appropriately in unforeseen situations. It is often connected with intellectual acuity for the situation and the necessary intellectual capability to cope with newly arising challenges.

In the recent years, computational advantages allowed for the rebirth of machine learning, in particular based on data-driven approaches (Witten et al. 2016). The usefulness of such approaches and the success of data science related approaches are clearly documented in many publications as well as solutions, as IBM's famous Watson (High 2012). However, when it comes to CPS, these approaches may not necessarily be the best choice. Many alternative methods supporting computational intelligence, such as collectively described in Steinbrecher (2016), are benefiting from the same computational advantages and provide traceable and explainable solutions. Furthermore, many of the heuristics developed in the first wave of artificial intelligence have been refined over the years and provide today impressive solutions. It is worth mentioning that explainability itself is not well defined with the community. With a tighter focus on CPS, the terms assurance and traceability are often used as the two bounding examples to express the intention behind requiring this characteristic. Assurance is result oriented, asking for system to be able to explain who they ensure that the results of their rules are going to fall within given constraints, such as rules of engagements, safety concerns, etc. The recent Air Force study summarizes several concepts used by the community (Clark et al. 2013). Traceability is interested in how a result can be traced back to all the various decisions made and rules applied. It is very close to the requirements traceability asked for in systems engineering. A broader spectrum of related concepts has been recently compiled in Karlo Došilović et al. (2018).

CPS engineers must be aware of such alternative solutions that may fulfill their requirements better than those currently in the main stream.

1.5.2 Autonomy

Many of these characteristics also apply to enable adaptation and autonomy. As with intelligence, the definition of these terms is challenging.

An autonomous system performs the desired tasks and behaves well and adequate even in complex environments without continuous human guidance. Williams compiled an overview of definitions with focus on the various autonomy scales (2015). Like intelligence, autonomy is multi-faceted and can be observed via many modalities in various forms. The use of different levels is therefore common practice, in particular using multi-dimensional scales consisting of levels with descriptive indicators for each set of dimensions defining a particular facet of autonomy. Wiley concludes his synthesis with the definition of the following autonomy dimensions (Williams 2015).

- *Goals*: an autonomous agent has goals that drive its behavior.
- *Sensing*: an autonomous agent senses both its internal state and the external world by taking in information (e.g. electromagnetic waves, sound waves).
- *Interpreting*: an autonomous agent interprets information by translating raw inputs into a form usable for decision making.

- *Rationalizing*: an autonomous agent rationalizes information against its current internal state, external environment, and goals using a defined logic (e.g. optimization, random search, heuristic search), and generates courses of action to meet goals.
- *Decision making*: an autonomous agent selects courses of action to meet its goals.
- *Evaluating*: an autonomous agent evaluates the consequences of its actions in reference to goals and external constraints.
- *Adapting*: an autonomous agent adapts its internal state and functions of sensing, interpreting, rationalizing, decision making, and evaluating to improve its goal attainment.

When comparing the intelligence characteristics with the autonomy dimensions, the close proximity of both concepts immediately becomes eminent. Nearly all papers on the different degrees of autonomy reference the seminary work of Sheridan and his research into levels of autonomy (1992), who introduced the following 10 levels of increasing autonomy.

1) The computer offers no assistance, the human must do it all.
2) The computer offers a complete set of action alternatives.
3) The computer narrows the selection down to a few.
4) The computer suggests the best selection.
5) The computer executes the option upon human approval.
6) The computer allows the human a restricted time to veto before execution.
7) The computer automatically executes and informs the human.
8) The computer informs the human after execution.
9) The computer decides if to inform the human.
10) The computer acts completely on its own.

The Armed Forces used these categories to identify the following definition for classes of autonomous systems (Williams 2008).

- *Human operated systems* are fully controlled by humans. All activities result from human initialization, eventually based on provided sensor information.
- *Human assisted systems* perform activities in parallel with human inputs, augmenting the human's ability.
- *Human delegated systems* perform limited control activities. The human can overrule the system at any time.
- *Human supervised systems* conduct all activities needed to perform a given mission, but they inform the human consistently, including providing explanations for decisions.
- *Mixed initiative systems* are capable of human-machine teams and can take over given tasks independently.
- *Fully autonomous systems* require no human intervention or presence. They conduct all activities across all ranges of conditions.

It should be pointed out that in particular for CPS such systems are often integrated into a larger system of systems. An anti-break system conducts a well-defined function autonomously, assisting the human in the operation of a car. The auto-pilot of an airplane autonomously flies it under fully and well-define constraints, etc. The borders between the levels are therefore often fluent.

1.5.3 Adaption

The requirement to be adaptive is known from both concepts discussed so far. It describes the ability of a system to change to better fit into a changing environment. These changes can be behavioral as well as structural (Antonio Martn et al. 2009).

Structural changes modify the physical components of the CPS. Many CPS have so called actuators that are used to move components of the CPS into new positions, such as robotic arms, sensors, antennas, etc. In addition, many CPS have modular components that can be switched in case of need to support different environments, such as wheeled or tracked locomotion devices. An interesting featured not yet sufficiently researched is the applicability of new capabilities, such as 3D printing in the field. While we already utilize 3D printers for on-demand spare part production, in the future completely new components will be possible that will help the system to adapt to new challenges.

From the computational perspective, the behavioral adaption is interesting, which addresses mainly the cyber component of the CPS. These computational components provide functions that can be modified, optimized, or completely replaced by an alternative set. The aspect of learning and the methods discussed in Steinbrecher (2016) will be applied here. The aspects of building computational adaptive systems were introduced by Holland in (1992) and mainly embraced for the computational support of social studies. Some of the currently utilized methods are described in detail in Antonio Martn et al. (2009).

Like already observed for intelligence and autonomy, there is no generally accepted definition for adaption, the list of characteristics of adaptive systems is open as well. Nonetheless, the literature agrees on some aspects, as they are covered in the following enumeration. One of the reasons is the high degree of interdisciplinary of the field, as many different user domains take advantage of the computational progress and the ease of available tools.

- Adaptive systems are composed of individual agents that interact with each other, may compete for common resources, and that all follow usually well-defined decision and action processes.
- The macro-behavior of the system results from the interaction of these agents. In complex adaptive systems, the resulting behavior of the system can usually

not directly be derived from the well-understood behavior of the single agents within the systems.

- To enable adaption, feedback loops are a necessary element of the highly dynamic interaction processes. The agents have to be able to learn from this feedback information.
- Cooperation and specialization of agents are also characteristics often mentioned in the literature. The persuasion of a common goals under the constraint of limited resources are driving forces without having to program the behavior into the processes of the agents explicitly.

Currently, the CPS community does not yet take enough advantage of the rich body of knowledge regarding such methods and solutions from the discipline of M&S. As discussed in Tolk et al. (2018), the reason for this may be the lack of awareness of the methods available, but also of the complexity of the underlying problems.

However, as discussed in Mittal et al. (2018), the use of simulation methods is good – if not best – practice when coping with complex environment that often does not provide immediate feedback for actions, but exposes effect of effects and temporarily shifted feedback, often only after several decision cycles after the originating action. The highly non-linear nature of relations in complex system's constituent sub-systems is another challenge. Adaptability in complex environments will require capabilities as provided by M&S. As pointed out in Tolk (2015), there is a close connection between CPS and the agent metaphor. Within the virtual environment, intelligent software agents exhibit the same characteristics as CPS in their environment. As such, software agents are not only good candidates for co-simulation approaches to evaluate the scalability of control approaches, they can also serve to get insights into possible emergence and can serve as a test bed for new rule sets and supporting solutions, or they can be used for the optimization of existing ones. Obviously, the agent-based simulation approaches can and should be augmented by other paradigms and approaches, leading to hybrid approaches as initially discussed in Tolk et al. (2018).

What is furthermore of interest for the application of M&S methods to cope with the complexity of CPS is that they are naturally embedded into to life environment. The continuous feedback from their situated environment to allow them to adapt or optimize their decisions, as discussed above, has been the topic of applied M&S research as well. Of particular interest is the research on Dynamic Data Driven Application Systems, which were made popular in earlier US Air Force research (Darema 2004) and only recently re-introduced as a research topic that can take advantage of the developments in computer technology in Biswas et al. (2018). In the hybrid simulation world, the same principle is referred to as symbiotic simulation (Onggo et al. 2018). All these approaches have in common that simulation solutions, such as CPS cyber components, are regarded as embedded solutions within the situated environment, using feedback loops for control, and utilizing data science methods to create actionable observation. Again, many techniques used for years in the training community, such as captured under the LVC paradigm (Hodson and Hill 2014), should be closely evaluated regarding their ability to contribute to the management of complexity of CPS over the whole lifecycle.

1.6 Conclusion

CPS are complex hybrid systems that span multiple sectors of the society at deployment level. They employ both discrete and continuous systems as they interact with the continuous physical world and the discrete digital world that involve humans as well. The constituent systems may exhibit features of an intelligent, adaptive, or autonmous agent in varying degrees. With IoT in infancy and projected to increase manifold, the CPS when deployed on the World Wide Web (Internet) introduce a lot of risk in the deployed critical infrastructure (such as Energy, Water, Transportation, etc.) which they indirectly use. Consequently, the IoT and CPS of tomorrow is fraught with many challenges and their growth may get stifled as the stakeholders that manage and operate the underlying critical infrastructures become aware of the risks involved. To make matters even more complicated, CPS may have multiple modalities which put the existing CPS in a new operational environment. Without sufficient T&E and V&V, it cannot be predicted if the CPS would produce reliable behavior and would not cause new emergent stresses to the underlying infrastructures leading to cascaded failures. The IT and OT communities that operate various underlying infrastructures may not be cognizant of the larger IoT context as it may fall outside their business operations. Consequently, the responsibility rests solely on CPS engineers to develop a robust CPS with expected modalities and definition of the CPS operations. However, there does not exist an experimentation test-bed that allows CPS T&E in relation to the operational context (over the underlying critical infrastructure).

The use of agent-based development and testing platforms for CPS has been subject of several publications that make the clear connection between intelligent, adaptive, and autonomous CPS and their "digital twin" in form of intelligent software agents within a situated virtual environment that replicates the nature of the physical environment in which the CPS is acting in. One example is given in Sanislav and Miclea (2012), more will follow in the chapters of this book. However, while this approach is sufficient for developing and testing computational functionality in well-defined missions and environments, the complex environment of

CPS often require the human in the loop for testing. The many recent advantages in the domain of LVC methods, such as published in Hodson and Hill (2014) for the military domain, must be taken into consideration as well.

M&S technology provides tractable methods for analysis and experimentation. M&S allows the exploration of the emergent behavior when dealing with complex systems engineering provided adequate attention is paid to simulation engineering and no emergent behaviors arise as a result of error approximation due to selection of a wrong modeling method or error propagation through wrong simulation integration that does weaves the space-time-series incorrectly. M&S is integral part of complex systems engineering and must be utilized through the following mechanisms for maximum impact:

- Live, Virtual, and Constructive (LVC) environments
- Systems engineering testing, evaluation, validation, and verification
- Operations, distribution, and communications

The application of M&S technology to CPS engineering is a non-trivial effort. It requires development of a computational infrastructure that brings together a hybrid model employing various domain simulators in a co-simulation environment for CPS contextual use. The co-simulation environment must be integrated with the state-of-the-art simulation technology employing hybrid modeling, cloud-computing, DevOps, parallel and distributed execution, and abstract time implementation. This will allow simulation experiment studies to be conducted in fast mode with reduced time-to-results. Over the decades, M&S has been a strong partner in developing intelligent, adaptive, and autonomous systems. The communities in these areas acknowledge, use, and advance the M&S technologies themselves. This chapter has discussed the state-of-art in M&S and how the aspects of intelligence, adaptation, and autonomy guide the specifications of CPS in a broader context. The intelligent, adaptive, and autonomous systems communities have begun to co-simulation technologies in piece-wise manner wherein two domains may be brought together through simulators. CPS expands the co-simulation concept many fold as we require an M&S environment with a swappable "CPS context" to experiment with various modalities across multiple critical infrastructures.

The chapters provided in this book have been contributed by invited recognized experts in their contributing fields, which has not been limited to the M&S domain. We asked experts on co-simulation and modeling formalisms as well as CPS practitioners and experts interested in the social aspects of CPS integration to support us in writing this book. The resulting compendium should offer contributions as well as research ideas to bring our communities closer together to tackle the big challenges of complexity in the CPS domain as well as showing the many facets of research and viewpoints that should be considered in this process.

Acknowledgments

The work presented in this paper was partly supported by the MITRE Innovation Program. The views, opinions, and/or findings contained in this paper are those of The MITRE Corporation and should not be construed as an official government position, policy, or decision, unless designated by other documentation. Approved for Public Release; Distribution Unlimited. Public Release Case Number 19-0424.

References

Alberts, D. S., Huber, R. K., & Moffat, J. (2010). *NATO net-enabled capability command and control maturity model (N2C2M2)*. Technical report, Command and Control Research and Technology Program.

Antonio Martn, J., de Lope, J., & Maravall, D. (2009). Adaptation, anticipation and rationality in natural and artificial systems: Computational paradigms mimicking nature. *Natural Computing, 8*(4), 757.

Baheti, R., & Gill, H. (2011). Cyber-physical systems. *The Impact of Control Technology, 12*(1), 161–166.

Biswas, A., Hunter, M., & Fujimoto, R. (2018). Energy efficient middleware for dynamic data driven application systems. In M. Rabe, A. A. Juan, N. Mustafee, A. Skoogh, S. Jain, & B. Johansson (Eds.), *Winter Simulation Conference* (pp. 628–639). IEEE.

Clark, M., Koutsoukos, X., Porter, J., Kumar, R., Pappas, G., Sokolsky, O., ... Pike, L. (2013). *A study on run time assurance for complex cyber physical systems*. Technical report. Aerospace Systems Directorate, Air Force Research Lab, Wright-Patterson Air Force Base.

Darema, F. (2004). Dynamic data driven applications systems: A new paradigm for application simulations and measurements. In M. Bubak, G. D. van Albada, P. M. A. Sloot, & J. Dongarra (Eds.), *Computational Science – ICCS 2004. ICCS 2004*, Lecture Notes in Computer Science (Vol. *3038*, pp. 662–669). Berlin, Heidelberg: Springer.

Douglass, S., & Mittal, S. (2013). A Framework for Modeling and Simulation of the Artificial. In A. Tolk (Ed.), *Ontology, Epistemology, and Teleology for Modeling and Simulation*, Intelligent Systems Reference Library (Vol. *44*). Berlin, Heidelberg: Springer.

Håkansson, A., & Hartung, R. (2014). An infrastructure for individualised and intelligent decision-making and negotiation in cyber-physical systems. *Procedia Computer Science, 35*, 822–831.

High, R. (2012). *The Era of Cognitive Systems: An Inside Look at IBM Watson and How It Works*. Armonk, NY: IBM Corporation, Redbooks.

Hodson, D. D., & Hill, R. R. (2014). The art and science of live, virtual, and constructive simulation for test and analysis. *Journal of Defense Modeling and Simulation, 11*(2), 77–89.

Holland, J. H. (1992). Complex adaptive systems. *Daedalus, 121*, 17–30.

Jones, K. H. (2014). Engineering antifragile systems: A change in design philosophy. *Proceedings of the 1st International Workshop: From Dependable to Resilient, from Resilient to Antifragile Ambients and Systems.*

Karlo Došilović, F., Brčić, M., & Hlupić, N. (2018). Explainable artificial intelligence: A survey. *Proceedings of 2018 41st International Convention on Information and Communication Technology, Electronics and Microelectronics (MIPRO)* (pp. 210–215). IEEE.

Kruse, R., Borgelt, C., Braune, C., Mostaghim, S., & Steinbrecher, M. (2016). *Computational Intelligence: A Methodological Introduction.* New York: Springer.

Langley, P. (2017). Progress and challenges in research on cognitive architectures. *Proceedings of Thirty-first AAAI Conference on Artificial Intelligence.*

Lee, K. H., Hong, J. H., & Kim, T. G. (2015). System of systems approach to formal modeling of CPS for simulation-based analysis. *ETRI Journal, 37*, 175–185.

Mittal, S. (2013). Emergence in stigmergic and complex adaptive systems: A formal discrete event systems perspective. *Journal of Cognitive Systems Research, 21*, 22–39.

Mittal, S. (2014). Model engineering for cyber complex adaptive systems. In *European modeling and simulation symposium.*

Mittal, S., Diallo, S., & Tolk, A. (2018). *Emergent Behavior in Complex Systems Engineering: A Modeling and Simulation Approach*, volume 4 of *Steven Institute Series on Complex Enterprise Systems* (pp. 4). Newark: Wiley.

Mittal, S., Doyle, M., & Portrey, A. (2015). Human-in-the-loop modeling in system of systems M&S: Applications to live, virtual and constructive (LVC) distributed mission operations (DMO) training. In L. B. Rainey & A. Tolk (Eds.), *Modeling and Simulation Support for System of Systems Engineering Applications.* Hoboken, NJ: Wiley.

Mittal, S., & Rainey, L. B. (2015). Harnessing emergent behavior: The design and control of emergent behavior in system of systems engineering. In *Proceedings of Summer Computer Simulation Conference.*

Mittal, S., & Risco-Martin, J. L. (2013). *Netcentric System of Systems Engineering with DEVS Unified Process.* Boca Raton, FL: CRC Press.

Mittal, S., & Risco-Martin, J. L. (2016). DEVSML Studio: A framework for integrating domain-specific languages for discrete and continuous hybrid systems into DEVS-Based M&S environment. In *Proceedings of Summer Computer Simulation Conference.*

Mittal, S., & Risco-Martín, J. L. (2017a). *Simulation-Based Complex Adaptive Systems* (pp. 127–150). Cham: Springer International.

Mittal, S., & Risco-Martin, J. L. (2017b). DEVSML 3.0 stack: Rapid deployment of DEVS farm in distributed cloud environments using microservices and containers. In *Proceedings of the 2017 Spring Simulation Multi-Conference (SpringSim'17)*.

Mittal, S., Ruth, M., Pratt, A., Lunacek, M., Krishnamurthy, D., & Jones, W. (2015). A system-of-systems approach for integrated energy systems modeling and simulation. In *Proceedings of Summer Computer Simulation Conference*. SCS.

Mittal, S., & Zeigler, B. P. (2014a). Modeling attention-switching in resource-constrained complex intelligent dynamical system (RCIDS). In *Proceedings of Symposium on Theory of Modeling and Simulation/DEVS, Spring Simulation Conference*.

Mittal, S., & Zeigler, B. P. (2014b). Context and attention in activity-based intelligent systems. In *Proceedings of Activity-based Modeling and Simulation (ACTIMS14), ITM Web of Conferences* (vol. 3).

Mittal, S., & Zeigler, B. P. (2017). The practice of modeling and simulation in cyber environments. In A. Tolk & T. Oren (Eds.), *The Profession of Modeling and Simulation* (pp. 223–264). Hoboken, NJ: Wiley.

Nicholas Taleb, N. (2012). *Antifragile: Things That Gain from Disorder* (Vol. 3). New York: Random House Incorporated.

Nutaro, J., Kuruganti, P. T., Shankar, M., Miller, L., & Mulle, S. (2008). Integrated modeling of the electric grid, communications, and control. *International Journal of Energy Sector Management, 2*, 420–438.

Onggo, B. S., Mustafee, N., Juan, A. A., Molloy, O., & Smart, A. (2018). Symbiotic simulation system: Hybrid systems model meets big data analytics. In M. Rabe, A. A. Juan, N. Mustafee, A. Skoogh, S. Jain, & B. Johansson (Eds.), *Winter Simulation Conference* (pp. 1358–1369). New York: IEEE.

Pratt, A., Ruth, M., Krishnamurthy, D., Sparn, B., Lunacek, M., Jones, W., ... Marks, J. (2017). Hardware-in-the-loop simulation of a distribution system with air conditioners under model predictive control. In *Power & Energy Society General Meeting* (pp. 1–5).

Pratt, M. R. A., Lunacek, M., Mittal, S., Wu, H., & Jones, W. (2015). Effects of home energy management systems on distribution utilities and feeders under various market structures. In *Proceedings of the 23rd International Conference and Exhibition on Electricity Distribution*.

Rajkumar, R., Lee, I., Sha, L., & Stankovic, J. (2010). Cyber-physical systems: The next computing revolution. In *Proceedings of the 47th Design Automation Conference (DAC), 2010*, Anaheim, CA, (pp. 731–736). New York: IEEE.

Sanislav, T., & Miclea, L. (2012). Cyber-physical systems-concept, challenges and research areas. *Journal of Control Engineering and Applied Informatics, 14*(2), 28–33.

Sheridan, T. B. (1992). *Telerobotics, Automation, and Human Supervisory Control.* Cambridge, MA: MIT Press.

Thule, C., Broman, D., Larcen, P. G., Gomes, C., & Vangheluwe, H. (2008). Co-simulation: A survey. *ACM Computing Surveys, 51*, 420–438.

Tolk, A. (2015). Merging two worlds: Agent-based simulation methods for autonomous systems. In A. P. Williams & P. D. Scharre (Eds.), *Autonomous Systems: Issues for Defence Policymakers* (pp. 291–317). Norfolk, VA: NATO Allied Command Transformation.

Tolk, A., Adams, K. M., & Keating, C. B. (2011). Towards intelligence-based systems engineering and system of systems engineering. In A. Tolk & L. C. Jain (Eds.), *Intelligence-Based Systems Engineering* (pp. 1–22). Berlin, Heidelberg: Springer–Verlag.

Tolk, A., Page, E., & Mittal, S. (2018). Hybrid simulation for cyber physical systems: State of the art and a literature review. In *Proceedings of Annual Simulation Symposium, Spring Simulation Multi-Conference* (pp. 122–133).

Traoré, M. K., Zacharewicz, G., Duboz, R., & Zeigler, B. (2018). Modeling and simulation framework for value-based healthcare systems. *Simulation, 95*(6), 481–497.

Turing, A. M. (1950). Computing machinery and intelligence. *Mind, 59*(236), 433–460.

Vangheluwe, H. (2000). Devs as a common denominator for multi-formalism hybrid systems modelling. In *IEEE International Symposium on Computer-Aided Control System Design* (pp. 129–134).

Williams, A. (2015). Defining autonomy in systems: Challenges and solutions. In *Autonomous Systems: Issues for Defence Policymakers* (pp. 27–62). Norfolk, VA: NATO Allied Command Transformation.

Williams, R. (2008). *Autonomous systems overview.* Technical report, BAE Systems.

Witten, I. H., Frank, E., Hall, M. A., & Pal, C. J. (2016). *Data Mining: Practical Machine Learning Tools and Techniques.* Cambridge, MA: Morgan Kaufmann.

Zeigler, B. P., Praehofer, H., & Kim, T. G. (2000). *Theory of Modeling and Simulation. Integrating Discrete Event and Continuous Complex Dynamic Systems* (2nd ed.). San Diego, San Francisco, New York, Boston, London, Sydney, Tokyo: Academic Press.

2

Challenges in the Operation and Design of Intelligent Cyber-Physical Systems

Sebastian Castro, Pieter J. Mosterman, Akshay H. Rajhans, and Roberto G. Valenti

MathWorks, Natick, MA, USA

2.1 Introduction

Cyber-physical systems (CPS) are computer-controlled physical systems that deploy computational – or *cyber* – elements to sense, control, and operate in a physical environment. Given the rapid strides made by technology advances in computation and communication, these systems are becoming smart and interconnected. Advances in computation have enabled the introduction of artificial intelligence into such systems – sometimes called *intelligent cyber-physical systems* (Müller 2017) or simply *intelligent physical systems* (Koditschek et al. 2015) – that are not just able to sense, understand, and manipulate the physical environment around them, but can also learn and improve over time. Advances in communication have made distributed architectures of intelligent physical systems possible to the extent that individual autonomous systems can operate together in a collaborative fashion.

Modern society is expected to experience a transformational impact of intelligent CPS in a range of sectors. Compelling examples can be found in the utilities, transportation, and manufacturing sectors.

- In the utilities sector, safe and reliable power grids are CPS applications that exemplify *smart energy*, which enables the integration of renewable energy sources that may be intermittent, such as wind energy, by exploiting weather forecasting.
- In the transportation sector, CPS applications are instances of *smart mobility* and include: on land, connected autonomous vehicles that reduce fatalities and

Complexity Challenges in Cyber Physical Systems: Using Modeling and Simulation (M&S) to Support Intelligence, Adaptation and Autonomy, First Edition. Edited by Saurabh Mittal and Andreas Tolk.
© 2020 John Wiley & Sons, Inc. Published 2020 by John Wiley & Sons, Inc.

optimize for congestion; in water, vehicles that optimize surface shipping and improve underwater exploration; and in the air, unmanned aerial vehicles (often referred to as drones) that improve and enable complex search and rescue operations.

- In the manufacturing sector the *Industry 4.0* paradigm increases economics, efficiency, safety, reliability, and throughput of industrial production as part of *smart manufacturing*.[1] This also relates to the *digital twin* notion that uses computational simulation of (oftentimes expensive) physical assets on the one hand to predict and detect faults and on the other hand, in advanced cases, to replace physical functionality with simulated behavior.

Motivated by the global societal-scale relevance of CPS, this chapter intends to outline important design, test, and operation challenges for development and deployment of such systems. Faced with the breadth of a topic of such vast magnitude, systems are classified based on levels of increasingly sophisticated capabilities along dimensions that compare with levels of cognition and communication in humans.

The developed classification provides structure to the chapter content, which is organized as follows. Section 2.2 introduces an example from the smart mobility domain to motivate the overall discussion. Section 2.3 draws parallels between the evolution of physical and cognitive faculties in humans and engineered systems as a basis for a feature classification presented in Section 2.4. With focus on the communicating and collaborating classes, challenges in operating intelligent CPS are outlined in Section 2.5 followed by challenges for the design and testing of intelligent CPS in Section 2.6. Section 2.7 summarizes and concludes the discussion.

2.2 Connected Autonomous Vehicles

To motivate the discussion in this chapter, let us consider the case of connected autonomous vehicles as an exemplar application of intelligent CPS. Human driving is inefficient and error-prone. By one estimate, traffic congestion cost the US economy $87 billion in lost productivity in just the year 2018.[2] As per the National Highway Traffic Safety Administration (NHTSA), more than 37 000 lives were lost in driving accidents in the United States in the year 2017 alone.[3] It is widely believed that the introduction of connectivity and autonomy in automobiles will

---

1 arminstitute.org
2 https:// www. cnbc. com/2019/02/11/ americas-87-billion-traffic-jam-ranks-boston-and-dc-as-worst-in-us.html
3 www.nhtsa.gov

improve driving safety and efficiency, for example by sensing real-time information about the immediate surroundings to make safer driving decisions and by connecting to and leveraging aggregated historical traffic congestion information to make smarter routing decisions, respectively. A key to achieving these objectives is the aspirational goal of developing engineered systems that are able to navigate traffic situations and driving scenarios like and better than humans.

In order to improve the situational and self-awareness of the individual vehicles, automobiles of today and tomorrow must be instrumented by sensors of various modalities, such as cameras, RADAR (radio detection and ranging) and LIDAR (light detection and ranging) elements, global positioning systems (GPS), and inertial measurement units (IMUs). Based on such rich sensory input, advances in software algorithms – such as computer vision algorithms for camera data – are necessary for classifying the raw sensor data and assigning semantic meaning to it, such as road, obstacles, and other agents in the environment. Availability of large amounts of prerecorded data is necessary for offline training of deep neural networks that are then deployed on hardware for real-time online semantic inference. Semantic inferences from various sensory inputs must be combined via sensor fusion algorithms to eliminate shortcomings and potential blind spots of any one sensing mechanism, thereby increasing the reliability.

Processed data must be leveraged in order to build accurate maps of the environment where the vehicle can localize itself and perform planning to safely and effectively navigate traversable areas. Hierarchical planning algorithms must break down an overall goal of reaching a desired destination into a series of steps, such as merges and lane changes, that are further achieved by low-level control tasks such as trajectory optimization and path planning. The extent to which these tasks can be performed by an automobile without the help of a human determine the various *Levels of Driving Automation* identified by the Society of Automotive Engineers (SAE) J3016™ standard.[4]

Connectivity between a network of vehicles and between vehicles and the infrastructure is opening up cooperative and collaborative operation as a new mechanism of operation. Dedicated Short-Range Communication (DSRC) protocols such as those specified in the IEEE 802.11p standard[5] are necessary for real-time sharing of information in vehicle-to-vehicle (V2V) and vehicle-to-infrastructure (V2I) communication modalities (Christian 2015). Raw data, semantic information, and even learning outcomes can now be shared with a fleet of vehicles, so if one car learns a better way to drive, the entire fleet benefits from it.[6]

---

4 https://www.sae.org/standards/content/j3016_201806
5 https://standards.ieee.org/standard/802_11p-2010.html
6 https://www.recode.net/2016/9/12/12889358/tesla-autopilot-data-fleet-learning

Smart instrumented intersections such as the variants of Cooperative Intersection Collision Avoidance Systems (CICAS):

- the Stop-Sign Assist (SSA) variant when crossing a high-speed through traffic on rural highways at dangerous stop-sign-controlled intersections (Becic et al. 2013), and
- the Signalized Left-Turn Assist (SLTA) variant when making a left turn in front of oncoming vehicles (Misener 2010),

have been studied as mechanisms to aid human drivers in making better split-second decisions. Even in these infrastructure-based Advanced Driver Assistance Systems (ADAS) incarnations that merely make suggestions to human drivers, there are important heterogeneity, architecture, and verification challenges (Rajhans et al. 2011; Rajhans and Krogh 2012, 2013; Rajhans 2013; Rajhans et al. 2014).

With increasing autonomy and connectivity, as the SSA functionality transfers from roadside to in-vehicle (Becic et al. 2012) and the SLTA functionality is being replaced by connected vehicles that communicate and carry out on-the-fly collision avoidance policies (Zhang et al. 2018), the burden of decision making is increasingly being transferred from humans to the connected autonomous vehicles. In process, what once were CICAS suggestions are now morphing into actions with actual consequences. Attaining correct behavior is even more challenging, yet at the same time of paramount importance from a safety point of view.

From an efficiency point of view, meta-level information such as city-level traffic patterns can be used to devise a globally optimal traffic routing strategy to optimize fuel efficiency and traffic throughput, thereby avoiding the so-called *price of anarchy* (Zhang et al. 2018) caused by following modern-day routing strategies that are individually optimal but socially suboptimal.

These exemplar challenges and opportunities can be extrapolated to other application domains in CPS. Indeed the need for sensing, perception, decision making, planning, and control and execution to effectively operate in an environment is universal across various intelligent CPS domains. Collaboration between ensembles of smart agents and humans can also be seen across other CPS application domains such as search and rescue (Mosterman et al. 2014a, 2014b; Zander et al. 2015).

2.3 The Evolution of Physical and Cognitive Faculties in Humans

In order to assess challenges for building intelligent CPS applications that tackle complex challenges such as the ones encountered in connected autonomous driving, it is insightful to consider how humans evolved as intelligent beings who carry out complex tasks in a collaborative society. While a comparison between humans and engineered systems has been made in the literature for specific application domains, such as smart manufacturing (Dias-Ferreira 2016), security and resilience (Azab and Eltoweissy 2012), and multi-objective Pareto optimization and tradeoff analysis (Keller 2018), the objective of this chapter is to create a feature classification for intelligent CPS. Specifically, parallels are drawn between the evolution of physical and cognitive faculties of humans with that of engineered systems to build such a feature classification.

2.3.1 Energy Efficiency and Physical Manipulation

In the early stages of evolution when *Homo erectus* emerged, great advances were made in physical capabilities in order to adapt to the environment. Bipedalism freed up hands and arms while the opposing thumb and padded fingers created for improved manipulation abilities. The ankle and elasticity in tendons enabled energy efficiency that supported exceptional endurance. As additional improvement in the manipulation faculty, *Homo sapiens* developed a shoulder joint that is distinctly different from other species (Roach et al. 2013). This shoulder joint specifically supports the unique ability to throw projectiles with great precision.

2.3.2 Cognition

Multi-modal sensory processing helped create rich information about the environment. Cognitive abilities to orient in large spaces and plan actions enabled effective use of environment, flora, and fauna. Moreover, the cognitive ability to imitate conspecifics helped share techniques for tool creation and form a culture of tool use that was at the genesis of evolution into modern human (MacWhinney 2005).

While evidence of tool use shows incremental advances over the course of evolution up till *Homo sapiens*, processing abilities changed more significantly (MacWhinney 2005; Ralph 2013). Brain size reached a modern volume but perhaps more important are changes in brain architecture to support the cognitive ability that enabled taking the perspective of conspecifics and considering them actors as oneself. Additionally, episodic memory developed that supported the storage of rich sensory information and allowed elaborate planning.

2.3.3 Language and Communication

The next step in cognitive processing abilities enabled the use of symbols, a unique quality of *Homo sapiens* (Holloway 1981). Symbols as syntactic reference may have developed along with speech, which was enabled first by the anatomical components necessary for rich vocalization. Combined with phonological composition, this provides a practically infinite vocabulary to associate symbols with concepts and notions. The ability to handle elaborate plans supports grammar that requires keeping track of notions expressed in a sentence while keeping track of multiple perspectives not just of different actors but also in future, in past, and at locations other than the current.

With the ability to communicate in a rich language of symbols, larger social structures became possible, which, in turn, supported a much improved culture of knowledge sharing and retention.

2.3.4 From Natural to Technical

The human faculties that evolved as energy efficient and sophisticated manipulation, cognition, and communication in a symbolic language can be grafted onto the three foundational pillars of CPS: control, computation, and communication (Kim and Kumar 2012). Control enables efficient energy use in and sophisticated manipulation of the physical world. Computation enables machine intelligence to perform tasks that would have traditionally required human cognition. Communication of data between machines achieves the same tasks as communication in terms of symbols between humans.

Control as manipulation of physical objects and quantities in an energy efficient manner has come to leverage increasing computational power and communication availability. As such, control can be considered functionality that is layered on top of a compute and communication stack. Hence, the classification of CPS ensembles that follows considers two dimensions. The first dimension considers advances in computation as they relate to advances in cognition from the perspective of control functionality. The second dimension extends the stages of cognition with corresponding communication capability and identifies classes of control functionality that the communication enables.

2.4 The Landscape of Intelligent Cyber-Physical Systems

Given the long history of technology, this chapter presents a structure of technical systems that have emerged as part of the digital revolution. In particular, the cognitive processing faculty of humans is used as a structuring mechanism with communication as a key technology superimposed. The structure enables an outlook as to where technology might move next.

2.4.1 A Classification of Engineered System Ensembles

A classification of engineered systems is developed based on the increasing cognitive abilities in organisms (Minsky 2006; Samsonovich & De Jong 2005). Table 2.1

Table 2.1 Behavior at various levels of cognition.

Type	Capability
Reflective	Evaluation and assessment behavior
Reasoned	Deliberate, planned behavior
Reactive	Learned, adaptive behavior
Reflexive	Instinctive behavior, reflexes

lists behavior types that involve an increasing level of cognition. The least demanding is behavior of a *reflexive* nature. Examples of this are the dilation of a pupil based on light and the beating of a heart. Behavior that is of a *reactive* nature is learned and adaptable behavior that can be practiced such as reactively kicking a ball to hit a target. Behavior that is *reasoned* involves planning actions to achieve a goal, for example, navigating traffic to enter or exit a highway. Finally, behavior of a *reflective* nature relies on awareness of the organism and allows setting goals based on an assessment of benefit, for example, adopting a healthy lifestyle.

In addition to the behavior of individuals that require different levels of cognitive ability, key in the evolution of *Homo erectus and later Homo sapiens* is their ability to communicate with conspecifics (MacWhinney 2005; Tattersall 2014). While primitive forms of communication are based on direct reference, the ability to assume the perspective of conspecifics was fundamental to more sophisticated communication such as in the form of mimesis (including chant, dance, repeated gesturing, and partonomy). This form of communication provided *Homo erectus* with a powerful advantage over contemporary species and allowed spreading all throughout Africa and onto other continents. If the power of communication is of such terrific magnitude, the value of language based on symbolism and endowed with sophisticated concepts such as deixis, as practiced by *Homo sapiens*, is difficult to overstate.

In the technical world, communication technology is quickly becoming the hallmark of engineered systems. Table 2.2 shows a classification of engineered systems based on the increasingly sophisticated levels of behavior of individual systems with communication among systems in an ensemble superimposed. The "Individual" row characterizes the behavior of individual engineered systems:

Table 2.2 A classification of behavior in operation.

Configuration	Behavior			
	Reflexive	Reactive	Reasoned	Reflective
Individual	Automatic	Adaptive	Autonomous	Aware
Ensemble	Distributed	Connected	Collaborative	Coallied

- Reflexive behavior is preprogrammed and directly responds to stimuli from the environment. This compares with automatic control architectures where measurements of physical quantities are input to a fixed control law that directly determines an actuation response.
- Reactive behavior is learned, for example by practice or training, and requires an interpretation of the stimuli from the environment. This compares with adaptive systems where different control setpoints or parameters are used based on observations (e.g. an intelligent thermostat may learn preferred temperatures based on historical choices by users).
- Reasoned behavior plans how to reach a goal, which corresponds to a fundamental component of autonomous behavior. Developing plans relies on models of the system itself as well as the observed entities in its surrounding that affect the planning.
- Reflective behavior sets goals while evaluating advantages and drawbacks. This requires an engineered system to be aware of itself, its ambitions, its proclivities, etc., and corresponds to behavior not currently known as generally available in engineered systems.

Turning to the communication aspect, the behaviors of ensembles of engineered systems are characterized in the "Ensemble" row. The corollary of automatic control is *distributed control*, where multiple automatic control loops rely on sharing measurement data among them. In the vein of the intelligent intersection example of Section 2.2, traffic lights that are synchronized are an example of control that is distributed across a number of individual control loops. Input to such a distributed system may be the arrival of vehicles as detected by an inductive loop in the road surface.

A *centralized* intelligent intersection around vehicle-to-infrastructure (V2I) communication is an example of a *connected control* system in that the coordinator (as part of the infrastructure) creates a situational awareness and then adjusts ensemble setpoints. For example, the coordinator interprets (or it is communicated) whether a vehicle is a luxury car or a tractor trailer combination. Setpoints for their speed are then computed according to the respective dynamics of the vehicles. These setpoints are communicated to the different vehicles that then adjust in response. As another example, Waze[7] is a smartphone app that allows users to share traffic data such as collisions that might have happened at certain geolocations. In addition, Waze analyzes and interprets smartphone data to determine traffic congestion. If, in a future scenario, vehicles are enabled to automatically respond to such data, an adaptive Waze vehicle becomes another example of *connnected control*. In this scenario, when traffic congestion is detected and shared

---

7 www.waze.com

with vehicles elsewhere, these connected vehicles may adapt the route to their destination.

A *decentralized* intelligent intersection based on vehicle-to-vehicle (V2V) communication is an example of *collaborating control* where autonomous systems form an ensemble and share rich information to support overall planning behavior such as individual plans and planning considerations. In contrast with the V2I intelligent intersection, vehicles may not rely on predefined control laws at that intersection with setpoints determined by the coordinator. Instead the different vehicles act autonomously and as they approach an intersection they must reason together about their speed, in which lane to drive, distance to intersection, vehicle dynamics, possible urgency, etc. to come up with an overall plan for the ensemble. This decentralized control orchestrates the movement of each vehicle to best achieve the individual plans of all vehicles using the intersection at that given time period. Returning to Waze, another example of collaborating systems would be a scenario that involves the various vehicles adapting their planning given new information. This could involve changing the order of an overall plan, modifying times of arrival or departure, speed, or deciding on different stops along the way to minimize vehicles contending for the same road segments at certain times. Yet another example of collaborative systems is a smart emergency response system (Mosterman et al. 2014a, 2014b; Zander et al. 2015), where autonomous ground vehicles may plan how to arrive at select locations where they set up a depot. Autonomous rotorcraft may then plan sorties to deliver provisions in the field while minimizing overall time of delivery.

Aware systems in a *coallied control* ensemble would share information about internal states such as intent or assessments, though no examples of such operational ensembles are known.

2.4.2 The Lifecycle of Engineered System Ensembles

Traditional Model-Based Design has been studied following a "V"-shaped approach (e.g. Mosterman et al. 2004) as drawn in Figure 2.1a, where models are

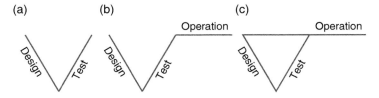

(a) (b) (c)

Figure 2.1 Paradigm shift from the traditional Model-Based Design "V" to a Model-Based Design-Test-Operation triad. (a) Traditional "V" (b) emerging "square root sign" and (c) future cyclic triad.

used to design and test the implementation that is deployed. However, intelligent CPS require a paradigm shift in this traditional viewpoint. With new applications such as Industry 4.0 and industrial internet of things, modeling, and simulation are also being used in operation after systems are deployed. For example, a computational model – a so-called "digital twin" – of an expensive physical asset is simulated during operation for prognostics and predictive maintenance purposes. This emerging paradigm can be thought of as a "square root sign" rather than a "V". In the more sophisticated intelligent CPS that constitute dynamically reconfigurable ensembles, the operation facet connects back to design and test for runtime reconfiguration. This reconfigurability morphs the "square root" into a cyclic triad as is depicted in Figure 2.1c.

It is worth noting that this triad shares some similarities with the CPS trustworthiness framework developed by the US National Institute for Standards and Technology (NIST), which considers conceptualization, realization, and assurance as three sides in a triangular (prism) representation (Figure 8 in CPS Public Working Group (2017)). Though the terms conceptualization, realization, and assurance are related to design, operation, and test, the NIST framework considers models in a limited context as the output of the conceptualization phase and an input to the realization phase. In contrast, the view of this chapter holds that modeling and simulation play an equally important role in all three sides of the triad.

The two-dimensional representation from Figure 2.1c can be projected in three dimensions by adding the various increasingly sophisticated behavior classes from Table 2.2 as the third dimension. This forms a prism in three

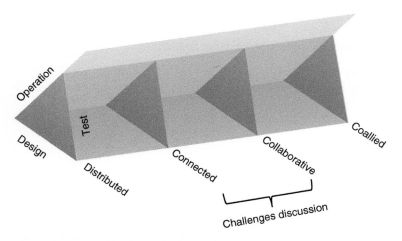

Figure 2.2 Intelligent CPS Ensemble Design-Test-Operation Prism.

dimensions, as is depicted in Figure 2.2 with the behavior classes in the ensemble configuration – distributed, connected, collaborative, and coallied – annotated next to each triad.

There are several challenges associated with the development of intelligent CPS from each of these classes. This chapter focuses the discussion around challenges for systems that are open by communicating with other systems in ensembles. The study of the ensemble dimension is further narrowed by concentrating on systems that are currently under most scrutiny: (i) connected systems and (ii) collaborative systems. The remaining classes in the ensemble dimension are either well understood (distributed control systems) or they require significant research to conceive of and develop technology that is still very much in its infancy (coalitions of systems). Challenges for these systems are discussed in the context of (i) machines being connected, (ii) machines being collaborative, (iii) design of such connected and collaborative ensembles of machines, and (iv) testing such ensembles at runtime as they assemble and dissolve. The challenges are first discussed from the perspective of systems in operation in Section 2.5 and then the design and test perspective in Section 2.6. Related work (Mosterman and Zander 2016) details a corresponding needs analysis.

2.5 Challenges in System Operation

During operation, communication among intelligent CPS in an ensemble opens up opportunities for new functionality, even more so when sophisticated collaboration is supported. Future enabling technology that is needed, the current state, and challenges are discussed for connected and collaborative operation.

2.5.1 Connected Operation

Systems that are part of ensembles where connections with other systems are established during operation have a need for wireless communication, data sharing, and service utilization.

2.5.1.1 Wireless Communication
High performance wireless communication will allow reliable configuration of flexible system configurations for features with varying quality of service. There are two key challenges.

- The communication protocol stack must be physically aware and configurable while being compatible with the internet protocol (IP), for example, IEEE 802.15.4e (Palattella et al. 2013). Such a protocol stack supports real-time services of graded quality with a low energy footprint and enables including (precise) time and location information in communicated data. Useful approaches are the modeling of building blocks that comprise communication protocols as well as modeling of the performance characteristics across electronics hardware targets.

- Precise timing and synchronization (e.g. the precision time protocol IEEE 1588 (Cooklev et al. 2007)) must be supported in a distributed and wirelessly connected environment. Two strands of advances that are important are: (i) physical layer based timing and synchronization architectures, which benefit from modeling of the physical radio frequency (RF) layer as well as the antenna and (ii) scheduling of periodic and aperiodic events with reliable execution times (e.g., Zhang et al. 2010), which builds on advances in scheduler configuration, dynamic scheduling with guarantees, and support for mixed synchronous and asynchronous behavior.

2.5.1.2 Data Sharing

Advanced data sharing will allow distributed information resources to be effectively exploited and enable system features that were not considered *a priori* at the sharing sources. Where in an offline approach system integration would be responsible for synchronization of data streams (Muller 2007), in online connectivity scenarios this must be resolved by construction. There are two specific challenges to address.

- The functionality must provide support for multirate architectures where the methodologies concern the synchronization of data from incongruent sources. Solution aspects to consider include communication modeling, systematic (and automatic) analysis of double buffering schemes, timing properties of software, and clock recovery. An increasing use of models for system integration is imperative (Mosterman et al. 2005).
- To derive value for the system features it must be possible to reliably extract corresponding (unambiguous) information from the communicated data. Approaches in support of this represent information as high-level models with well-defined metamodels and ontologies with model import/export under version control, automatic generation of metamodels (e.g. from model libraries), and sharing and comparing of model concepts. The RoboEarth network is an example of concrete advances in that internet connections enable robots to generate, share, and reuse data (Waibel et al. 2011).

2.5.1.3 Service Utilization

Systems that are dynamically assembled post deployment will be endowed with the capability to purpose available functionality in service of specific (singular) needs. Three challenges are listed.

- Service-based approaches must operate as real-time embedded services in a physical environment. Advances stem from work on real-time middleware and service-oriented architectures with physical capabilities that must address key technical issues such as service discovery response time (latency, averages, time-out) (Douglas 2007) and request for services in different modalities. Middleware that is real-time capable ranges from a real-time version of the High Level Architecture (HLA), and a real-time Common Object Request Broker Architecture (CORBA), to the Data Distribution Service (DDS), and the Robot Operating System (ROS) (Krishnamurthy et al. 2004; McLean et al. 2004; Schmidt et al. 2008; Quigley et al. 2009; Pérez and Gutiérrez 2014).
- Service discovery must increase in logic capabilities (be "smart"). Possible solutions may be the use of service ontologies for service provider matching that rely on taxonomies for similarity and transformability matching (e.g. Song et al. 2010). Capabilities for type similarity checking and conversion as well as semantics definition are key.
- Information sharing must be enabled in heterogeneous system ensembles. A language and ontology infrastructure, for example, the Ontology Web Language (OWL) (Horrocks and Patel-Schneider 2011) used for describing semantic web services, may serve as an underlying technology to support translation and transformation. Additional technology to build on includes the reliable generation of models and (implementation) code from models.

2.5.2 Collaborative Operation

Three needs for enabling system ensembles to operate in a collaborative manner include runtime system adaptation, emerging behavior design, and functionality sharing.

2.5.2.1 Runtime System Adaptation

The ability for safe and reliable system adaptation at runtime enables a system in an ensemble to exploit functionality that is exogenous, implemented by other systems, for efficient, economical, and resilient operation.

- The main challenge is reasoning and planning adaptation of an ensemble of systems, which builds on a number of technologies: (i) introspection of the systems in an ensemble to determine the system state, configuration (possibly using runtime variants), and available services (possibly based on a middleware service description specification); (ii) handling of ensemble (in)consistency with a level of fidelity that is sufficient for runtime needs, which may build on traceability between representations, possibly across transformations (e.g. models@run.time Bencomo et al. 2014); and (iii) online (re)calibration of models (e.g. Huang et al. 2010) to continuously ensure accuracy, which may use collected data, along with statistical, optimization, or machine learning tools, to modify the parameters or structure of software artifacts within the system.

2.5.2.2 Emerging Behavior Design

Robust methods to design emerging behavior allow for the systematic design of systems that are part of an ensemble such that the ensemble as a whole realizes desired behavior in an optimal sense.

- The overall challenge is about collaborative planning, guidance, and control with a number of methods to build on: (i) analysis methods across loosely coupled architectures are key, especially for embedded operation (e.g. globally asynchronous/locally synchronous, GALS, architectures (Steven 2005)), which spans event-driven control, discrete-event modeling and analysis, and uncertainty modeling; (ii) planning and synthesis of distributed control functionality on concurrent resources is potentially core and involves concurrency and platform modeling, functionality decomposition, and service composition (e.g. the Towers of Hanoi as a CPS) (Mosterman et al. 2013; Mosterman and Zander 2015); and (iii) formal methods to ensure conformance and be applicable to collaborative problems with concurrency semantics while enabling property proving with performance models, all by retaining the rigor of formality yet in an accessible manner (e.g. design refinement in Conway's Game of Life (Sanders and Smith 2012)).
- In addition, behavior can be learned (Hofer 2017). By tracking the performance of collaborative systems based on the decisions made by the system – for example, allocating tasks, resources, or operating modes – techniques such as online optimization and reinforcement learning (Kober et al. 2013) can use this historical data to automatically refine the strategy and rules of a complex decision-making system, thus giving rise to novel behavior patterns that may better solve an existing environment or even adapt to an evolving environment.

2.5.2.3 Functionality Sharing

Sharing functionality among systems in an ensemble, not only by making the functionality available but also by meaningful exchange of information and meta information about the functionality will allow the creation of novel system features post deployment. There are two challenges to highlight.

- Using functionality for multiple (different) purposes post deployment, which may build on a number of advances: (i) generation of models for a particular task by property identification (e.g. property based model slicing) and model behavior selection (e.g. behavioral analysis and functionality mining such as determining the requirements for a design from its behavior); (ii) hardware resource sharing by creating a dispatch architecture (e.g. a real-time virtual machine Gu and Zhao 2012) and following a platform-based design approach (Balarin et al. 2009; Marco Di Natale 2012); (iii) performance characterization (e.g. Liu and Feng Zhao 2005) via performance models and measures (e.g. critical path analysis and code performance reporting and advise); and (iv) online calibration (e.g. Levinson and Thrun 2013) based on objective and performance criteria, which is supported by adaptive filtering, distortion modeling, and automatic groundtruthing (baselining).

- Interaction between features (e.g. Mosterman and Zander 2016) that leverage the shared functionality to find potential (re)solutions in assumption formalization and dependency effect analysis. Specific technologies include model slicing based on properties or assumptions, tracing between sources and destinations for behavioral anomalies, and mapping formulated assumptions about functionality to behavior.

2.6 Challenges in System Design and Test

The operational opportunities for ensembles of intelligent CPS build on corresponding advances in their design and test. A key aspect is the cyclic nature of the design, test, and operation stages as illustrated in Figure 2.1a. Future technology that is needed is compared against the current state with challenges to arrive at the future state.

2.6.1 Design

In the design of connected and collaborating ensembles of systems, two main needs are the ability for virtual system integration and for design artifact sharing.

2.6.1.1 Virtual System Integration

The ability to realistically integrate systems in a virtual sense will enable the confident design of systems as part of a reliable system ensemble that configures during operation. There are three challenges to elaborate on.

- Obtaining proper models in design, which potentially builds on the generation of models with necessary detail given a selected property of interest. This can involve structural model changes, operating point selection and linearization, implementation model generation, and so forth. A number of approaches to highlight include selecting model detail based on properties of interest (Ferris 1995; Mosterman and Biswas 2000; Stein 1995), counterexample guided refinement (Clarke et al. 2000), and requirements guided abstraction selection (Jiang et al. 2014).

- System-level design and analysis of a heterogeneous ensemble by using models requires advances in (i) connecting, combining, and integrating models represented in different formalisms, potentially at the behavior level via cosimulation or a shared simulation API or at a shared semantics level by code generation and (ii) efficient simulation models that can be used across dynamics and execution semantics, which involves an array of potentially interacting solver configurations for continuous-time, discrete-time, and discrete-event behaviors. These challenges are subject of study in the field of computer automated multiparadigm modeling (Mosterman and Vangheluwe 2000) where combining dynamic semantics (Mosterman and Biswas 2002) and execution semantics (Mosterman 2007) are important advances.
- Connectivity among models, software, and hardware corresponding to different vendors and end manufacturers can build on efforts in creating open tool platforms with trusted interfaces for communication across synchronized and coordinated models, software, and hardware devices. Some underlying technologies include data streaming, target connectivity support, standardized communication protocols (e.g. TCP, UDP), and real-time simulation (Popovici and Mosterman 2012).

2.6.1.2 Design Artifact Sharing

Being able to securely and reliably share design artifacts across design efforts for a system ensemble will allow convenient, efficient, and consistent collaboration between stakeholders in design and ultimately throughout the system lifecycle. Two challenges are presented.

- Given the different organizations (vendors, end manufacturers, and others) that are invariably involved, tool coupling among these disparate organizations may be addressed by building on (e.g. Open Services for Lifecycle Collaboration, OSLC (Seceleanu and Sapienza 2013)): (i) support for traceability across semantic and technology adaptation, for example, based on a service API with change notification to establish relations across abstractions, formalisms, and transformations, all while honoring intellectual property protection, and (ii) information extraction from protected intellectual property (e.g. by obfuscation or encryption) and use of trusted compilers.
- Supporting manifold views and tools that are essential in design, especially for system ensembles can advance based on core technologies such as (i) configurable view projections (e.g. Atkinson et al. 2010) that are tool specific and support model generation, pattern extraction or slicing, and XML interexchange, and (ii) use of consistent semantics across tools by modeling the execution engines (e.g. Mosterman and Zander 2011) that are combinations of code libraries (e.g. numerical integration, root finding, and algebraic equation

solving) with a broad spectrum of optimization so that semantic analysis of an execution engine as a dynamic system is enabled.

2.6.2 Test

To test behavior of system ensembles that rely on interfaces for runtime configuration, it is key to develop support for runtime system adaption, collaborative functionality testing, and hardware resources sharing.

2.6.2.1 Runtime Adaptable System Testing

Testing of systems in a runtime adaptable configuration will allow confidently exploiting functionality across the overall ensemble, as was mentioned in the context of collaborative operation. A challenge specific to testing is discussed here.

- Testing complex functionality on a deployed system is critical, yet challenging for systems that operate embedded in a physical environment. Often, a full model of the system's interaction with the environment is not available or impractical to implement given the available computational resources. *Surrogate models* (Viana et al. 2010) can provide an approximation of the actual process that is accurate enough and computationally feasible for online testing and design optimization procedures. Specific technologies include variable and dimensionality reduction through methods such as sensitivity analysis and principal component analysis, using data to generate response surfaces, and fitting low-fidelity models ranging from polynomials to artificial neural networks.

2.6.2.2 Collaborative Functionality Testing

The ability to test collaborative functionality will enable assurance of collaboration quality on shared resources while being able to identify and automatically mitigate root causes of failure in a distributed environment. Two challenges are expounded on.

- Systematic test suite generation and automated test evaluation is especially challenging for collaborative functionality. Solution approaches possibly build on model-based test generation from requirements while preserving the context of a dynamic ensemble configuration. Specific technologies (e.g. Arrieta et al. 2014; Zander 2009) include coverage based automatic test generation, variants-based testing, and closed-loop testing.
- Reproducible test results under minimum uncertainty could leverage two key advances: (i) setting of initial conditions and injecting fault data, especially when using a service architecture, with specific technologies such as system state restoration, stateless services, and test fixture generation (e.g. Leitner et al.

2013), and (ii) temporal and spatial partitioning to isolate functionality for a specific system architecture under investigation, which includes time partition testing and functionality extraction.

2.6.2.3 Hardware Resource Sharing

Robust, safe, and secure support for hardware resource sharing will allow contracting out system resources within an ensemble and will support a balanced use of external resources for resiliency and runtime cost optimization. Two challenges are discussed.

- Determination of key test cases for different implementations builds on characterization of computational architectures, for example, by use of static analysis methods (such as abstract interpretation), automatic test generation, and detailed models of hardware architecture behavior implementations (e.g. to know of "corner cases").
- Safety of heterogeneous system ensembles is critical and requires advances in: (i) modeling the semantics of time (e.g. Zander et al. 2011) to support safety monitoring components (e.g. watchdogs and mitigators (Kaiser et al. 2003, 2010)), which includes time represented as harmonic periods (e.g. integer time), synchronous behavior (e.g. a single clock or discrete time), simultaneous behavior (iterations of change), dense (e.g. a rational time base), or continuous (e.g. variable step numerical solutions) and (ii) dynamically mixing safety integrity levels (SiL) and the use of certification kits for components of mixed SiL, which supports matching software with hardware.

2.7 Conclusions

Cyber-physical systems (CPS) are rapidly increasing in complexity owing to the progress in computation and communication technology. As these systems become connected and autonomous, the complexity of their design, test, and operation presents several key challenges. To outline these challenges in a structured manner, a classification scheme is developed along the behavior and configuration dimensions drawing a parallel between human faculties and engineered systems capabilities. Challenges throughout the lifecycles of these intelligent systems are discussed in terms of design, test, and operation, with a focus on connected and collaborative behavior in an ensemble configuration. The class of ensembles with distributed control has been subject of much research and is relatively well understood. In contrast, at the opposite end of the classification spectrum, there is no known scholarly content on operations of the class of ensembles with coallied systems that are self-aware and so create self-aware ensembles.

The use of computational models is key in addressing the various design, test, and operation challenges. While the traditional role of models during the design phase is well understood, models are also needed for successful real-time operation, for example in predictive maintenance and prognostics and health monitoring applications. For connected and collaborative ensembles that may include several instances of dynamic reconfigurations at runtime, several online iterations of design and test would be necessary along with each instance of reconfiguration (along the "development and operations" or DevOps paradigm (Jabbari et al. 2016), albeit with differences). Here, models would provide an effective – and perhaps the only – way to carry out the necessary run-time evaluations as the systems themselves are already deployed and in operation.

The goal of the chapter is to provide a broad overview, yet the discussion is by no means exhaustive. Despite the number and scale of the challenges, the potential of intelligent CPS to fundamentally transform human lives is enormous. Continuing to push the frontiers of CPS will require close collaboration between the research and industry communities, hardware and software vendors, and multiple disciplines in science, engineering, technology, and more.

References

Arrieta, A., Sagardui, G., & Etxeberria, L. (2014). A model-based testing methodology for the systematic validation of highly configurable cyber-physical systems. In *The Sixth International Conference on Advances in System Testing and Validation Lifecycle* (pp. 66–72). IARIA XPS Press.

Atkinson, C., Stoll, D., & Bostan, P. (2010). Orthographic software modeling: A practical approach to view-based development. In L. A. Maciaszek, C. González-Pérez, & S. Jablonski (Eds.), *Evaluation of Novel Approaches to Software Engineering*, volume 69 of *Communications in Computer and Information Science* (pp. 206–219). Berlin, Heidelberg: Springer.

Azab, M., & Eltoweissy, M. (2012, November). Bio-inspired evolutionary sensory system for cyber-physical system defense. In *2012 IEEE Conference on Technologies for Homeland Security (HST)* (pp. 79–86).

Balarin, F., DAngelo, M., Davare, A., Densmore, D., Meyerowitz, T., Passerone, R., ... Zhu, Q. (2009). Platform-based design and frameworks: Metropolis and metro ii. In G. Nicolescu & P. J. Mosterman (Eds.), *Model-Based Design for Embedded Systems, Computational Analysis, Synthesis, and Design of Dynamic Systems* (pp. 259–322). Boca Raton, FL: CRC Press. ISBN: 9781420067842

Becic, E., & Manser, M. (2013). *Cooperative intersection collision avoidance system stop sign assist traffic-based FOT: MNDOT contract number 98691 deliverable for task 4.1 final report*. Technical report, Minnesota Department of Transportation.

Becic, E., Manser, M. P., Creaser, J. I., & Donath, M. (2012). Intersection crossing assist system: Transition from a road-side to an in-vehicle system. *Transportation Research Part F: Traffic Psychology and Behaviour, 15*(5), 544–555.

Bencomo, N., France, R. B., Cheng, B. H., & Aßmann, U. (Eds.) (2014). *Models@run. time,* volume 8378 of Lecture Notes in Computer Science (LNCS). Berlin, Germany: Springer.

Clarke, E. M., Grumberg, O., Jha, S., Lu, Y., & Veith, H. (2000, July 15–19). Counterexample-guided abstraction refinement. In *Proceedings of the 12th International Conference on Computer Aided Verification, CAV 2000* (pp. 154–169). Chicago, IL, USA.

Cooklev, T., Eidson, J. C., & Pakdaman, A. (2007). An implementation of IEEE 1588 over IEEE 802.11b for synchronization of wireless local area network nodes. *IEEE Transactions on Instrumentation and Measurement, 56*(5), 1632–1639.

CPS Public Working Group. (2017, June). *Framework for cyber-physical systems: Volume 1, overview.* Technical report, National Institute of Standards and Technology. NIST Special Publication 1500-201.

Di Natale, M. (2012). Specification and simulation of automotive functionality using AUTOSAR. In K. Popovici & P. J. Mosterman (Eds.), *Real-time Simulation Technologies: Principles, Methodologies, and Applications, Computational Analysis, Synthesis, and Design of Dynamic Systems* (pp. 523–548). Boca Raton, FL: CRC Press. ISBN: 9781439846650

Dias-Ferreira, J. (2016). *Bio-inspired self-organising architecture for cyber-physical manufacturing systems* (PhD thesis), The Royal Institute of Technology (KTH), Stockholm, Sweden.

Ferris, J. B., & Stein, J. L. (January 1995). Development of proper models of hybrid systems: A bond graph formulation. In F. E. Cellier & J. J. Granda (Eds.), *1995 International Conference on Bond Graph Modeling and Simulation (ICBGM '95), Number 1 in Simulation* (pp. 43–48). Las Vegas: Society for Computer Simulation, Simulation Councils, Inc. Volume 27

Gu, Z., & Zhao, Q. (2012). A state-of-the-art survey on real-time issues in embedded systems virtualization. *Journal of Software Engineering and Applications, 5*(4), 277–290.

Hofer, L. (2017). *Decision-making algorithms for autonomous robots* (PhD thesis), Université de Bordeaux.

Hoffert, J., Jiang, S., & Schmidt, D. C. (2007). A taxonomy of discovery services and gap analysis for ultra-large scale systems. In *Proceedings of the 45th Annual Southeast Regional Conference,* ACM-SE 45 (pp. 355–361). New York, NY, USA: ACM.

Holloway, R. L. (1981). Culture, symbols, and human brain evolution: A synthesis. *Dialectical Anthropology, 5,* 287–303.

Holloway, R. L. (2013). The evolution of the hominid brain. In W. Henke & I. Tattersall (Eds.), *Handbook of Paleoanthropology* (pp. 1–23). Berlin, Heidelberg: Springer.

Horrocks, I., & Patel-Schneider, P. F. (2011). Knowledge representation and reasoning on the semantic web: Owl. In J. Domingue, D. Fensel, & J. A. Hendler (Eds.), *Handbook of Semantic Web Technologies* (pp. 365–398). Berlin, Germany: Springer.

Huang, Y., Seck, M. D., & Verbraeck, A. (2010). Towards automated model calibration and validation in rail transit simulation. In P. M. A. Sloot, G. D. van Albada, & J. Dongarra (Eds.), *Proceedings of the International Conference on Computational Science, ICCS, Amsterdam, the Netherlands* (pp. 1253–1259). Elsevier.

Jabbari, R., Ali, N., Petersen, K., & Tanveer, B. (2016, April). What is devops? A systematic mapping study on definitions and practices. In *Proceedings of the Scientific Workshop of XP 2016* (pp. 1–11).

Jiang, Z., Mosterman, P.J., & Mangharam, R. (2014, December). *Requirement-guided model refinement*. Technical Report MLAB-70, University of Pennsylvania.

Kaiser, B., Klaas, V., Schulz, S., Herbst, C., & Lascych, P. (2010, September). Integrating system modelling with safety activities. In *Proceedings of the 29th International Conference on Computer Safety, Reliability, and Security, SAFECOMP 2010* (pp. 452–465). Vienna, Austria.

Kaiser, B., Liggesmeyer, P., & Mäckel, O. (2003). A new component concept for fault trees. In *Proceedings of the 8th Australian Workshop on Safety Critical Systems and Software: Volume 33*, SCS '03, Darlinghurst, Australia, 37–46. Australian Computer Society, Inc.

Keller, K. L. (2018). Leveraging biologically inspired models for cyberphysical systems analysis. *IEEE Systems Journal, 12*(4), 3597–3607.

Kim, K., & Kumar, P. R. (2012). Cyberphysical systems: A perspective at the centennial. *Proceedings of the IEEE, 100*(Special Centennial Issue), 1287–1308.

Kober, J., Bagnell, J. A., & Peters, J. (2013). Reinforcement learning in robotics: A survey. *The International Journal of Robotics Research, 32*(11), 1238–1274.

Koditschek, D. E., Kumar, V., & Lee, D. D. (2015). Future directions of intelligent physical systems: A workshop on the foundations of intelligent sensing, action and learning (FISAL). Available at: https://basicresearch.defense.gov/Portals/61/Documents/future-directions/4_FISAL.pdf

Krishnamurthy, Y., Gill, C., Schmidt, D. C., Pyarali, I., Mgeta, L., Zhang, Y., & Torri, S. (2004, May 25–28). The design and implementation of real-time CORBA 2.0: Dynamic scheduling in TAO. In *Proceedings of the 10th IEEE Real-time Technology and Application Symposium (RTAS '04)* (pp. 121–129). Toronto, Ontario, Canada.

Leitner, P., Schulte, S., Dustdar, S., Pill, I., Schulz, M., & Wotawa, F. (2013, May 26). The dark side of SOA testing: Towards testing contemporary soas based on criticality metrics. In *Proceedings of the 5th International ICSE Workshop on*

Principles of Engineering Service-Oriented Systems, PESOS 2013 (pp. 45–53). San Francisco, CA, USA.

Levinson, J, & Thrun, S. (2013, June). Automatic online calibration of cameras and lasers. In *Proceedings of Robotics: Science and Systems*, Berlin, Germany.

Liu, J., & Zhao, F. (2005, February). *Towards service-oriented networked embedded computing*. Technical Report MSR-TR-2005-28, Microsoft Research.

MacWhinney, B. (2005). Language evolution and human development. In B. J. Ellis & D. F. Bjorklund (Eds.), *Origins of the Social Mind: Evolutionary Psychology and Child Development* (pp. 383–410). New York, NY: Guilford Press.

McLean, T., Fujimoto, R. M., & Brad Fitzgibbons, J. (2004). Middleware for real-time distributed simulations. *Concurrency and Computation: Practice and Experience*, *16*(15), 1483–1501.

Miller, S. P., Whalen, M. W., OBrien, D., Heimdahl, M. P., & Joshi, A. (2005, September). *A methodology for the design and verification of globally asynchronous/ locally synchronous architectures*. Technical Report NASA/CR-2005-213912, National Aeronautics and Space Administration (NASA), Langley Research Center.

Minsky, M. (2006). *The Emotion Machine: Commonsense Thinking, Artificial Intelligence, and the Future of the Human Mind*. New York, NY: Simon & Schuster.

Misener, J. A. (2010). *Cooperative intersection collision avoidance system (cicas): Signalized left turn assist and traffic signal adaptation*. Technical report, University of California, Berkeley.

Mosterman, P. J., & Biswas, G. (2000, March 23–25). Towards procedures for systematically deriving hybrid models of complex systems. In *Proceedings of the Third International Workshop on Hybrid Systems: Computation and Control, HSCC 2000* (pp. 324–337). Pittsburgh, PA, USA.

Mosterman, P. J., Ghidella, J., & Friedman, J. (2005, July). Model-based design for system integration. In *Proceedings of the Second CDEN International Conference on Design Education, Innovation, and Practice* (pp. CD–ROM: TB–3–1 through TB–3–10). Kananaskis, Alberta.

Mosterman, P. J., Prabhu, S., & Erkkinen, T. (2004, July). An industrial embedded control system design process. In *Proceedings of the Inaugural CDEN Design Conference (CDEN'04)*, Montreal, Canada. CD-ROM: 02B6.

Mosterman, P. J., & Vangheluwe, H. (2000, September). Computer automated multi-paradigm modeling in control system design. In *Proceedings of the IEEE International Symposium on Computer-Aided Control System Design* (pp. 65–70). Anchorage, Alaska.

Mosterman, P. J., & Zander, J. (2011, September). Advancing model-based design by modeling approximations of computational semantics. In *Proceedings of the 4th International Workshop on Equation-Based Object-Oriented Modeling Languages and Tools* (pp. 3–7). Zürich, Switzerland. keynote paper.

Mosterman, P. J., & Zander, J. (2015, January). GitHub Repository: Towers of Hanoi in MATLAB/Simulink for Industry 4.0. doi: 10.5281/zenodo.13977.

Mosterman, P. J., Zander, J., & Han, Z. (2013, April). The towers of Hanoi as a cyber-physical system education case study. In *Proceedings of the First Workshop on Cyber-Physical Systems Education*, Philadelphia, PA.

Mosterman, P. J. (2007). Hybrid dynamic systems: Modeling and execution. In P. A. Fishwick (Ed.), *Handbook of Dynamic System Modeling* (pp. 15-1–15-26). Boca Raton, FL: CRC Press.

Mosterman, P. J., & Biswas, G. (2002). A hybrid modeling and simulation methodology for dynamic physical systems. *Simulation, 78*(1), 5–17.

Mosterman, P. J., Escobar Sanabria, D., Bilgin, E., Zhang, K., & Zander, J. (2014b). Automating humanitarian missions with a heterogeneous fleet of vehicles. *Annual Reviews in Control, 38*(2), 259–270.

Mosterman, P. J., Sanabria, D. E., Bilgin, E., Zhang, K., & Zander, J. (2014a). A heterogeneous fleet of vehicles for automated humanitarian missions. *Computing in Science and Engineering, 12*, 90–95.

Mosterman, P. J., & Zander, J. (2016). Cyber-physical systems challenges: A needs analysis for collaborating embedded software systems. *Software and Systems Modeling, 15*(1), 5–16.

Mosterman, P. J., & Zander, J. (2016). Industry 4.0 as a cyber-physical system study. *Software and Systems Modeling, 15*(1), 17–29.

Muller, G. (2007, June). Coping with system integration challenges in large complex environments. In *The Seventeenth International Symposium of the International Council on Systems Engineering INCOSE* paper ID: 7.1.4.

Müller, H. A. (2017). The rise of intelligent cyber-physical systems. *Computer, 50*(12), 7–9.

Palattella, M. R., Accettura, N., Vilajosana, X., Watteyne, T., Grieco, L. A., Boggia, G., & Dohler, M. (2013). Standardized protocol stack for the internet of (important) things. *IEEE Communications Surveys and Tutorials, 15*(3), 1389–1406.

Pérez, H., & Gutiérrez, J. J. (2014). A survey on standards for real-time distribution middleware. *ACM Computing Surveys, 46*(4), 49:1–49:39.

Popovici, K., & Mosterman, P. J. (Eds.) (2012). *Real-time Simulation Technologies: Principles, Methodologies, and Applications. Computational Analysis, Synthesis, and Design of Dynamic Systems*. Boca Raton, FL: CRC Press.

Quigley, M., Conley, K., Gerkey, B. P., Faust, J., Foote, T., Leibs, J., ... Ng, A. Y. (2009). ROS: An open-source robot operating system. In *ICRA Workshop on Open Source Software*.

Rajhans, A., Bhave, A., Loos, S., Krogh, B. H., Platzer, A., & Garlan, D. (2011, December). Using parameters in architectural views to support heterogeneous design and verification. In *2011 50th IEEE Conference on Decision and Control and European Control Conference* (pp. 2705–2710).

Rajhans, A., & Krogh, B. H. (2012). Heterogeneous verification of cyber-physical systems using behavior relations. In *Proceedings of the 15th ACM International Conference on Hybrid Systems: Computation and Control*, HSCC '12 (pp. 35–44). New York, NY, USA: ACM.

Rajhans, A., & Krogh, B. H. (2013). Compositional heterogeneous abstraction. In *Proceedings of the 16th International Conference on Hybrid Systems: Computation and Control*, HSCC '13 (pp. 253–262). New York, NY, USA: 253–262ACM.

Rajhans, A. H. (2013). *Multi-model heterogeneous verification of cyber-physical systems* (PhD thesis). Carnegie Mellon University.

Rajhans, A., Bhave, A., Ruchkin, I., Krogh, B. H., Garlan, D., Platzer, A., & Schmerl, B. (2014). Supporting heterogeneity in cyber-physical systems architectures. *IEEE Transactions on Automatic Control, 59*(12), 3178–3193.

Richard, C. M., Morgan, J. F., Bacon, L. P., Graving, J. S., Divekar, G., & Lichty, M. G. (2015) *Multiple sources of safety information from v2v and v2i: Redundancy, decision making, and trustsafety message design report.* Technical report, Office of Safety Research and Development, Federal Highway Administration. Available at https://www.fhwa.dot.gov/publications/research/safety/15007/15007.pdf.

Roach, N. T., Venkadesan, M., Rainbow, M. J., & Lieberman, D. E. (2013). Elastic energy storage in the shoulder and the evolution of highspeed throwing in *Homo. Nature, 498*(7455), 483–486.

Samsonovich, A. V., & De Jong, K. A. (2005). Designing a self-aware neuromorphic hybrid. In K. R. Thórisson, H. H. Vilhjálmsson, & S. Marsella (Eds.), *AAAI-05 Workshop on Modular Construction of Human-Like Intelligence* (pp. 71–78). Menlo Park, CA: AAAI Press.

Sanders, J. W., & Smith, G. (2012). Emergence and refinement. *Formal Aspects of Computing, 24*(1), 45–65.

Schmidt, D. C., Corsaro, A., & Hag, H. (2008, May). Addressing the challenges of tactical information management in net-centric systems with DDS. CrossTalk special issue on Distributed Software Development.

Seceleanu, T., & Sapienza, G. (2013). A tool integration framework for sustainable embedded systems development. *IEEE Computer, 46*(11), 68–71.

Song, Z., Cárdenas, A. A., & Masuoka, R. (2010). Semantic middleware for the internet of things. In F. Michahelles, & J. Mitsugi (Eds), *Proceedings of the 2010 Internet of Things (IOT)*, Tokyo, Japan: IEEE. doi: 10.1109/IOT.2010.5678448

Stein, J. L., & Louca, L. S. (1995, January). A component-based modeling approach for system design: Theory and implementation. In F. E. Cellier & J. J. Granda (Eds.), *1995 International Conference on Bond Graph Modeling and Simulation (ICBGM '95), Number 1 in Simulation, Las Vegas* (Vol. 27, pp. 109–115). Society for Computer Simulation, Simulation Councils, Inc.

Tattersall, I. (2014). An evolutionary context for the emergence of language. *Language Science, 46*(Part B), 199–206.

Viana, F. A. C., Gogu, C., & Haftka, R. (2010). Making the most out of surrogate models: Tricks of the trade. *Proceedings of the ASME Design Engineering Technical Conference*, *1*, 587–598.

Waibel, M., Beetz, M., D'Andrea, R., Janssen, R., Tenorth, M., Civera, J., ... van de Molengraft, R. (2011). RoboEarth: A World Wide Web for robots. *Robotics and Automation Magazine*, *18*(2), 69–82.

Zander, J. (2009). *Model-based testing of real-time embedded systems in the automotive domain* (PhD Thesis). Technical University Berlin.

Zander, J., Mosterman, P. J., Hamon, G., & Denckla, B. (2011, September). On the structure of time in computational semantics of a variable-step solver for hybrid behavior analysis. In *Proceedings of the 18th IFAC World Congress*, Milan, Italy.

Zander, J., Mosterman, P. J., Padir, T., Wan, Y., & Fu, S. (2015). Cyber-physical systems can make emergency response smart. *Procedia Engineering*, *107*, 312–318.

Zhang, J., Pourazarm, S., Cassandras, C. G., & Paschalidis, I. C. (2018). The price of anarchy in transportation networks: Data-driven evaluation and reduction strategies. *Proceedings of the IEEE*, *106*(4), 538–553.

Zhang, Y., Cassandras, C. G., Li, W., & Mosterman, P. J. (2018). A discrete-event and hybrid simulation framework based on simevents for intelligent transportation system analysis. In *14th IFAC Workshop on Discrete Event Systems (WODES)* (pp. 323–328).

Zhang, Y., Gill, C. D., & Lu, C. (2010). Configurable middleware for distributed real-time systems with aperiodic and periodic tasks. *IEEE Transactions on Parallel Distributed Systems*, *21*(3), 393–404.

3

NATO Use of Modeling and Simulation to Evolve Autonomous Systems

Jan Mazal[1], Agostino Bruzzone[2], Michele Turi[1], Marco Biagini[1], Fabio Corona[1], and Jason Jones[1]

[1] *NATO Modelling & Simulation Center of Excellence (M&S COE), Italy*
[2] *Genoa University, Genoa, Italy*

3.1 Introduction

The idea of a synthetic and autonomous creature with performance and "mission" effectiveness comparable to human performance has existed since ancient times, and lives on in several legends such as the Golem[1] (a story that was born in ancient Egypt, and later inspired national legends in various centuries). Militaries started to purse this issue more seriously about a century ago, but technology did not permit more than rudimentary attempts with tele-operated features. Nicola Tesla was one of the first who achieved serious results, introducing the first functional demonstrator[2] enabling the installation of "radio-control" in various systems. During the Second World War, we saw significant advances in routine process automation, a key enabler to entering the space domain. But even by the 1980s, state of the art technology was more about advanced automation, and far from simulating high-level reasoning necessary for significant breakthroughs in the Autonomous Systems area.

Several key milestones in this field came in the mid-1990s, where a demonstration of a machine performance in a "world-class" chess match introduced the technological potential of complex reasoning (Deep Blue vs Garry Kasparov, 1997).[3]

---

1 https://en.wikipedia.org/wiki/Golem
2 https://en.wikipedia.org/wiki/Radio_control
3 https://en.wikipedia.org/wiki/Deep_Blue_versus_Garry_Kasparov

Complexity Challenges in Cyber Physical Systems: Using Modeling and Simulation (M&S) to Support Intelligence, Adaptation and Autonomy, First Edition. Edited by Saurabh Mittal and Andreas Tolk.
© 2020 John Wiley & Sons, Inc. Published 2020 by John Wiley & Sons, Inc.

Even though this experiment brought more expectation then serious results, it spurred enthusiasm and investment in Artificial Intelligence (AI) research, the backbone of autonomous systems area.

Today's bleeding edge technologies are introducing new computational and operational capabilities in robotic systems, extending the value of the word "autonomy" to provide the crucial element. Even in the recent past there were many things that seemed promising but were unable to reach the desired capabilities, as seen with the original ROV (Remotely Operated Vehicle) AQM-34V, which was planned to be used in operations (Bruzzone et al. 2018a; Williams and Scharre 2014). It should be stated, that some achievements are comparable to results produced with today's technologies: in the nineties some tests demonstrated notable capabilities of RPV (Remotely Piloted Vehicle) interceptors compared to an F-14 piloted by experienced pilots (Larm 1996). These tests led to some serious doubts about the future of human pilots. These and other experiments have indicated that machines could outperform the human in the decision-making domain, and now this future is even closer.

Since the first flights of the forerunners of today's UAVs (Unmanned Aerial Vehicle), it took over a decade to bring about the first cases of operational use of lethal force by autonomous aircraft, and two decades were required to observe the first UAV-to-UAV refueling event between two Global Hawks (Quick 2012). This is a clear demonstration of the fact that these early attempts at Autonomous Systems evolved over decades into operational systems with somewhat limited experimentation of their potential on other uses (e.g. support, airborne early warning, etc.). Simulation allows the creation of virtual worlds where it is possible to investigate these elements and to evaluate alternatives and potential solutions (Bruzzone 2016; Bruzzone et al. 2016c).

An additional consideration in the operational use of autonomous systems are issues not only in terms of doctrine and technological solutions, but also with respect to the legal, ethical, and practical issues concerning lethal force (Bruzzone et al. 2018a). These considerations are summarized by the acronym LAWS (Lethal Autonomous Weapon Systems) and represent a very critical challenge to new generations of systems operating across the different domains: UAV, UGV, USV (Unmanned Surface Vehicle), UUV (Unmanned Underwater Vehicle), or AUV (Autonomous Underwater Vehicle).

It is also necessary to realize that rapid technological progress in selected areas, particularly in Defense,[4] initiated the rise of opposition communities, especially

---

4 Just today, there exists a set of available (combat) technology which significantly outperforms humans in close combat activities, which are mainly about "real-time issues" (meaning who shoots first, wins). Equally important are the economic aspects: like other industrial products, autonomous systems could theoretically be manufactured very quickly (within minutes or seconds). An incomparably lower price than investment in human soldiers which includes not only the training but also decades of "social" investment. This leads to the fact that the dominance of Autonomous Systems on the future battlefield is inevitable, and is just a question of time.

in the context of autonomous technology linked to a fully automated weapons system. This has been a topic of several important international events in recent years. One of the more publicly known activities pursuing a ban on autonomous weapons is the "Campaign to Stop Killer Robots," an international coalition of non-governmental organizations.

Recognizing a gap in this domain and the potential for M&S to assist NATO and national research and understanding of these important Autonomous Systems aspects, the NATO M&S COE and other NATO partners are utilizing M&S for Autonomous System research and experimentation. The NATO Modeling and Simulation Centre of Excellence (M&S COE) is working in education and information sharing, specifically with the MESAS – Modeling and Simulation for Autonomous Systems event and in experimentation with the R2CD2 – Research on Robotics Concept and Capability Development project, both of which will be described in this chapter. Similarly the University of Genoa, a partner to the NATO M&S COE, is studying autonomous systems using M&S environments to experiment with Dual-Use systems, systems that have both military and civilian applications.

3.2 Autonomous Systems in NATO

Since the 1990s the progress and perception of the Autonomous Systems field has differed significantly within each individual NATO nation. And at the NATO level, autonomous systems were not seriously incorporated into considerations and studies at the political and strategic level until relatively late, within the last decade. Two of NATO's important activities from these times are discussed in following sections.

3.2.1 NATO RTO/SAS-097: Robots Underpinning Future NATO Operations

The work of the NATO Research Task Group (RTO) for Systems Analysis and Studies (SAS)-097[5] took place from January 2012 to January 2015 and included members from 10 NATO countries.[6] In the final report of SAS-097, robots were defined as physically-embodied, artificially intelligent autonomous devices which can sense their environment and act within it to achieve some goals. Their study

---

5 Final report – https://www.sto.nato.int/publications/Pages/default.aspx, (search for SAS-097).
6 In 2014 the RTO (Research and Technology Organization) was renamed to the NATO STO (Science and Technology Organization).

highlights that robotics is the discipline aimed at creating autonomous machines and as a multidisciplinary area, robotics integrates the outcomes of several scientific and technological fields. The group focused its effort on the issue of deploying robots and integrating them in a military context and describing the gaps between robotic technology and its operational context.

The group concentrated on the most important concept opening new horizons for robots: autonomy. The report mentions that the autonomous machine sector has been boosted in the last decade by the attempt to create a self-driving car. The group noted that both key car manufacturers and informatics industrial giants like Apple and Google have been investing tremendous amount of money in autonomous cars, with universities following these challenges as well. The progress in this field is highly dynamic, changing substantially even during the brief life-time of SAS-097. In any case fundamental facts arising from this effort were found to be:

- The defense sector will benefit from the achievements in self-driving cars.
- Robots have become the reality in defense, on the ground, in the air, in water, and in space. Robots will likely constitute one of the key changing factors in future warfare. The related technology is present, especially safe and reliable remote-control.
- Robotics is well accommodated/established in research and technology areas.
- There exists a wide pool of robotic experts at these times, but the future will require much higher ratio of this specialty within a population.
- There is a great potential to attract, educate, and train a new generation of roboticists.
- Most industries already highly automate their manufacturing processes and the military is a step behind, underspending in robotic research/development, applications, and operational benefits.

Final results and the objectives/achievements of the study were:

- *Analyzed the gap between operational requirements and technological possibilities (relationship to the NATO Long-Term Capability Requirement and Local Target Areas).* The group conducted a trend analysis in Autonomous Systems (AxS) in the areas of Control, Sensors and Platform, completing an analysis of Operational Requirements, analyzing the EU Perspective and completing analysis of research into Human–Robot Cooperation.
- *Provided experimentation support for robotics concept development and testing.* The group conducted joint experiments between the Czech University of Defense and United States Army Tank Automotive Research, Development and Engineering Center (TARDEC), worked on multipurpose platform development – Project TAROS, and supported real mission deployments and joint exercises with

National entities and engaged with the academic community by participating in numerous academic conferences and publishing journal articles.

- *Created and supervised bi-directional working links to the European Commission R&D activities in dual-use of robotics. Leading member of the SAS-097,* Czech Technical University (CTU) becomes a member of euRobotics non-profit working group and participated in the euRobotics forum in Rovereto, Italy, in 2014 and Vienna, Austria, in 2015. Contributed to the Multi-Annual Roadmap (MAR) document and explored potential of dual-use of robotics in the fields of search and rescue robotics, human–robot interaction and robotic manipulation with soft materials.
- *Opened possibilities for new robotics research motivated by military needs and funded by third parties.* Established in 2012 the Center of Advanced Field Robotics (CAFR): Institution founded by four Czech universities and one industrial partner. The aim of the CAFR is to bring together organizations in the Czech Republic that are engaged in research and development in the field of advanced robotics and autonomous systems. CAFR is focused on applied research, primarily in the security, industry, and military domains.

3.2.2   MCDC: Autonomous Systems (2013–2014)

MCDC[7] campaigns have a number of focus areas designed and coordinated under one common theme. The theme for MCDC (2013–2014) was "Combined Operational Access." One of the seven Focus Areas was dedicated to studying human, operational, legal, and technological implications of introducing autonomy to military systems. This Autonomous Systems Focus Area (abbreviated as AxS FA, where the "x" is treated as a variable for "air," "ground," "surface," etc.) was proposed and led by NATO Allied Command Transformation (ACT) Headquarters, Norfolk, VA.

NATO ACT's decision to lead the MCDC AxS Focus Area came after identifying that while NATO and some Allies were already successfully testing prototype systems with autonomous capabilities, they had no formal guidance or understanding of implications of this new technology. And even native English speaking scientists were confusing the terms *autonomous, autonomic, autonomical, automated, automatic,* or *unmanned* and *remotely controlled.* This was further complicated by adding the prefix "semi-."

---

7 The Multinational Capability Development Campaign (MCDC) is a series of two-year-long campaigns coordinated by the United States Joint Staff J7. The program is a follow-on to the Multinational Experiment (MNE) series initiated by the former United States Joint Forces Command in 2001. MCDC Campaigns are designed to enhance the force's operational effectiveness in joint, interagency, multinational, and coalition operations by undertaking projects dedicated to multinational development of new military capabilities.

A very important aspect of AxS from NATO's perspective is the fact that some autonomous platforms may soon have capabilities (agility, stealth, endurance, precision, shared awareness, and swarm operations) which can challenge existing military equipment and capabilities. Therefore, NATO must urgently improve its knowledge of autonomous systems and begin preparing to face the challenge of countering such systems in near future.

The intention of the MCDC AxS FA working group was to improve the awareness and understanding of military implications of this rapidly advancing new technology and establish common understanding and foundation for the development of AxS capabilities and doctrine. The final product of this effort was a document presented to MCDC and NATO Nations' senior leadership to help capability planners, commanders, industry or academia understand why, where, when, and how the military could, and may have to, effectively employ autonomy in systems during future armed conflicts, humanitarian assistance, and disaster relief operations.

Finally the main achievements and "output" of MCDC AxS FA were two documents, one a policy guidance and the other a book on issues for defense policy:

- *Policy Guidance* – Autonomy in Defense Systems/Role of Autonomous Systems in Gaining Operational Access, a study for policy makers consolidated from five studies:
    - Definitional study (led by HQ ACT[8]) focusing on the meaning of autonomy;
    - Legal study (led by Switzerland), which examined legal issues mainly concerning weapon systems with autonomous capability;
    - Human factors and ethics study (led by the United States and HQ ACT), which explored future ethical, organizational, and psychological implications;
    - Military operations study (led by HQ ACT), describing operational benefits and challenges;
    - Technology study (led by the Czech Republic), which summarized key technological developments and challenges.

    These complete study findings and records from the various workshops and seminars are published separately in an MCDC Autonomous Systems Proceedings report, available from the MCDC Secretariat. As was mentioned before, this policy guidance document of about 30 pages is focused on top government and military authorities, to raise awareness of the importance of autonomy in future defense capabilities, and its potential quick evolvement by adversaries.
- *Autonomous Systems – Issues for Defence Policymakers* is a book issued as a second volume of Allied Command Transformation's *Innovation in Capability Development* series. It extends a previous document and within 4 topical parts

---

8 NATO Headquarters Allied Commander Transformation.

and 13 chapters, discusses in depth terminology, autonomy levels, ethical, legal, policy, operational and technical aspects accompanying Autonomous Systems deployed in operational environment.

From the fundamental point of view, MCDC AxS came up with several important statements, which could shape the evolution of autonomous Systems within NATO:

- *From the legal perspective:* International law does not prohibit or restrict the delegation of military functions to autonomous systems, provided that such systems are capable of being used in full compliance with applicable international law. The principles of international law governing State responsibility and individual criminal responsibility appear to adequately regulate the responsibility and consequences of harmful acts resulting from the use of autonomous systems.
- *From the ethical perspective:* Generally, open and public discourse should be stimulated about the ethical aspects of the development, proliferation, and use of autonomous technology. There should exist a transparency about the ethical benefits and concerns associated with autonomous technology. There should exist explanation how, from an ethical perspective, autonomous technology is different than other technological advances. This includes considering the ethical permissibility of autonomous systems targeting humans and defining levels of responsibility for the intended and unintended consequences of tasks performed by autonomous systems; however, it should not neglect nonlethal tasks performed autonomously.
- *From the operational perspective:* There must be ensured that any claimed benefits of increasing autonomy is accompanied with appropriate analysis of trade-offs and risks. Discussion should emphasize autonomy as a capability of systems in general, rather than a feature of predominately unmanned platforms. There is insufficient evidence to claim cost-saving as a generic benefit; each system must be evaluated in terms of whole life-cycle costs and compared against multiple baselines.

NATO's latest effort in this field, the NATO/ACT Autonomy Program, was started early in 2018 and indicates NATO's perception of future warfare challenges. It is necessary to understand that for NATO, Ethical, Legal, Political, and other considerations accompanying Autonomous Systems seriously complicates its operational deployment. As a secondary effect, this restricts investment in that technology, slowing down research, development, and experimentation.

3.2.3 NATO M&S COE Efforts in Autonomous Systems and the Cyber Domain

A reliance of Autonomous Systems on cyberspace creates a domain overlap, providing a parallel focus to both areas for the NATO M&S COE. In 2016 cyberspace was recognized by NATO as a separate operational domain.[9] Cyber training and exercises are growing fast, and NATO's cyber community of interest is growing too, engaging more and more with other affected communities. This section illustrates the exploration of many possible aspects related to the innovation and integration of the modeling and simulation discipline with Cyber Physical Systems, in particular robotics systems like UAxS (Unmanned Autonomous Multidomain System). Their continuous evolution and use in tactical scenarios makes them vulnerable to possible cyber threats, particularly through the communication infrastructure that these systems need for information exchange. A holistic approach of the infrastructures to be protected is necessary to identify software and hardware components that compose UAxS (Madan et al. 2016). Furthermore, cyber risk analysis should be used to prioritize identified threats and Cyber Operations should be modeled and simulated to provide support to risk analysis activities.

One of the most significant of the NATO M&S COE's research and development efforts in Autonomous Systems is the R2CD2[10] project. The objective was to create a simulation environment dedicated to the implementation of the UAxS communication infrastructure and is capable of implementing cyber-associated countermeasures. That environment, called the UAxS Cyberspace Arena (UCA) (Biagini and Corona 2016), is based on an integrated simulation environment that is able to provide capabilities to support the evaluation and experimentation of UAxS tactical communication networks. The R2CD2 project provides a first embryonic implementation of the UCA architecture, an architecture and simulation environment that can be reused for developing related Cyber Services (Biagini et al. 2017).

The R2CD2 project supports the development of Innovative Unmanned Autonomous Multi-domain Systems capabilities, in particular it has been used for:

- NATO's Allied Command Transformation (ACT) task on autonomy, providing a platform for proof-of-concept experiments for multiple ACT projects and reuse of concepts developed by ACT in the Countering Autonomous System project.
- NATO Science and Technology Organization (STO) research group "Operationalization of Standardized C2-Simulation Interoperability," studying the standardization of the C2SIM Interoperability language and the development and implementation of a C2SIM extension for UAxS.
- NATO experimentation on interoperability, using the R2CD2 platform for tests on C2SIM interoperability involving simulated UAxS and Command and Control systems.

The R2CD2 project answers the need to develop requirements for new military capabilities that are able to both employ and counter robotic systems

---

9 https://www.nato.int/cps/en/natohq/topics_78170.htm
10 Research on Robotics Concept and Capability Development (R2CD2).

(Biagini et al. 2017). This project leverages M&S to conduct experimentation on new systems, weapons, doctrine, training and logistics, in a cost-efficient way. The NATO M&S COE concentrates on five primary areas of investigation, taking into consideration an urban operational environment in the near, mid, and future terms:

- the interaction between human troops and Autonomous Systems;
- the military Command and Control (C2) of robotic units;
- technical and tactical procedures for UAxS;
- development of functional requirements of new robotic platforms;
- countering UAxS.

For such areas, an M&S-based, scalable and modular platform was built on an open standard architecture by the NATO M&S COE in collaboration with Industry and Academia. This platform was based on selected constructive simulators to execute scenarios on the military employment of autonomous systems in order to be an innovative tool for proof-of-concept activities on robotic capabilities supporting NATO. M&S technology has allowed the reuse of models, prototypes, systems, and studies, developed for different projects and saving precious resources. Currently, the R2CD2 project platform is being further expanded to include new UAxS models, behaviors, and countermeasures in order to potentially address and solve issues regarding:

- Detection and identification of enemy forces utilizing UAxS and sensors;
- Situational Awareness augmented by Artificial Intelligence (AI) Decision Support;
- Defense against UAxS applying new capabilities and related TTPs (Tactics, Techniques and Procedures).

The M&S platform developed within the R2CD2 project makes use of the Level of Autonomy (LOA) work developed during NATO ACT's autonomous systems countermeasures project (C-UAxS) for behavior and human–robot interactions. The terrain for the project is a mega-city model built by the NATO M&S COE for use in NATO ACT's Urbanization Project (UP). And a new C2SIM interoperability language standard was included in the platform architecture to provide interoperability between C2 systems and simulated robotic entities (Biagini et al. 2018).

The C2SIM extension to UAxS needs to be continuously enriched and aligned to the last developments of the SISO[11] C2SIM Product Development Group. In this way, the R2CD2 platform will be ready to participate in further experimentation on C2SIM usability in a distributed simulation environment and contribute to C2SIM standardization in the area of robotics assets management. The objective

---

11 Simulation Interoperability Standards Organization.

state of the R2CD2 platform is to provide a core database of air and ground unmanned autonomous platforms that are able to perform missions with different levels of autonomy, according to coded TTPs (Biagini et al. 2017).

Depending on their capabilities, an UAxS could send information from the battlefield based on sensors it has that contribute to an overall Common Operational Picture (COP). A set of kinetic and non-kinetic countermeasures are also planned to demonstrate the potential for the platform as a tool for developing requirements for new counter-UAxS capabilities. The platform will also provide a decision support capability for UAxS shared between a separated module and coded Artificial Intelligence (AI) algorithms for the UAxS behaviors in each simulator.

Another effort is to deploy the UAxS simulation environment and its associated services in a cloud environment, following NATO's Modelling and Simulation as a Service (MSaaS) paradigm. The NATO M&S COE used the R2CD2 prototype to contribute to the development of the C2SIM interoperability language, for C2SIM standardization and the proposal of a NATO STANAG (Standardization Agreement). In particular, this will provide the opportunity to test the C2SIM to UAxS extension for Command and Control of robotic units. And ultimately further work to integrate real military C2 systems using the C2SIM interface.

The NATO M&S COE is planning further investigations into the Cyber-Physical Systems domain, to include cyber countermeasures to autonomous systems (Biagini and Corona 2018). An additional communications and networks simulation is planned to be included in the architecture of the R2CD2 project for traffic generation and experimentation on communication protocols, procedures, cyber effects, and countermeasures in the electromagnetic spectrum.

3.3 Modeling and Simulation for Autonomous Systems Conference (MESAS)

MESAS is a scientific conference that brings together government, industry, and academic experts from the M&S and Autonomous Systems fields. Initiated by NATO, the conference program, execution, and published proceedings are managed by the NATO M&S COE. At the beginning of 2012, NATO Allied Command Transformation invited all associated Centres of Excellence (organizations of subject matter experts) to identify their possible expertise involving the integration of Autonomous Systems in operational activities. In response to this request, the NATO M&S COE became actively involved in this effort through the study group mentioned earlier in this chapter, SAS 097: "Robotics Underpinning Future NATO Operations."

This work, and subsequent participation in the Multinational Capability Development Campaign mentioned in Section 3.2.2, led to the idea of creating a scientific workshop dedicated to autonomous systems. The idea was further extended by electing to focus on modeling and simulation for autonomous systems. This led to the launch of the MESAS (Modeling and Simulation for Autonomous Systems) Conference under the leadership of the NATO Modelling and Simulation Centre of Excellence.

The idea was to bring together communities involving modeling and simulation, mathematics, AI, robotics, mechanical engineering, electronics and others, with the aim to collect new ideas for concept development and experimentation in this domain. Since its inception, the event brings together recognized experts from these different scientific areas from around the world. MESAS enables an effective information exchange and cooperation between industry, academic institutions, and military.

This event promotes the potential gaps in military areas to the scientific communities, and quickly brings academia's focus to military applications associated with Autonomous Systems in the operational environment. One of the key benefits of the MESAS conference organization for NATO and the M&S COE is the ability to keep abreast of the state of the art in the Autonomous Systems area, without additional or duplicate analysis efforts.

The MESAS conferences have seen the presentation and publishing of several thousands of pages of scientific and engineering work related to M&S and the AS field. Each edition of MESAS is a peer-reviewed, scientific conference, with proceedings indexed and available in scientific databases. MESAS conferences are supported by many sponsors from the scientific and industrial associations, including: IEEE, MIMOS, Afcea, IEEE-RAS, Simulation Team, ONRG, Finmeccanica (now Leonardo), Selex, Czech Technical University in Prague, University of Palermo, Future Forces Forum, and many others.

This section highlights some of the key outcomes from past MESAS.

3.3.1 MESAS 2014

The first MESAS conference took place in Rome, Italy as a MCDC AxS supporting activity. This first event saw great presence from the military, industry, and academic communities. This first event included a technical committee of 17 international experts, and the conference proceedings (MESAS 2014) include 32 selected papers.

The conference was segmented in eight topical areas, presented concurrently in several conference streams, particularly:

- Unmanned Aerial Vehicle: this section included computer vision approaches in UAV navigation and simulation approaches and services for UAV mission planning.
- Distributed Simulation: dedicated to HLA standards, interoperability and experimentation issues, and also the cyber domain.
- Robotic Systems: M&S technology and approaches in robot simulation and autonomous function validation, motion planning and architecture development.

- Military Applications: several topics within various operational domains, including: autonomous swarms in maritime operations, terrain impact on AS ground maneuver and operational capability development issues.
- M&S Validation: focused on simulation tools for algorithm development, validation, and sensors and specific systems modeling.
- Human–Machine Communication: dedicated to issues of machine learning and human machine interface in the context of autonomous systems.
- Simulation and Algorithms: presented enhancement of M&S tools and innovations in Simultaneous Localization and Mapping (SLAM) algorithms and "traveling salesman" problem solutions.

As stated, one of the main achievements of MESAS 2014 was the successful start of an event that brings together the military, academic, and industrial communities to exchange information and discuss important topics in the Autonomous Systems field. From the scientific perspective, MESAS 2014 brought many results from applied research area and operational topics.

This first edition of MESAS was strongly supported by many experts across international and Italian organizations (both public institutions and private companies). Participation was very good in terms of nations, with attendees from Canada, Colombia, Czech Republic, Germany, Italy, Spain, and the United Kingdom (MESAS 2014). Among the topic presented were works dealing with surveillance within a smart city and threats related to potential attacks to cyber physical systems. Several papers addressed issues related to modular robotics, covering a wide spectrum of alternatives from solid engineering solutions to snake configurations. The proposed papers addressed multi-domain cases covering air, ground, and marine environments, and demonstrations of innovative systems were done within a small, but very interactive exhibition area. Along with operational systems, the show presented advanced solutions, including an interactive, virtual simulation of off-shore platform protection carried out by autonomous and traditional systems and collaborative examples using small UGVs.

The crucial topics of man–machine communications, the use of High Level Architecture standards for distributed M&S, and innovative optimization techniques for autonomous system path planning based on AI solutions and swarm intelligence (Stodola et al. 2014). Interesting applications of collaborative use of autonomous systems were proposed in reference to surveillance as well as odor plume tracking.

3.3.2 MESAS 2015

The MESAS Conference 2015 was held in Prague, in conjunction with ITEC (Industry, Training, and Education Conference), Europe's premiere M&S conference. The aim of this second conference was to explore the role of modeling and simulation in the development of systems with autonomous capabilities and in its operationalization to support multi-national interoperability. The plenary session and way-ahead discussion panel included six invited experts and key note speakers. The MESAS 15 technical committee consisted of 14 international distinguished experts and the conference proceedings (MESAS 2015) include 18 selected papers.

The MESAS 2015 proceedings content can be characterized in three topical segments:

- State of the Art and Future of AS (five papers): M&S in context of Ontology for AxS development, Improving AxS Interoperability and UAS Integration.
- Experimental M&S Frameworks for AS (three papers): Interoperability Issues for Multi-robot Systems, Applications and Experiments with AxS Platforms and Data Analysis and Reporting.
- Methods and Algorithms for Autonomous Systems (10 papers): A group of papers discussing a range of topics, including: path planning for various platforms, environments and conditions; next stages of virtual or augmented reality solutions for human–machine interface; further simulation approaches and framework for robotic systems development; performance testing; calibration and validation.

MESAS 2015 continued the previous conference effort and also benefited from being held in conjunction with ITEC. MESAS was also growing in terms of international cooperation in its organization and the topics of high relevance for the simulation community regarding autonomous systems. This edition of the conference addressed the crucial aspects of interoperability among systems and modern solutions based on autonomous systems. Several participants addressed algorithms for improving capabilities, autonomy in particular, while providing discussion on testing systems in virtual frameworks and integrating real components (e.g. software in the loop).

3.3.3 MESAS 2016

MESAS Conference in 2016 returned to Rome and focused on the following topics:

- Human–Machine Integration, Interaction, and Interfaces: Challenges in Human–Robot Teaming, Ethical and Philosophical Aspects, Human–Machine Interface, Machine Learning and Modeling of Visual Communication.
- Autonomous Systems and MS Frameworks and Architectures: Security of Autonomous Systems, M&S for Counter UxS, Interoperability, M&S of Operational Environment, M&S in Cyber Domain

- Autonomous Systems Principles and Algorithms: Body/Physics Modeling, Image Recognition, Multi-agent Surveillance, Data and Algorithms, Decision-Making, Path Planning, and Simultaneous Localization and Mapping (SLAM).
- Unmanned Aerial Vehicles and Remotely Piloted Aircraft Systems: Disasters/ Emergency Management, Communications, Simulation Environment, M&S performance measurement.
- Modeling and Simulation Application: Route Planning, Robot Navigation and Mapping, Object Manipulation and M&S for Decision Support.

The MESAS 2016 technical committee consisted of 27 international experts and the published conference proceedings (MESAS 2016) included 32 selected papers.

MESAS 2016 copied the success of previous sessions and took part in parallel with the NATO Modelling and Simulation Group Business Meeting, a biannual gathering of NATO and National M&S experts and policy makers. From a scientific point of view, the conference also discussed the philosophical and ethical considerations about operational AxS performance.

Curtis Blais discussed the mission performance of fully Autonomous Systems, warning about early optimism while pointing to the challenges in developing improved models of human systems, robotic systems, and human–robot teams in combat simulations. He presented examples posed in the context of the Combined Arms Analysis Tool for the Twenty-First Century (COMBATXXI), saying: "We have come to a point in combat modeling where we can no longer be satisfied with simplistic models of human performance. We must be able to investigate the complex interplay that will occur among humans and unmanned systems."

Further ethical and philosophical considerations came from the Mark Coeckelbergh and Michael Funk, dealing with human–machine cooperation problems such as security of data flows, speed/distance, and knowledge in cooperative human–autonomous system configurations. They identified ethical problems and conflicts with regard to security, speed, and cooperation, and also pointed to cultural differences with regard to knowledge in ethics and related military technologies. It is important to take these problems into account at an early stage of system development.

Several papers were dedicated to human–machine interaction and application of M&S in that process. Miroslav Kulich and his team dedicated effort to the interesting problem of machines learning human behavior in order to maximize exploration performance. University of Bundeswehr demonstrated a relatively innovative a concept of HMI with promising applications results using a computer vision approach to communicating with UAVs using human gestures.

MESAS 2016 papers dealt with architectural concepts of AxS from several perspectives: security components, software and hardware design, countering AxS, control systems, etc., as well as topics on various M&S tools, standardization and interoperability issues.

There was a strong presentation about algorithms, which introduced a chain of improvements on the various approaches used in autonomous navigation and object recognition. Martin Dorfler with Libor Preucil presented on robust recognition using combined image descriptors performing on terrain images. Gonzalo Perez-Paina and his team introduced three-dimensional UAV pose estimation enriched by computer vision processing an image stream from an on-board camera.

Additionally articles were presented dealing with path planning problems, including Mateo Ragaglia whose team presented a solution based on rapidly exploring random tree (RRT) for multiple cooperative agents. Basak Sakcak with his colleague also used RRT in human-like path planning for specific purposes. Jan Mazal and his colleagues presented an algorithm for 3D path planning in complex operational conditions, while Ove Kreison and Toomas Ruuben also dealt with path planning in operational conditions, demonstrating how a sniper threat could be factored into maneuver planning optimization.

The proceedings also included papers dealing with application issues demonstrating implementation of AS in a M&S environment in selected scenarios (Agostino Bruzzone), Tactical Decision Support (P. Stodola), Satellite Communication and Navigation (Giancarlo Cosenza and colleagues) and UAV as a Service, dealing with the cloud framework for operational applications of coordinated commercial UAVs.

3.3.4 MESAS 2017

The MESAS Conference 2017 was again held in Rome, with a focus on:

- M&S of Intelligent Systems – AI, R&D, and Applications
- Autonomous Systems in Context of Future Warfare and Security
- Future Challenges and Opportunities of Advanced M&S Technology

The MESAS 2017 technical committee consisted of over 50 international experts and the published conference proceedings (MESAS 2017) included 32 selected papers.

The event brought many important topics and problem solutions, specifically in the area of 3D exploration and self-localization. Several approaches and experimental results were presented, in particular the presentation Autonomous 3D Exploration of Large Areas, by Anna Mannucci and her team, which demonstrated an innovative, coordinated approach to 3D exploration of large uncluttered areas by a team of UAVs with constrained payloads. Their work on a 3D exploration strategy combining local and global information in 4D showed that effective coordination is critical for exploration in a feasible amount of time with limits on computational power. Also presented were several improvements in navigation and localization systems, like a robotic navigation system for an unknown environment by Simone

Nardi, who presented a novel platform for Autonomous navigation using a system designed as an open framework with sets of interconnected modules that are applicable to different robotic platforms.

Other interesting achievements were presented by Viktor Walter and his team in the area of onboard self-localization processing and analyzing optical flow from a bottom camera, and an integrated approach to autonomous environment modeling written by Miroslav Kulich. The underwater dimension was coved by Filippo Arrichiello and his team in an article focused on Dynamic Modelling of a Streamer of Hydrophones Towed with an AUV.

As in the past, in MESAS 2017 there were several significant efforts on path optimization problems. Path planning for a formation of mobile robots by Estefanía Pereyra, was based on the standard Dijkstra's Algorithm looking for the optimal path for a formation of robots, while allowing the possibility of split and merge. The algorithm explores a graphical representation of the environment, computing for each node the cost of moving a number of robots and their corresponding paths. A paper written by Petr Stodola – Route Optimization for Cooperative Aerial Reconnaissance, describes a high-level route planner for a UAV swarm in order to carry out optimal and feasible reconnaissance operation. Basak Sakcak, Luca Bascetta, and Gianni Ferretti presented a paper titled "An Exact Optimal Kinodynamic Planner Based on Homotopy Class Constraints" which proposes a kinodynamic planning algorithm for robotic vehicles with bounds on actuation. The described approach is based on identification of homotopy classes to decompose the global obstacle avoidance problem into several simpler sub-problems. And finally a very interesting paper and presentation by Davide Vignotto, Federico Morelli, Daniele Fontanelli, focusing on the Modelling of a Group of Social Agents Monitored by UAVs. In this paper, a UAV team has to control the trajectories of a group of agents having an internal behavioral logic, such as a group of human beings moving in a shared environment or the movements of a flock of animals. Their paper proposes an effective model for the group of agents, inspired by the Social Force Model and a distributed estimation and control algorithm for controlling the UAVs.

Additional focus included M&S for Performance Evaluation, Reliability, Maintenance and Data Fusion, which found interesting results in standardization, data fusion, failure detection, and System of Systems framework development. The presentation of A Simulation Program for Performance Prediction of an AUV with Different Propulsion System Configurations captured a lot of attention, dealing with a series of primary maneuver standards for underwater vehicles and examining different propulsion system configurations. The remainder of MESAS 2017 was largely dedicated to operational M&S applications, in robot control and Scenario Modelling, with a separate topic dedicated to Autonomous Systems in operations and one presentation covering the somewhat underestimated Education and Training area in the context of Autonomous Systems application and cooperation.

MESAS 2017 clearly indicated that the application portfolio of Autonomous Systems is growing and thus the need for proper implementation of M&S tools, approaches, standards, and methodology. MESAS 2017 is incrementally enriching the M&S scientific knowledge portfolio and performs well as an integration platform for "multi-domain" communities around the world.

3.4 Autonomous Systems: Future Challenges and Opportunities

Nowadays it is possible to use simulation and AI to evaluate advanced and collaborative use of autonomous systems over complex scenarios (Bruzzone et al. 2016c). And at the same time new challenges are emerging due to the potential high density of autonomous systems in the future and their vulnerability to multiple threats (Haas and Fischer 2017). These aspects do not apply to only military systems, but also with civilian drones, which have great developmental potential for institutions, businesses, and private operators (Luppicini & So 2016; Yayla and Speckhard 2017; Bruzzone et al. 2018a; Zwickle et al. 2018). In such a future state, autonomous systems should be highly reliable and their use by NATO should include the capability to operate in urban and other complex environments. To achieve such results, it necessary to develop experience and extend their operational range in a way that traditional military systems have yet to achieve (Box 1979). One solution to this challenge is to derive benefits from dual-use autonomous systems and to train and exercise in this framework with modeling and simulation. Recently NATO developed dual-use capabilities of robotic systems that are theoretically capable of supporting national and international agencies (e.g. Civil Protection, Humanitarian Organizations, Fire Fighting Services, Law Enforcement Agencies, Coast Guards, etc.). In future the situations will become even more intense, with a complicated structure involving many autonomous systems including those belonging to private individuals or organizations. The use of Modeling & Simulation will be crucial to reinforce current capabilities and it will be critical to further develop them to face future challenges and threats.

It is important to also mention a potential threat that risks being underestimated: the unexpected reactions and possibly unpleasant emergent behaviors resulting from autonomous systems controlled by AI in complex mission environments (Bruzzone et al. 2018a). Indeed it is quite surprising to realize that while Isaac Asimov half a century ago identified the risk of delegating AI to control sophisticated robotic systems as potentially dangerous to humans, in the last two decades these aspects have been almost totally neglected in the development of lethal systems with harsh limitations to operating over high density civilian areas (Asimov 1950; Cellan-Jones 2014; Stewart 2016). From this point of view, in the future the necessary and "natural" development of much more complex collaborative tasks among different autonomous systems will move humans out of the loop, or better, over the loop (Magrassi 2013).

So it is evident that increasing autonomy in robotic systems could result into a large variety of possible behavior, and one not always corresponding positively for safety and the good of our Nations. In fact it is evident that AI perceives reality in a different way than humans, and as such they could react to it in unpredictable or aggressive ways. We could even see a future with AI evolving into *good shepherds,* driving decisions based on a comprehensive view that surpasses individual capacity and using autonomous cyber-physical systems as their "fingers"; one could consider also the possibility of developing *AI driven conspiracies* as the evolution of some kind of super Artificial Intelligence *"equipped"* with friendly or hostile behavior. Until a few years ago this was mostly sci-fi, but we should consider that autonomy could produce dangers and threats on micro levels, or other negative events simply by confusing priorities, improper sensor processing or failures to properly finalize scenario awareness (Levin 2018). Scientific literature considers some of these elements, and for sure new predictive and reactive capabilities should be addressed in some way without losing a pragmatic point of view (Duderstadt 2005; Blackmore 2006; Barrat 2013).

Based on these considerations it is evident that it will be more and more necessary to test extensively over simulation and in complex scenarios these new solutions, often involving multi-elements interoperating with many different systems. Respecting these challenges, we observed recently multiple attempts to define rules that should be adopted with respect to robot cognitive processes, such as in the case of the Engineering and Physical Research Council and the Arts and Humanities Research Council of Great Britain (epsrc.ukri.org) and in scientific papers (Vincent 2016). Obviously the key technology to evaluate these risks and make a quantitative assessment of related threats should be based on Modeling and Simulation (Bruzzone 2018a).

3.4.1   Dual-Use as Key for Reliability and Sustainability

As anticipated the sustainability of autonomous systems with respect to future mission environments and threats requires developing very reliable and flexible solutions. Technologies are currently evolving in this direction providing great support and leading to a downsizing of the systems and cost reductions. It is evident that this is strongly related to newly available capabilities that allow extending the use of autonomous systems over a wide spectrum of applications. In this sense, military use is evolving relatively slowly in respect to civil applications and solutions. For these reasons it is reasonable to consider the dual-use modes of using autonomous systems in both military and civilian aspects as key for success (Sandvik and Lohne 2014; Bruzzone and Massei 2017).

In this sense, just looking at the reliability of civilian UAV systems shows an increase in low cost systems, particularly when compared to dedicated military solutions, especially in the mini and micro classes. This could correspond to soon having a civilian device that could have significant potential for being extended to military and homeland security sectors, bringing competition to respected, dedicated systems. In addition to small devices, there are also systems belonging to the small- and medium-size classes evolving, and we expect to soon observe the proliferation of new systems covering different applications areas (e.g. precision agriculture, media coverage, logistics, inspections, and controls).

These phenomena are evolving in all domains, including marine and land, so due to the high density and complexity of future environments, we expect to observe that new generations of high performance solutions based on bleeding edge technologies will able to largely surpass the current capabilities of existing, dedicated military systems. Based on these considerations, it is evident that future NATO Autonomous Systems would benefit from incorporating the new civilian emerging capabilities and developments for use in challenging civil operations. In such cases, the use of dual-use autonomous systems would provide more sustainable development and maintenance of systems that could otherwise prove too expensive or too limited in future environments. It is vital to note that operations in civilian areas will require very high reliability to avoid accidents and eliminate risks.

Obviously dual-use represents a major opportunity and advantage for NATO, allowing improvements to reliability and interoperability for intensive activities across multiple domains. These aspects are even more crucial considering the potential vulnerability of autonomous systems to new threats, in particular cyberspace (Javaid et al. 2012; Bruzzone et al. 2013; Hartmann and Steup 2013; Kwon et al. 2013; Hartmann and Giles 2016). The future is expected to propose scenarios where autonomous and intelligent systems will be used over all domains in strong interaction with other traditional assets in support of operations. In this context, the use of modeling and simulation plays a vital role in the design, development, testing and evaluation of future systems. Simulation supports the evaluation of different alternatives and supports choosing the most effective solutions for dealing with problems that often largely surpass the limits of traditional military operations (Bruzzone 2018b; Bruzzone and Di Bella 2018): most current missions are about Search & Rescue, Intelligence, Reconnaissance, Interdiction almost always carried out of domestic urbanized areas.

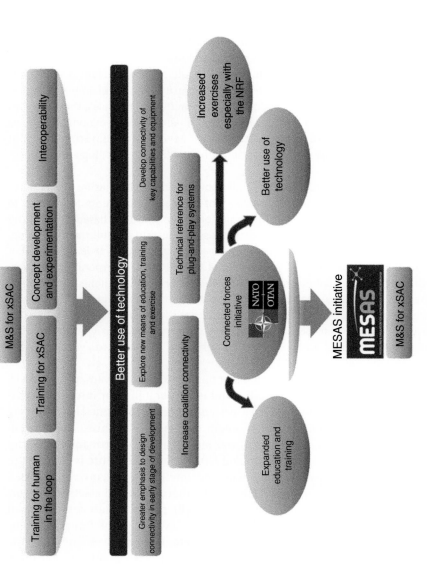

Figure 3.1 Initial MESAS Objective starting in 2014, Systems with Autonomous Capabilities are abbreviated with the acronym xSAC, where the x indicates the operational domain: ground, maritime, air, space, and cyber.

3.4.2 Dual-Use Capabilities in New Scenarios

In fact several challenges for NATO are dealing with new emerging scenarios; among these it is well known that Hybrid Scenarios as well as Critical Protection of Infrastructure are cases where the use of autonomous systems represents crucial elements. In the future it will be necessary to develop capabilities for large and complex cases where autonomous systems could be active on both sides and overlap with intensive use of civil systems over urban areas. These mission environments will require highly reliable and efficient solutions, characterized by flexibility and versatility. Broad operational application of dual-use systems requires the development of new and emerging capabilities, and extensive evaluation in high-density, urban areas where suspect players as well defensive and offensive assets are operating across the military and social domains, including cyber and social media (Bruzzone et al. 2016a, 2017). From this point of view an interesting example has been demonstrated in NATO by T-REX (Threat network simulation for REactive eXperience). In this case, a scenario was proposed for a small desert region with a few towns and a city where energy, water, communication, and oil represent major strategic assets and related critical infrastructures that are fundamental to sustain population. T-REX allowed the evaluation of different kinds of threats and potential attacks combining media, cyber and real assets, including autonomous systems, used to compromise Critical Infrastructures. In the case of Hybrid Warfare, these operations are often carried out cloaked, within the regular city life and operations. The autonomous systems in this example are crucial to protect, prevent, and mitigate effects from attacks on critical resources. In fact, T-REX proposes exactly these kinds of issues by combining the use of multi domain autonomous systems on both sides as well as cyber and hybrid warfare targeting of critical infrastructure. For instance, the simulation allows setting different layer parameters, with the cyber layer proposed in the Figure 3.2, to evaluate the vulnerability with respect to combined attacks that involve in this case a swarm of micro UAV, while medium size UAV and UGV are used to support Joint Intelligence, Surveillance, and Reconnaissance (JISR) and identify physical threats as well as those in cyber space (Bruzzone et al. 2016d).

3.4.3 Autonomous Systems in Emergency Management Enabling New Capabilities for NATO

Another crucial field to develop innovative integrated solutions to augment NATO capabilities is the complex scenario of disaster relief and emergency management, an area that has been extensively addressed in NATO M&S COE's MESAS conference (Bruzzone et al. 2016b, 2018a).

Figure 3.2 T-Rex Simulator reproducing combined Cyber and Drone Attack to Critical Infrastructure.

In fact, operating in an emergency management situation requires capabilities similar to operational mission environments, and there is equally an opportunity for simulation to support the testing of new technologies, policies, doctrines, and capabilities. In this framework, the use of interoperable models is obviously crucial (Bruzzone and Massei 2017). A good example in this sense is proposed in the Figure 3.3 where an advanced simulator, Immersive Disaster Relief and Autonomous System Simulation (IDRASS), is operating within an immersive, interactive, interoperable environment called SPIDER (Simulation Practical Immersive Dynamic Environment for Reengineering) at the Simulation Team Labs. Simulation Team scientists have extensive experience using autonomous systems and innovative simulations for disaster relief operations and emergency management (Bruzzone et al. 1996, 2016b).

This study reviews and evaluates operations involving different kinds of UAVs and UGV (Unmanned Ground Vehicles) within a Chemical, Biological, Radiological, and Nuclear (CBRN) scenario. In this way different hypothesis about policies and the rules of engagement (ROE) are evaluated with respect to MoM (Measures of Merits) and risks (Bruzzone et al. 2016a).

Simulation Team developed and used advanced simulation standards to address specific simulation solutions for different applications, including: training; policy definition and capability assessment for accidents related to military operations; CBRN; and Homeland Security and Industrial Scenarios.

The Immersive Disaster Relief and Autonomous System Simulator, mentioned earlier in this section, was presented at MESAS in 2016, and includes extensive use of autonomous systems for detection, assessment, cordoning, triage, and recovery in different kinds of CBRN Scenarios. The system was applied to Industrial, Urban, and Nuclear cases, and allows testing the benefits of operational roles for autonomous

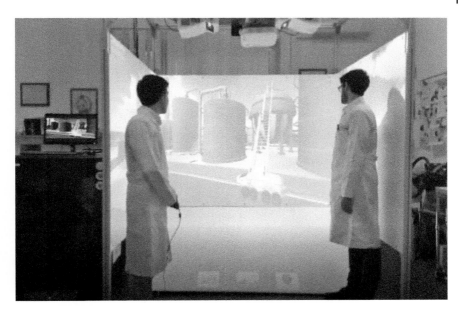

Figure 3.3 IDRASS Simulation Solution within SPIDER Virtual Environment applied to test collaborative procedures among autonomous systems over different domains in case of CBRN.

systems in outdoor and indoor operations (Bruzzone and Massei 2017). The MESAS presentation of this innovative MS2G (Modeling, interoperability Simulation & Serious Games) allowed validation of the potential of simulation in dual-use mission environments with respect to crisis management (Bruzzone 2016). In particular, Immersive Disaster Relief and Autonomous System Simulator was validated operating in scenarios dealing with hazardous material spills to support Education and Training. Indeed it adopts the innovative paradigm defined by MS2G that combines simulation's High-Level Architecture (HLA) concept with the engaging approach typical of Serious Games (Bruzzone 2018a). Thanks to the MS2G approach, Immersive Disaster Relief and Autonomous System Simulator is guaranteed to be flexible, supporting multiple operational modes such as running as a standalone system, operating federated with other models or systems via the HLA protocol and integrated with the IOT (Internet of Things) for live simulation.

3.5 Conclusion

Rapid technological developments in Modeling and Simulation, Artificial Intelligence, Robotics, and Advanced Weapons Systems brings new approaches, opportunities, and threats to a wide range of operational and industrial applications. These fields are leading to significant changes in the way we perceive our social and operational environment, and how we will measure human performance in the future. It is now more clear that these advances in operational capabilities, opportunities, and mission requirements will dramatically change in the future. Expectations are somewhat mixed with concerns about the level of understanding of this ongoing technological revolution, especially in context of advances in AI, a field which some technologists fear could get out of our control.

On the other hand, technological advances in the field of autonomous systems (AI/Robotics) will likely select the winner in future military conflicts. It is logical, therefore, that NATO and Nations cannot disregard autonomous systems evolvement, because of the risk to security from slowing development in this highly critical area.

The NATO M&S COE seeks out the best way ahead in the area of M&S for Autonomous Systems and M&S for Cyber Physical Systems to investigate how M&S can be applied to improve integration of AxS into operational environments. MESAS is one such way to do this, creating a Community of Interest focused on M&S in support of Autonomous Systems. The MESAS conference has proven to be a significant opportunity for high level discussion and debate on critical topics in the autonomous systems field mentioned in this chapter and others that have yet to be identified.

This vision is further extended when the impacts on society, education and training, command and control, operational effectiveness, and future combat concepts are considered. The future of integration and interoperability is reliant upon the exchange of ideas, visions and fresh perspectives, as well as experience, knowhow and frank and open dialog among all stakeholders. MESAS fits perfectly into the mission of the M&S COE, providing M&S tools and concepts to NATO and Nations through collaboration with military, industry, academia, and other institutions. MESAS and other efforts by NATO and its Nations provide an opportunity for the establishment of a Community of Interest around M&S and AS that will continue enhancing future development.

References

Asimov, I. (1950). *I, Robot*. New York: Gnome Press.

Barrat, J. (2013). *Our Final Invention: Artificial Intelligence and the End of the Human Era*. New York: Thomas Dunne Book, Macmillam.

Biagini, M., & Corona, F. (2016). Modelling & Simulation Architecture Supporting NATO Counter Unmanned Autonomous System Concept Development. In J. Hodicky (Ed.), *MESAS 2016*, LNCS (Vol. *9991*, pp. 118–127). Rome, Italy: Springer.

Biagini, M., & Corona, F. (2018). M&S-Based Robot Swarms Prototype. In J. Mazal (Ed.), *M&S for Autonomous Systems (MESAS 2018) Conference*. Brno: Springer.

Biagini, M., Corona, F., & Casar, J. (2017a). Operational Scenario Modelling Supporting Unmanned Autonomous System Concept Development. In J. Mazal (Ed.), *MESAS 2017*, LNCS (Vol. *10756*, pp. 253–267). Rome, Italy: Springer.

Biagini, M., Corona, F., Innocenti, F., & Marcovaldi, S. (2018). *C2SIM Extension to Unmanned Autonomous Systems (UAXS): Process for Requirements and Implementation*. Roma: NATO Modelling and Simulation Centre of Excellence.

Biagini, M., Corona, F., Wolski, M., & Shade, U. (2017b, November 6–8). Conceptual Scenario Supporting Extension of C2SIM to Autonomous Systems. In *22nd International Command and Control Research and Technology Symposium (ICCRTS)*, Los Angeles, CA, USA.

Biagini, M., Scaccianoce, A., Corona, F., Forconi, S., Byrum, F., Fowler, O., & Sidoran, J. L. (2017c, April 9–13). Modelling and Simulation Supporting Unmanned Autonomous Systems (UAxS) Concept Development and Experimentation. In *Proceedings of SPIE, Disruptive Technologies in Sensors and Sensor Systems, 102060N*, Anaheim, CA, USA.

Blackmore, S. (2006). *Conversations on Consciousness*. Oxford, UK: Oxford University Press.

Box, G. E. P. (1979). Robustness in the Strategy of Scientific Model Building. In R. L. Launer & G. N. Wilkinson (Eds.), *Robustness in Statistics*. New York: Academic Press.

Bruzzone, A. G. (2016). New Challenges & Missions for Autonomous Systems operating in Multiple Domains within Cyber and Hybrid Warfare Scenarios. Invited Speech at Future Forces, Prague, Czech Rep.

Bruzzone, A. G. (2018a, September). MS2G as Pillar for Developing Strategic Engineering as a New Discipline for Complex Problem Solving. Keynote Speech, Proceeding I3M, Budapest.

Bruzzone, A. G. (2018b, September). Strategic Engineering: How Simulation could Educate and Train the Strategists of Third Millennium. In *Proceedings of CAX Forum*, Sofia.

Bruzzone, A. G., & Di Bella, P. (2018). Tempus Fugit: Time as the Main Parameter for the Strategic Engineering of MOOTW. In *Proceedings of WAMS*, Praha, CZ, October.

Bruzzone, A. G., Franzinetti, G., Massei, M., Di Matteo, R., & Kutej, L. (2018). LAWS: Latent Demand for Simulation of Lethal Autonomous Weapon Systems. In *Proceedings of MESAS*, Praha, CZ, October.

Bruzzone, A. G., Giribone, P., & Mosca, R. (1996). Simulation of Hazardous Material Fallout for Emergency Management During Accidents. *Simulation*, *66*(6), 343–355.

Bruzzone, A. G., Longo, F., Agresta, M., Di Matteo, R., & Maglione, G. L. (2016c). Autonomous Systems for Operations in Critical Environments. *International Journal of Simulation and Process Modelling (IJSPM)*, *1*, 11.

Bruzzone, A. G., Longo, F., Massei, M., Nicoletti, L., Agresta, M., Di Matteo, R., ... Antonio, P. A. (2016b, June 15–16). Disasters and Emergency Management in

Chemical and Industrial Plants: Drones simulation for education & training. In *Proceedings of MESAS*, Rome.

Bruzzone, A. G., & Massei, M. (2017). Simulation-Based Military Training. In S. Mittal, U. Durak, & T. Oren (Eds.), *Guide to Simulation-Based Disciplines* (pp. 315–361). Springer.

Bruzzone, A. G., Massei, M., Longo, F., Cayirci, E., di Bella, P., Maglione, G.L., & Di Matteo, R. (2016d, October 17–21). Simulation Models for Hybrid Warfare and Population Simulation. In *Proceedings of NATO Symposium on Ready for the Predictable, Prepared for the Unexpected, M&S for Collective Defence in Hybrid Environments and Hybrid Conflicts*, Bucharest.

Bruzzone, A. G., Massei, M., Maglione, G. L., Di Matteo, R., & Franzinetti, G. (2016a, September). Simulation of Manned & Autonomous Systems for Critical Infrastructure Protection. In *Proceedings of DHSS*, Larnaca, Cypurs.

Bruzzone, A. G., Massei, M., Mazal, J., Di Matteo, R., Agresta, G.L., & Maglione, G. L. (2017, September). Simulation of Autonomous Systems Collaborating in Industrial Plants for Multiple Tasks. In *Proceedings of SESDE*, Barcelona, Spain.

Bruzzone, A. G., Merani, D., Massei, M., Tremori, A., Bartolucci, C., & Ferrando, A. (2013, September 25–27). Modeling Cyber Warfare in Heterogeneous Networks for Protection of Infrastructures and Operations. In *Proceedings of European Modeling and Simulation Symposium*, Athens, Greece.

Cellan-Jones, R. (2014). Stephen Hawking warns artificial intelligence could end mankind. *BBC News*, 2.

Duderstadt, J. J. (2005). *A Roadmap to Michigan's Future: Meeting the Challenge of a Global Knowledge-Driven Economy*. Washington, DC: National Academy Press.

Haas, M. C., & Fischer, S. C. (2017). The Evolution of Targeted Killing Practices: Autonomous Weapons, Future Conflict, and the International Order. *Contemporary Security Policy*, *38*(2), 281–306.

Hartmann, K., & Giles, K. (2016, May). UAV Exploitation: A New Domain for Cyber Power. In *Proceedings of the 8th International Conference on Cyber Conflict (CyCon)*, (pp. 205–221). IEEE.

Hartmann, K., & Steup, C. (2013, June). The vulnerability of UAVs to Cyber Attacks-An Approach to the Risk Assessment. In*Proceedings of the 5th International Conference on Cyber Conflict (CyCon)* (pp. 1–23). IEEE.

Javaid, A. Y., Sun, W., Devabhaktuni, V. K., & Alam, M. (2012, Nevember). Cyber Security Threat Analysis and Modeling of an Unmanned Aerial Vehicle System. In *Proceedings of the IEEE Conference on Technologies for Homeland Security (HST)* (pp. 585–590). IEEE.

Kwon, C., Liu, W., & Hwang, I. (2013). *Security analysis for cyber-physical systems against stealthy deception attacks*. American Control Conference (ACC) (pp. 3344–3349). IEEE.

Larm, D. (1996, June). *Expendable Remotely Piloted Vehicles for Strategic Offensive Airpower Roles.* Thesis at School of Advanced Airpower Studies, Maxwell Air Force Base, Alabama.

Levin, S. (2018, March 22) Uber crash shows 'catastrophic failure' of self-driving technology, experts say. *The Guardian.*

Luppicini, R., & So, A. (2016). A Technoethical Review of Commercial Drone Use in the Context of Governance, Ethics, and Privacy. *Technology in Society*, *46*, 109–119.

Madan, B. B., Banik, M., Wu, B. C., & Bein, D. (2016, October 16–21). Intrusion Tolerant Multi-cloud Storage. In *Proceedings of IEEE Conference on Smart Cloud* (pp. 262–268). Bangalore, India.

Magrassi, C. (2013, May 22–24). Education and Training: Delivering Cost Effective Readiness for Tomorrow's Operations. Keynote Speech at ITEC, Rome.

MESAS. (2014, May 5–6). *Modelling and Simulation for Autonomous Systems.* First International Workshop, Rome, Italy, Revised Selected Papers, ISBN 978-3-319-13823-7

MESAS. (2015, April 29–30). *Modelling and Simulation for Autonomous Systems.* Second International Workshop, Prague, Czech Republic, Revised Selected Papers, ISBN 978-3-319-22383-4

MESAS (2016). *Modelling and Simulation for Autonomous Systems.* Rome: Springer International. ISSN 0302-9743. ISBN 978-3-319-47604-9

MESAS. (2017, October 24–26). Modelling and Simulation for Autonomous Systems. In *Proceedings of the 4th International Conference*, Rome, Italy, Revised Selected Papers, ISBN 978-3-319-76072-8

Multinational Capability Development Campaign (MCDC). (2013–2014), Focus Area "Role of Autonomous Systems in Gaining Operational Access", Policy Guidance: Autonomy in Defense Systems, Supreme Allied Commander Transformation HQ, Norfolk, United States, 29 October 2014 Electronic copy. Available at: http://ssrn.com/abstract=2524515

Quick, D. (2012, October 9). Global Hawk UAVs fly in close formation as part of aerial refueling program. *New Atlas Magazine.*

Sandvik, K. B., & Lohne, K. (2014). The Rise of the Humanitarian Drone: Giving Content to an Emerging Concept. *Millennium*, *43*(1), 145–164.

Stewart, P. (2016). Drone Danger: Remedies for Damage by Civilian Remotely Piloted Aircraft to Persons or Property on the Ground in Australia.

Stodola, P., Mazal, J., & Podhorec, M. (2014, May 5–6). Improving the Ant Colony Optimization Algorithm for the Multi-Depot Vehicle Routing Problem and Its Application. In *Proceedings of the MESAS*, Rome.

Vincent, J. (2016, June 29). Satya Nadella's rules for AI are more boring (and relevant) than Asimov's Three Laws. Retrieved from The Verge.

Williams, A. P., & Scharre, P. D. (Eds.) (2014). *Autonomous Systems: Issues for Defense Policy Makers*. Norfolk, VA: Capability Engineering and Innovation Division, Headquarters Supreme Allied Commander Transformation. ISBN 9789284501939

Yayla, A. S., & Speckhard, A. (2017). The Potential Threats Posed by ISIS's Use of Weaponized Air Drones and How to Fight Back. *Huffington Post*, https://www.huffingtonpost.com/entry/the-potential-threats-posed-by-isiss-use-of-weaponized_us_58b654b3e4b0e5fdf6197894

Zwickle, A., Farber, H. B., & Hamm, J. A. (2018). Comparing Public Concern and Support for Drone Regulation to the Current Legal Framework. *Behavioral Sciences and the Law*, *37*(1), 109–124.

Part II

Modeling Support to CPS Engineering

4

Multi-Perspective Modeling and Holistic Simulation

A System-Thinking Approach to Very Complex Systems Analysis

Mamadou K. Traoré

IMS UMR CNRS, University of Bordeaux, Bordeaux, France

4.1 Introduction

Cyber Physical Systems (CPS) are gaining tremendous interest in Modeling and Simulation (M&S), probably because of the level of complexity they exhibit (Baheti and Gill 2011) which calls for innovative analysis approaches. CPS are systems with embedded software, which record data using sensors, affect physical processes using actuators, are connected with other systems/objects in digital communication networks, and interact with their environment and human in various interfaces. As such, they are an enabling technology, i.e., they enable numerous innovative applications, including but not limited to logistics and intelligent transportation, mobile energy provision and consumption, healthcare and assisted living, and factory of the future. Their benefits span from new ways of avoiding accidents, to new strategies for optimized use of energy resources, consumers-based production planning, reduction of environmental pollution, fast and remote care, improved route guidance, and many more.

CPS engineering introduces complexity that changes the risk profile of the solution (Lee 2008), mainly because several requirements are often combined, each of which being a critical challenge taken in isolation. Some of the most salient requirements are:

- Decentralized collaboration of autonomous individuals: physical and software components are intertwined, each operating on different spatial and temporal scales, exhibiting distinct behavioral modalities, and interacting in ways that change with context. Together, they manifest (desired or/and undesired) emergent behavior;

- Openness: sometimes, CPS components and users interact in an ad hoc manner. One example is the dynamic planning of travel in line with the current situation, through the integration of up-to-date traffic information with air and rail schedule changes;
- Learning and resilience: in some cases, CPS elements must learn from a novel situation and adapt to it (such as the failure of a component, or a rare and impacting event happening in the environment), or coordinate with new systems that did not exist initially when the CPS was built;
- Intelligence: the ability to make decisions is central to CPS. This often requires self-awareness, as well as third party-awareness. It also entails the capability to reason and infer goal-oriented resolves;
- Extensive human–machine interaction: interactions between systems and individuals or groups who monitor and influence the CPS operations need a consistent support for seamless integration of cyber, physical, and human elements.

Model-based paradigm is recognized as a powerful approach to systems engineering in many disciplines (Tolk et al. 2018b), and Modeling & Simulation (M&S) provides a core mechanism to such an approach (Zeigler et al. 2018b). Here, we place our discussion under the umbrella of M&S. Hence, when we refer to a model (or models), we mean simulation model(s), even if in some cases, what is stated also holds for other types of model.

From a M&S point of view, CPS can arguably be seen as the most-advanced form of hybridization since they involve both computational and physical components, as shown in Figure 4.1 (where hybrid strategies occur at three levels: concepts, specifications, and operations levels) initially proposed in (Tolk et al. 2018a). At the concept level, fundamental notions (such as state, event, concurrency...) and their relationships are defined and formally captured by appropriate methods and formalisms. At the specification level, real-world systems/problems under study are expressed as models, using the concepts adopted. At the operations levels, virtual and physical engines execute the instructions abstractly expressed at the immediate upper level. Consequently, the heterogeneity of engines (respectively models and formalisms) dictates that hybrid approaches be defined/adopted to fully integrate the entities and concepts of interest. While the M&S community traditionally distinguishes between discrete and continuous phenomena as regard to central time-related concepts (translated respectively into DisM and ContM, respectively discrete and continuous models), qualitative and quantitative computational approaches, such as Operation Research or Artificial Intelligence methods, rather focus on problem-solving steps and mechanisms (translated into Alg, the problem-solving algorithm). As indicated by the legend of Figure 4.1, the literature has coined various terms to qualify the various possible hybridizations. For example, Sim + Int is often referred to as hybrid simulation, where "+" denotes a composition/mixing operation that can vary from loose to tight integration.

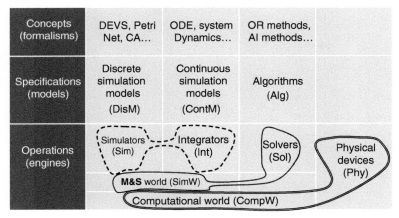

Concepts (formalisms)	DEVS, Petri Net, CA…	ODE, system Dynamics…	OR methods, AI methods…	
Specifications (models)	Discrete simulation models (DisM)	Continuous simulation models (ContM)	Algorithms (Alg)	
Operations (engines)	Simulators (Sim)	Integrators (Int)	Solvers (Sol)	Physical devices (Phy)

M&S world (SimW)

Computational world (CompW)

Legend

Sim+Int: often referred to as hybrid simulation
SimW+Sol: often referred to as combined simulation
CompW+Phy: often referred to as Cyber Physical System (CPS)
+ denotes a composition operation

Figure 4.1   Hybridization strategies in computational frameworks.

Similarly, SimW + Solv is often referred to as combined simulation, where SimW is Sim or Int or Sim + Int. The essence of CPS is the hybridization done at the operations level between computational engines (CompW, i.e., SimW, or Sol, or SimW + Sol) and physical components (Phy).

Hybridization is obviously stronger at the top level of Figure 4.1 (Concepts level) and weaker at the bottom level (Operations level). Consequently, taking CPS challenges from the Operations level to the Specifications level and the Concepts level has the potential to provide the same benefits as Model-Driven Engineering (MDE) aims at in Systems Engineering, that is:

- Precise capture of system information in an integrated way.
- Understanding and analysis of design and decision alternatives at model level rather than physical system level, allowing reasoning and symbolic manipulation.
- Incremental, automated model-driven verification for a significant part of the system under study.
- Continuous transition to system development processes (i.e. model continuity), with continuous traceability.

While CPS realize heterogeneous compositions at the operations level, their sound analysis requires frameworks that can support heterogeneous compositions at upper levels. Such frameworks can be envisioned under multiple rubrics, as illustrated by Figure 4.2:

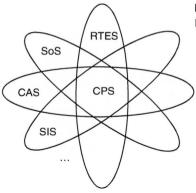

Figure 4.2 CPS from multiple engineering perspectives.

- Software Intensive Systems (SIS) engineering (Hölzl et al. 2008), in that software contributes key influences to the design, construction, deployment, and evolution of the system as a whole, and hence require properties such as reliability, safety, security, correctness, performance, availability, and dependability;
- System of Systems (SoS) engineering (Maier 1998), in that the optimization of CPS processes cannot be based on sub-optimization of the components, but must be holistically addressed;
- Networked Systems (NS) engineering (Tatikonda and Mitter 2004), in that CPS decentralized control is highly constrained by communication capabilities (such as the bandwidth) and Quality of Service (QoS);
- Real-Time Embedded Systems (RTES) engineering (Fan 2015), in that CPS major elements are embedded systems that must meet various timing and other constraints that are imposed on them by the real-time behavior of the external world to which they are interfaced, hence rising the challenge of their predictability;
- Complex Adaptive Systems (CAS) engineering (Miller and Page 2007), in that CPS are agent systems which elements must sometimes learn and adapt to new situations;
- And more.

Each of these rubrics has adequate formalisms and sometimes patterns of models to capture the system from a given perspective. Figure 4.2 is not exhaustive by no means, hence a deeper understanding of CPS at the Specifications level requires multiple levels of explanation be provided to capture these different perspectives and analyze the various aspects involved, while keeping a holistic understanding of the behavioral pattern of the overall system and its interaction with the surrounding environment. Each perspective provides a level of explanation that focuses on one of the major characteristics of the CPS under interest and abstract the others through simplifications captured as parameters of the model built under the current perspective. Rajhans et al. (2014) give illustrative examples, such as the heterogeneous models of a quadrotor from four different design domains, i.e., a signal-flow model of the system to study stability and performance analysis; an equation-based model to study the dynamic response of the system to external forces and torques; a process algebra model to study safety conditions; and a hardware model to study trade-offs between specifications and system-level performance.

An appropriate means to address CPS complexity could be a hybridization of approaches that would evidently provide useful knowledge from various angles on how such systems perform at the holistic level rather than focusing on specific problems in isolation for specific solutions. Multi-Perspective Modeling and Holistic Simulation (MPM&HS) has been introduced as a framework to achieve such a goal in the context of healthcare systems (Djitog et al. 2017, 2018; Traoré et al. 2018). Here, we envision its applicability to broader classes of complex systems, and more specifically to CPS. MPM&HS proposes different models of a system of interest be built from various perspectives to provide multiple levels of explanation, while the holistic understanding of the system is obtained by a parameter-based integration of these models. As such, MPM&HS models can be used to study a CPS of interest, or to be part of its decision-making (aka intelligence) process, or replace some of its components (in case of failure, or as a digital twin) or environment, or even be added to supply new features (for the CPS to adapt to a novel situation).

The remaining of the chapter is organized as follows. Section 4.2 discusses related work. Section 4.3 introduces the conceptual foundations to MPM&HS. Section 4.4 gives details of the multi-perspective modeling leg of the approach, while Section 4.5 does the same for the holistic simulation leg. The whole MPM&HS process is presented in Section 4.6, and Section 4.7 illustrates an application of the approach. Section 4.8 discusses key challenges of generalizing MPM&HS and suggest ways to tackle them. Finally, Section 4.9 concludes the chapter.

4.2   Related Works

Models heterogeneity and the need to compose them in larger modeling activities have been largely studied in literature, from different viewpoints, ranging from theoretical contributions to pragmatic approaches. Such contributions appear under different labels, such as multi-paradigm M&S (Vangheluwe et al. 2002), multi-modeling (Fishwick 1995), hybrid M&S (Tolk et al. 2018a), co-simulation (Camus et al. 2016; Mittal and Zeigler 2017), etc. Looking closer to Figure 4.1, we see that more details can be brought in terms of the various possibilities that exist as a simulation expert is trying to compose heterogeneous simulation components. Table 4.1 presents these details, while mirroring them with the well-recognized

Table 4.1 Heterogeneous composition challenges.

Level	Challenge	Example of situation	Mirror in LCIM
Concepts (formalisms)	Integrated perspectives (semantical consistency and compositional validity)	Disease spreading perspective + population dynamics perspective	Conceptual
Specifications (models)	Syntactic model (syntactic composability)	Perspective-specific DEVS model + perspective-specific System Dynamics model	Pragmatic
Operations (engines)	Operational semantics (simulation time management: co-simulation, federated simulation, hybrid simulation...)	DEVS simulator + System Dynamics integrator	Semantic
	Data management (middleware for interoperability: data exchange standard format and protocol)	Linux-Java-implemented DEVS code on + Windows-C++-implemented DEVS code	Syntactic
	Code/Event synchronization (Parallel and Distributed Simulation)	Executable code on computer 1 + executable code on computer 2	Technical

Levels of Conceptual Interoperability Model (LCIM) introduced by (Tolk and Muguira 2003) in the context of military application data interoperability and later improved and generalized to other domains (Tolk et al. 2009):

- At the operations level, we reveal three sub-levels. The first and very basic one corresponds to the situation where heterogeneity comes from the geographic distribution of the components to compose. The main composability challenge is the need for synchronizing all events generated by the composed codes such that the resulting trajectories of the entire whole are faithful with reality. The literature on Parallel and Distributed Simulation (Fujimoto 2000) has largely studied this concern. The second sub-level is where components are heterogeneous in terms of operating environment (such as different programming languages, or operating systems). The main concern there is to ensure data interoperability, i.e., a common format for data exchange and the associated protocol architecture and technologies. The third and last sub-level is where simulation components implementing the simulation algorithms (aka operational semantics) of various simulation modeling formalisms are mixed. The main concern is then how to consistently manage the simulated time, as these components can concurrently run and communicate through different techniques (such as co-simulation, federated simulation, or hybrid simulation techniques).

- At the specifications level, models using various syntaxes can be mixed (e.g., a DEVS-specified component mixed with a System Dynamics-specified component to model the spread of a disease). The challenge there is the syntactic composability of such specifications.
- At the concepts level, various perspectives can be mixed (as combining a disease spreading model with a population dynamics model into a holistic view). The main challenge there is to ensure the semantic consistency and validity of merging concepts from various perspectives.

Even if the vocabulary used is slightly different, the LCIM provides a sound conceptual background to our framework. The LCIM levels of interest for us are the following:

- Technical level, where systems have technical connection(s) and can exchange data.
- Syntactic level, where systems agree on the protocol to exchange the right forms of data in the right order, but the meaning of data elements is not established.
- Semantic level, where systems exchange terms that they can semantically parse.
- Pragmatic level, where systems are aware of the context and meaning of information being exchanged.
- Conceptual level, where systems are completely aware of each other's information, processes, contexts, and modeling assumptions.

The discussion done in this paper concerns the LCIM conceptual level. Tolk et al. (2013) stated that "*on this* (conceptual interoperability)*, we need a fully specified but implementation independent model*". This is exactly what MPM&HS allows to obtain. The topic of simulation models composability has also been addressed by (Petty and Weisel 2003) to define a formal theory of validity. Two forms of composability have been defined there: syntactic and semantic (also known as engineering and modeling). The work in this paper can be placed within the context of semantic composability. Close to our work, Seck and Honig (2012) introduced a generic multi-perspective modeling approach, which they formalized by adding to the DEVS system specification hierarchy (Zeigler 1976), a top layer to represent multi-perspective models. However, no perspective identification process is suggested. Zeigler et al. (2012) presented a methodology and modeling environment for simulating national health care based on multi-level modeling and families of models, which are also applicable to other types of systems.

We argue that our contribution is original in that none of the works mentioned in this section offers a systematic way to identify, address concurrently, and simulate the perspectives of a complex system in a holistic way as does our framework.

4.3 Conceptual Foundations to MPM&HS

MPM&HS relies on the fundamental principle that a complex system combines multiple perspectives (which we also call facets) and that every perspective influences every other one. Let us use the well-established theoretical framework of M&S defined in Zeigler (1976) to give a semi-formal view of what is MPM&HS. Zeigler (1976) describes four basic entities to the M&S enterprise, as shown by Figure 4.3: the system under study, the model, the experimental frame (EF), and the simulator. The system under study is represented as a source of behavioral data. The EF is the set of conditions under which the system is being observed, and is operationally formalized to capture the objectives of the study. The model is a set of rules or mathematical equations that give an abstract representation of the system, which is used to replicate its behavior. The simulator is the automaton that is able to execute the model's instructions. More theoretical aspects on EF can be found in (Zeigler 1984). Later, Zeigler et al. (2000) gave a thorough description of various algorithms for the simulator, including sequential, as well as parallel and distributed variants. Elaborating on concepts developed in Zeigler (1984), Traoré and Muzy (2006) suggested that during modeling activities, models always come with the specification of associated EFs. The specified EF is a model

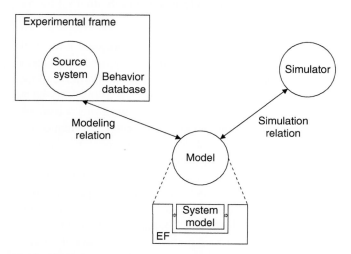

Figure 4.3 Basic entities in M&S and their relationships. *Source:* Adapted from Zeigler (1976).

component to be coupled with the model to produce the data of interest under specified conditions, as suggested by the bottom part of Figure 4.3.

The first leg of MPM&HS, namely Multi-Perspective Modeling (MPM), relies on the idea that different levels of explanation can be obtained from a given system by repeatedly studying the system under various perspectives (or facets), each of which expressed as an experimental frame, as depicted by Figure 4.4. Perspective-specific EFs are elaborated to provide answers to questions of very different nature about the same system. Consequently, each of them is coupled with a model elaborated from the corresponding perspective to derive results of interest in this perspective. Each perspective-specific model abstracts the influences due to other perspectives by means of parameters which values explicitly reflect implicit assumptions and simplifications done about these influences.

While in practice, perspective-specific models are executed in isolation, i.e., without recourse to the processes from other perspectives, all these models are related in reality, since they depict various abstractions of the same system. However, building a monolithic highly detailed mega-model that involves all inter-influencing factors is not viable. Therefore, the second leg of MPM&HS, namely Holistic Simulation (HS), suggests gluing the perspectives together by enabling live exchanges of information between models from different perspectives through integrators, as shown in Figure 4.5. That way, a holistic view is obtained, which encompasses isolated perspective-specific simulations and their mutual influences, without a drastic increase of complexity. Such an integration is done by dynamically feeding the parameters of a focused model in a given perspective with the outputs of models from other perspectives. When the perspective-specific models are holistically integrated that way, a holistic EF can be elaborated and questions that are transversal to different perspectives can then be addressed. This holistic EF is to be coupled with the resulting holistic model to derive results that cannot be accurately addressed in any of the perspective taken alone.

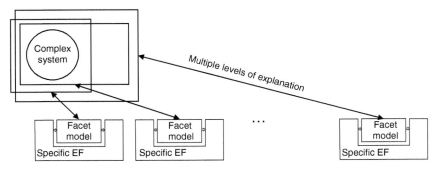

Figure 4.4 General principle of Multi-Perspective Modeling.

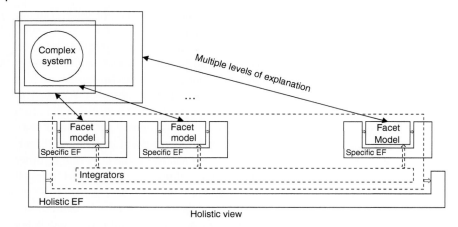

Figure 4.5 General principle of Holistic Simulation.

4.4 Multi-Perspective Modeling

Zeigler et al. (2018a) recently formulated fundamental requirements for M&S of modern complex systems as follows: "develop an organizational ontology that supports combinatorial model compositions, with major facets at the top level to ensure macro behavior and their refinement into meso and micro behaviors, and a large spectrum of models at the bottom level for combinatorial composition." We derive from this statement, a disciplined approach to building an ontology for the M&S of a domain of interest, to which we associate a modeling process, as detailed hereafter.

It is essential that the domain analysis ontology provides, at some general level, a formal way to capture all the knowledge that might be in the range of M&S of the domain for which it is likely to be used. Therefore, it must capitalize on the abstractions used for the simulation of the entire targeted domain, beyond aspect-specific modeling. Thus, we suggest a generic ontology that highlights key characteristic in such a way that the fundamental requirements formulated in Zeigler et al. (2018a) can always be met.

4.4.1 Generic Ontology for Complex Systems

MPM cannot be efficiently realized without a disciplined approach. Indeed, the identification of all possible perspectives, or at least the ones of major interest, is not obvious. To apply MPM to complex systems, we adopt a layered analysis approach, which is captured in Figure 4.6 using the System Entity Structure (SES) ontological framework (Zeigler 1984). SES enables a fundamental representation

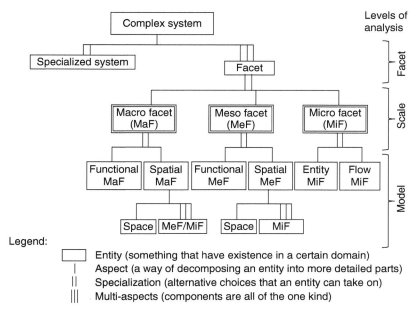

Figure 4.6   Generic Ontology for Complex Systems M&S (O4CS).

of a hierarchical modular model providing a design space via the elements of a system and their relationships in a hierarchical and axiomatic manner. It is a declarative knowledge representation scheme that characterizes the structure of a family of models in terms of decompositions, component taxonomies, and coupling specifications and constraints. In SES, entities (represented by boxes) are things that have existence in a certain domain. They can have variables, which can be assigned a value within a given range. An Aspect expresses a way of decomposing an object into more detailed parts. A Multi-Aspect is an aspect for which the components are all of one kind. A Specialization is a category or family of specific forms that a thing can assume. SES axioms are (Zeigler and Sarjoughian 2017): uniformity, strict hierarchy, alternating mode, valid brothers, attached variables, and inheritance. Uniformity forces that any two nodes with the same labels have isomorphic subtrees. Strict hierarchy prohibits a label from appearing more than once down any path of the tree. Alternating mode states that, if a node is an Entity, then the successor is either Aspect or Specialization, and vice versa. Valid brothers forbid having two brothers with the same label. Attached variables constrains that variable types attached to the same item shall have distinct names. Inheritance asserts that Specialization inherits all variables and Aspects from the parent Entity to the children Entities. Zeigler and Hammonds (2007) provide a formal set-theoretic characterization of the SES that shows how the axioms are satisfied.

The generic ontology is meant to be instantiated in the analysis of any new domain of interest in view of its M&S. Such an instantiation provides the domain-specific ontology that will drive the MPM&HS process of the targeted domain. As depicted by Figure 4.6, the following layers are defined (and detailed in Sections 4.4.2, 4.4.3, and 4.4.4):

- Facet level, where specializations of the class of systems that characterizes the domain of interest are highlighted, and cumulative aspects of a domain system are separated.
- Scale level, where major spatial and temporal scales are emphasized.
- Model level, where legacy models often originating from decades of theoretical findings are identified as reusable artifacts to be selected and integrated in new studies.

4.4.2 Facet Level

This level recognizes the whole complex system as a juxtaposition of multiple facets, while various specializations can be identified as possible instances of the same integrated set of facets in various specific contexts. For example, healthcare systems can be specialized into primary, secondary, ternary, and home care (Traoré et al. 2018), while transportation systems can be specialized into air, ground, rail, and aquatic transport, and military systems can be specialized into air, ground, and marine forces. An important element of this multi-facet (or multi-perspective) approach is that, while perspectives have mutual influence on each other, each perspective captures its received influences by means of parameters, which values explicitly reflect implicit assumptions and simplifications done about other perspectives influences. For example, let us consider a traffic regulating CPS, as made of two facets: one related to the dynamics of the traffic (dynamics facet), and the other related to the control of lights, barriers, etc. (control facet). When focusing on the dynamics facet, the corresponding model makes use of parameters (such as the arrival rate, the average speed, and the time of reaction of vehicles) that are assumptions and simplifications made on all processes belonging to the control facet. Similarly does the control facet, with parameters such as the spatial distribution of vehicles.

4.4.3 Scale Level

A characteristic feature of complex systems is the occurrence of interactions between heterogeneous components at different spatial and temporal scales. The hierarchy theory provides a guideline to model such complex systems, by emphasizing on the fact that at a given level of resolution, a system is composed of interacting lower-level components and is itself a component of a higher-level component (O'Neill et al. 1989). As such, it opens the way to scale-driven modeling methodologies (Wu and David 2002) with various interpretations of the notion of scale (Allen and Starr 1982; Marceau 1999; Dungan et al. 2002; Ratzé et al. 2007), and a major concern about scale transfer processes where inter-scale interactions must be properly described, as emphasized by Jelinski and Wu (1996) and Willekens (2005). Scale-driven modeling exhibits a hierarchical organization from differences in temporal and spatial scales between the phenomena of interest. Thresholds between scales are critical points along the scale continuum where a shift in the importance of variables influencing a process occurs. Traditional generic distinctions between scales include the partition into micro, meso, and micro levels (Blalock 1979), or strategic, operational, and tactical levels (Rainey and Tolk 2015), or $n-1$, n, and $n+1$ levels (Aumann 2007). All these efforts agree on the fact that there is a minimum requirement for a triadic view of causalities (Salthe 1985; Ulanowicz 1997).

While the hierarchy theory mainly focuses on a descriptive form, the concrete translation of its derivative scale-driven modeling methodology into formalized computational models is achieved within the MPM&HS framework at the scale level of the generic ontology. Indeed, scale-driven modeling provides a vertical stratification within each facet (or perspective), while multi-perspective modeling provides a horizontal stratification within the holistic analysis of the entire system. As a result, the generic ontology of Figure 4.6 exhibits macro, meso, and micro levels of abstraction within each facet, leading respectively to the generic MaF, MeF, and MiF models: a MaF is the model of a population; a MiF is the model of an individual; and a MeF is the model of a group of individuals (a level in-between individual and population). In the example of the traffic regulating CPS, the traffic facet can be described at the scale of the entire population of vehicles (a macro scale), or at the scale of individual vehicles (micro scale). The inter-scale transfer phenomena need to be described as well, such that what happens at the micro scales emerges at the macro scale and what is done at the macro scale impacts on the micro scales. The same considerations hold for the control facet.

4.4.4 Model Level

This level is where the abstractions that can be directly simulated are defined. We distinguish four types of model: entity models, flow models, functional models, and spatial models. This classification is different from the one proposed by Fishwick (1995) in his seminal work on multi-formalism modeling in that we focus on what abstractions are at the forefront rather than how the formalism used can be characterized (i.e. declarative, functional, constraints, and spatial, as defined by Fishwick). The following defines the types of model considered:

1) Entity models describe autonomous individuals with specific attributes and with or without goal-driven behavior;
2) Functional models are formulated as mathematical equations;
3) Spatial models are composed of individuals geographically located in a space model; and
4) Flow models capture scenarios an individual can undergo.

Consequently, the generic ontology (Figure 4.6) has in each facet entity and flow models at the micro level of abstraction, and functional and spatial models at the macro and meso levels of abstraction. One can notice that a spatial model at any macro level involves a space model that contains abstractions detailed at lower levels (i.e., meso and micro), and that similarly a spatial model at any meso level involves a space model that contains abstractions detailed at the micro level. For instance, in the case of the traffic regulating CPS, many legacy and conventional models exist for the traffic facet (Hoogendoorn and Bovy 2001), among which are functional MaF such as LWR models (Lighthill and Whitham 1955; May 1990), spatial MaF such as cellular automata-based models (Nagel and Schreckenberg 1992), or entity MiF such as agent-based models (Cicortas and Somosi 2005), or flow MiF such as car following models (Edie 1961; Brackstone and McDonald 1999). Similarly exist legacy and conventional models for the control facet (Di Steffano et al. 1967; Levine 1996; Hellerstein et al. 2004).

4.5 Holistic Simulation

As previously explained, perspectives integration enables live exchanges of information between models from different perspectives and hence offers a holistic view which encompasses isolated perspective-specific simulations and their mutual influences. Such an integration relates the outputs of some perspective-specific models to the parameters of others. Technically, this is realized by creating a coordinating model, which translates output received from models into new values for the parameters of other models. For instance, in the quadrotor system presented in Rajhans et al. (2014), the output of the signal-flow model of the system (stability facet) must be used during a simulation run to dynamically update the parameters of both the equation-based model (dynamic response facet), the process algebra model (safety facet), and the hardware model (performance facet). Similarly, the output of each of these remaining three models must be used to update the parameters of any of the others. Zeigler et al. (2019) gave a formal specification of such an integration in the context of DEVS M&S.

4.6 MPM&HS Process

The whole MPM&HS methodological process is driven by the SES-based generic ontology O4CS. As such, the MPM&HS process suggests a generic organization, where the family of alternative models displayed at the leaves of O4CS have to be implemented and saved in a model base (called MB4CS). Once the models are saved, they can be retrieved from their repository and reused to design complex systems. Legacy and conventional models (which we call white models) are models available in the literature, which are based on established theories, generic for a class of systems, and parameterized to be reusable in various similar situations. Examples of such models are the compartmental SIR model (Kermack and McKendrick 1927; Hethcote 2000) to model diffusion processes, or prey–predator differential equations (Voltera 1931; Hofbauer and Sigmund 1998) to model population dynamics, etc. Specific models (which we call gray models) are the ones implemented in the library by the user, either by adapting a white model to a specific system, or by building from scratch for his or her specific purposes (Figure 4.7).

The MPM&HS methodological process is depicted by Figure 4.8. The analysis of a specific domain is guided by the O4CS generic ontology and results in domain-specific ontology and model base. From intended objectives of a given study within that domain, the modeler develops multi-perspective models to meet these objectives, through the pruning of the domain-specific ontology and the selection of appropriate components in the domain-specific model base. Pruning is the process of extracting from the O4CS tree a specific system configuration, selecting particular subsets of Aspects, particular cardinalities of Multi-Aspects, and particular instances of Specializations, and assigning values to the variables. Pruning of the ontology results in a multi-perspective model, where each top facet model is a multi-scale coupled model that addresses intra scale transfer issues. Facet models are then integrated in a holistic model through well-defined integrators that address inter-perspective scale transfer issues. The holistic model can then be

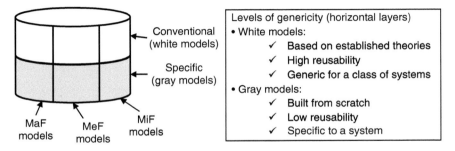

Figure 4.7 Generic model base for complex systems simulation (MB4CS).

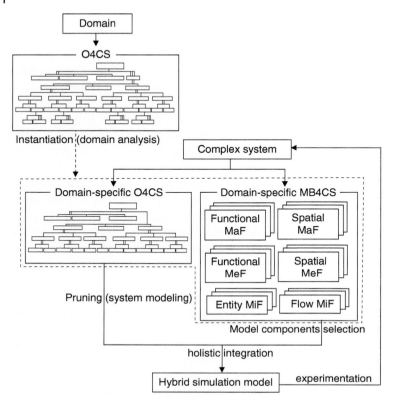

Figure 4.8 MPM&HS methodological process.

simulated under various scenarios to derive conclusions for the complex system of interest.

4.7 Application

A complete application of the MPM&HS approach has been realized for health-care systems analysis and design (Traoré et al. 2018). The instantiation of O4CS gives the ontology presented in Figure 4.9.

Analyses done at system, facet, scale, and model levels respectively, lead to a multi-layer hierarchy of models:

- At the Facet level, we distinguish healthcare systems specializations into primary, secondary, ternary, and home care. We also recognize three facets, i.e., "production facet", "consumption facet," and "coordination facet." The notion

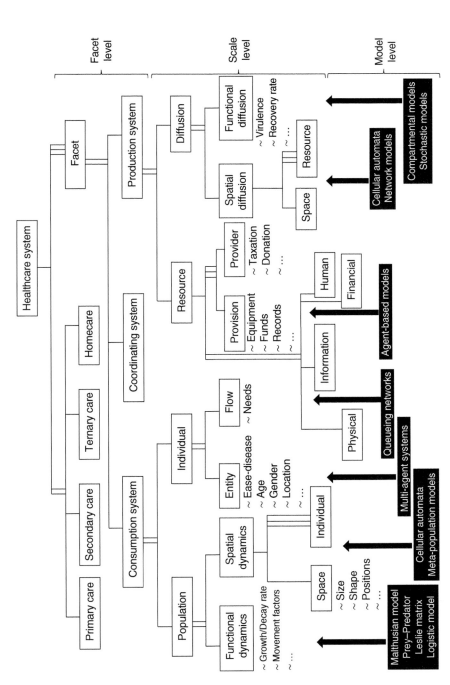

Figure 4.9 Ontology and conventional models for healthcare systems M&S.

of "Production" encompasses the traditional notion of "Supply" in that it involves not only the intentional supply of healthcare services needed, but all phenomena that produce positive (i.e. ease) and negative (i.e. disease) impacts on the system's stakeholders. Examples are vaccination and information diffusion to produce ease, and contamination and epidemics to produce disease (in the traditional supply-demand classification, epidemics are not specifically captured, since they are not seen as supply, nor as demand). Similarly, the notion of "Consumption" encompasses the notion of "Demand," as consumers (such as infected individuals) may not necessarily be intentionally in demand of that infection (though consumers of the infection). CPS in healthcare appear in the "coordination facet," as explained latter with Figure 4.11.

- At the Scale level, macro and micro views lead respectively to Population and Individual scales in the Consumption facet, and to Diffusion and Resource scales in the Production facet. The Population scale captures the dynamics of a population (i.e. growth, decay, and movements) in a functional or spatial way. The Individual scale captures the behavior of an individual (i.e. social, cultural, economic…) as an entity or a flow.
- At the Model level, conventional and legacy model (e.g. Malthusian model, Prey-Predator, etc.) are implemented in the model base.

Such a hierarchy leads to a framework under which healthcare systems simulation models are a combination of one or more facets, each related to Population Dynamics (PD), Individual Behavior (IB), Resource Allocation (RA), or Health Diffusion (HD). These facets cover the full set of healthcare M&S concerns as revealed by literature review, which, though interrelated, are often treated separately with the impact of other concerns on any one of them being approximated by parameters. For instance, when there is an epidemic in a community, it will naturally affect the provisions and allocations of the human and infrastructural healthcare resources in the health centers within the community. Technically, this means that the output of the epidemics model (HD facet) must be used to dynamically update the parameters of the RA facet model. Similarly, the dynamics of a population (PD facet) impacts on the spreading of a disease within that population (HD facet), by favoring it (e.g. through migrations), or by stopping it (e.g. migrations can also starve the infection process). Conversely, disease spreading impacts on population dynamics (e.g. by increasing migrations and deaths, or by decreasing births). In a recent study, we applied this framework to study the Ebola outbreak in Nigeria within the context of scarce health resources, vibrant population dynamics, and specific socio-cultural behavior of individuals (Djitog et al. 2017, 2018). While the simulation from each of these perspectives (i.e. allocation of health resources, population dynamics in Nigeria, the spreading of Ebola, and socio-economic behavior of individuals) abstracts realities concerning the rest of the perspectives, connecting different perspectives simultaneously takes into consideration all of them. The holistic simulation model we built is derived from the pruning of the ontology elaborated, as shown in Figure 4.10:

- Secondary care is selected as the type of care under consideration in step 1.
- An organization with both production and consumption facets is considered in step 2.
- A SEIRD model (a variant of the SIR model) is selected to model the Ebola spread in the major Nigerian city (i.e. Lagos) in step 3.
- A provision of health from Physical Resource is considered to capture health resource allocation in a Lagos hospital in step 4.
- This is modeled by a System Dynamics model in step 5.
- The population dynamics in Lagos is modeled with cellular automata in step 6.
- Individuals within the population with specific socio-economic behaviors (i.e. daily workers) are modeled as agents in step 7.
- The resulting coupled model corresponding to the pruned entity structure of Figure 4.10 integrates component models from the four facets identified, i.e., PD, IB, RA, and HD. The experimental frame built to experiment with the resulting holistic model allows us to see how all models impact on each other simultaneously, and in various scenarios of influence. The results of this study are reported in details in Traoré et al. (2018).

We have now extended this work to CPS in healthcare (CPSiH). There is not much cumulative experience on the M&S of CPSiH, therefore not much conventional or legacy models. However, in a review of the state-of-the-Art on the matter, Haque et al. (2014) characterize it as a system with the ability to observe patient conditions remotely and take actions, and suggest a generic graphical representation based on the literature. As such, CPSiH appears in our approach, as a coordinating facet of the healthcare system it is part of. This is of particular interest, as the design of such a CPS must integrate what happens at the population and individual levels as these are the providers and receivers of data on which the CPS operates, as well as at the resource level as this includes the sensing/acting devices, and the diffusion level as this is the reason of observation/actions done by the CPS.

The extension to the ontology of Figure 4.9 is depicted in Figure 4.11, where we consider two facets for CPSiH: one which is data-centered, and the other that is computation-centered. A CPSiH (and more generally, a CPS) relies on sensing, processing, and networking (Haque et al. 2014). From the data-centered facet, the study can focus on major issues such as data security, database management, and data availability. From the computation-centered facet, the focus can rather be on energy efficiency (for sensing/acting devices), processing efficiency and reliability (for servers and other computing features), storage capabilities and performances (for databases, data centers and cloud solutions), and communication efficiency

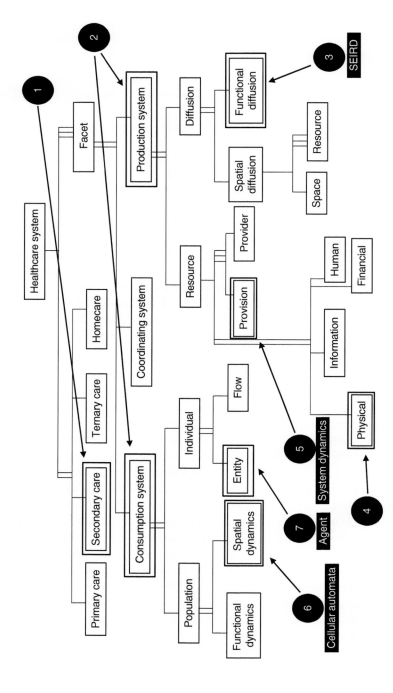

Figure 4.10 Pruning of the ontology for healthcare systems.

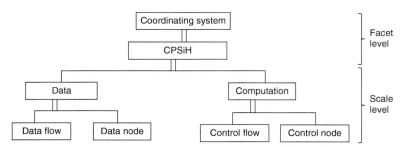

Figure 4.11 CPS in healthcare as coordinating systems.

and liability (for protocols). Macro and micro scales for data-centered (respectively computation-centered) facet lead respectively to Data Flow (respectively Control Flow) and Data Node (respectively Control Node) models. Only further experiences in M&S of CPSiH will allow to later populate these branches of the ontology at the model level. However, many network models (such as queueing models, cellular automata-based...) can already be seen as good potential candidates. From there, the MPM&HS process as presented and illustrated so far can be applied the same way to derive multi-perspective and holistic simulation models for CPSiH.

4.8 Discussion

While the MPM&HS fundamental principle looks appealing, the following two questions immediately arise:

1) What is the semantics behind the holistic integration of multi-perspective models?
2) How can we validate such an integration?

4.8.1 What Is the Semantics Behind the Holistic Integration of Multi-perspective Models?

A common practice is to consider simulation parameters constant throughout the different studies scenarios (Bard 1974). However, we have argued that the values of these parameters can change in concurrent simulation studies. Consequently, we have proposed an integration of the underling perspectives where independent simulation processes of disparate concerns of the same system exchange live updates of their influences on one another to bring results closer to the reality. A parameterized model is nothing more than a family of models having some common behavioral patterns but different structures. Each assignment of specific values to the components of the parameters vector corresponds to a single model of the family. Consequently, by giving live feedback to perspective-specific models through parameters, we realize dynamic change of structure of the models that receive these feedbacks. A general framework for dynamic change of structure is given in (Muzy and Zeigler 2014) from the DEVS perspective. If the DEVS formalism is not adopted when applying MPM&HS, there is a requirement that the working environment allows for dynamic change of structure in the course of a simulation experiment.

4.8.2 How Can the Holistic Integration Be Validated?

A key issue in developing multi-perspective models is the validity of the bridging components, i.e., the way parameters of a model are disaggregated using outputs of other models. Such an issue is directly correlated to the aggregation–disaggregation problem in Multi-Resolution Modeling (MRM). The concept of resolution in M&S refers to the level of detail of a model. Although a relative concept, it promotes the idea that a model (or an abstraction in the model) can be represented at different levels of detail. A greater insight about the phenomenology of a given process requires an increase of resolution. As such, a given parameter in a model means a decrease of the resolution of some external influencing process(es). In MRM, the dynamic change of resolution is called aggregation (from high-resolution entity/process to low-resolution entity/process) or disaggregation (the reverse operation), and the problem of linking simulations at different levels of resolution is known as the aggregation–disaggregation problem (Davis and Hillestad 1993; Reynolds et al. 1997; Davis and Bigelow 1998; Yilmaz and Oren 2004). The MPM&HS holistic integration approach is an aggregation/disaggregation technique, where low-resolution processes (i.e. parameters) aggregate with high-resolution processes (i.e. the feeding models). Since, live feedbacks are given in both directions between perspective-specific models, both models can be seen as disaggregation of each other's external processes. The aggregation/disaggregation operations are realized by the integrators components. As stated by Davis (1995), a recurring question is whether it is legitimate (and desirable) to disaggregate and aggregate processes during the course of a given simulation run. Translated within our framework, the question is the validity of the integrator models. One must notice that the parameters of a model may express external processes operating at a different time scale from the time scale of the model. Therefore, integrators are models where issues, such as the how and validity of scale transfer, are addressed. The correlation (whether linear, quadratic, polynomial, or a more complex relationship) between outputs of some models and parameters of others is either an a priori knowledge or need to be established. The establishment of such a knowledge may require a different simulation study be conducted, and interpolations be derived from several simulation experiments, quantities of data collected, and statistical analysis be performed (Duboz et al. 2003). More generally, the consideration of different perspectives means that as the modeler attempts to span a wider range of perspectives, there is a need to incorporate the effects of an increasing number of processes. It is therefore of equal importance to identify these perspectives and to manage the phenomenology and mathematics of the interactions between them.

4.9 Conclusion

Modern complex systems, like CPS, are hard to manage because of the heterogeneity that characterizes them. This renders their analysis and design more difficult, and requires multiple levels of explanation to be provided to achieve their various objectives, while keeping a holistic understanding of the behavioral pattern of the overall system and its interaction with the surrounding environment. As such, a hybridization of approaches that would evidently provide useful knowledge from various angles on how such systems perform at the holistic level rather than focusing on specific problems in isolation for specific solutions is an appropriate means to address their complexity. In this chapter, we argue that the MPM&HS approach can provide a helpful framework for that search.

In practice, M&S processes are often identified from a given perspective and are executed without recourse to the processes independently built from other perspectives. In reality, however, processes usually have mutual influences, which has a greater impact when it comes to complex systems like CPS, where *everything affects everything else*. However, building a monolithic highly detailed mega-model that involves all inter-influencing factors cannot be envisioned. To address this issue, MPM&HS suggests a stratification of the levels of abstraction into multiple perspectives and their integration into a common simulation framework. In each of the perspectives, models of different components of the system can be developed and coupled together. Concerns from other perspectives are abstracted as parameters. That way, each perspective can be seen as encompassing a family of questions that can be formulated through dedicated experimental frames. Consequently, the resulting top model within each perspective can be coupled with its experimental frame to run simulations and derive results. Live exchanges of information between models from different perspectives encompass these isolated perspective-specific simulations and their mutual influences, without a drastic increase of complexity. The resulting global model can be coupled with a holistic experimental frame to derive results that cannot be accurately addressed in any of the perspectives taken alone.

When CPS are specifically targeted, the ontology is useful to identify and implement in a library the building blocks of simulation software constructs that are necessary to model CPS perspectives (such as hardware, software, and network aspects). The holistic integration allows for addressing the complexity of scenarios in which these perspectives interfere. A key issue in applying the MPM&HS approach is the verification and validation of the multi-perspective holistic simulation models. Much remains to do on how to ensure the semantic consistency and validity of integrating heterogeneous models as well as their technical interoperability, as pointed out by Tolk et al. (2013). Such questions are part of our on-going research efforts.

References

Allen, T. H. F., & Starr, T. B. (1982). *Hierarchy: Perspectives for Ecological Complexity*. Chicago, IL: The University of Chicago Press.

Aumann, G. A. (2007). A Methodology for Developing Simulation Models of Complex Systems. *Ecological Modelling, 202*, 385–396.

Baheti, R., & Gill, H. (2011). Cyber-Physical Systems. In T. Samad & A. M. Annaswamy (Eds.), *The Impact of Control Technology* (pp. 161–166). New York: IEEE Control Systems Society.

Bard, Y. (1974). *Nonlinear Parameter Estimation*. New York: Academic Press.

Blalock, H. M. (1979). *Social Statistics*. New York: McGraw-Hill.

Brackstone, M., & McDonald, M. (1999). Car-Following: A Historical Review. *Transportation Research Part F: Traffic Psychology and Behaviour, 2*(4), 181–196.

Camus, B., Paris, T., Vaubourg, J., Presse, Y., & Bourjot, C. (2016). MECYSCO: A Multi-Agent DEVS Wrapping Platform for the Co-simulation of Complex Systems. *Research Report, LORIA*, UMR 7503, CNRS.

Cicortas, A., & Somosi, N. (2005). Multi-Agent System Model for Urban Traffic Simulation. In *Proceedings of the 2nd Romanian-Hungarian Joint Symposium on Applied Computational Intelligence* (pp. 107–120). Timisoara, Romania.

Davis, P. K. (1995). *Aggregation, Disaggregation, and the 3:1 Rule in Ground Combat*. Santa Monica, CA: RAND.

Davis, P. K., & Bigelow, J. H. (1998). Experiments in Multiresolution Modeling (MRM). *RAND Research Report MR-1004-DARPA*.

Davis, P. K., & Hillestad, R. (1993). Families of Models that Cross Levels of Resolution: Issues for Design, Calibration and Management. In G. W. Evans, et al. (Eds.), *Proceedings of the Winter Simulation Conference* (pp. 1003–1012). Piscataway, NJ: IEEE.

Di Steffano, J. J., Stubberud, A. R., & Williams, I. J. (1967). *Feedback and Control Systems*. Schaums Outline Series. New York: McGraw-Hill.

Djitog, I., Aliyu, H. O., & Traoré, M. K. (2017). Multi-Perspective Modeling of Healthcare Systems. *International Journal of Privacy and Health Information Management*, *5*(2), 1–20.

Djitog, I., Aliyu, H. O., & Traoré, M. K. (2018). A Model-Driven Framework for Multiparadigm Modeling and Holistic Simulation of Healthcare Systems. *SIMULATION: Transactions of the SCS – Special Issue on Hybrid Systems M&S*, *94*(3), 235–257.

Duboz, R., Ramat, E., & Preux, P. (2003). Scale Transfer Modeling: Using Emergent Computation for Coupling Ordinary Differential Equation System with a Reactive Agent Model. *Systems Analysis Modelling Simulation*, *43*(6), 793–814.

Dungan, J. L., Perry, J. N., Dale, M. R. T., Legendre, P., Citron-Pousty, S., Fortin, M. J., ... Rosenberg, M. S. (2002). A Balanced View of Scale in Spatial Statistical Analysis. *Ecography*, *25*, 626–640.

Edie, L. C. (1961). Car-Following and Steady-State Theory for Noncongested Traffic. *Operations Research*, *9*(1), 66–76.

Fan, X. (2015). *Real-Time Embedded Systems: Design Principles and Engineering Practices* (First ed.). Amsterdam: Elsevier. ISBN: 978-0128015070

Fishwick, P. A. (1995). *Simulation Model Design and Execution: Building Digital Worlds*. Englewood Cliffs, NJ: Prentice Hall.

Fujimoto, R. (2000). *Parallel and Distributed Simulation Systems*. New York: Wiley.

Haque, S. A., Aziz, S. M., & Rahman, M. (2014). Review of Cyber-Physical System in Healthcare. *International Journal of Distributed Sensor Networks*, *10*(4). 20 p. DOI:https://doi.org/10.1155/2014/217415

Hellerstein, J. L., Diao, Y., Parekh, S., & Tilbury, D. M. (2004). *Feedback Control of Computing Systems*. Hoboken, NJ: Wiley. ISBN:0-471-26637-X

Hethcote, H. (2000). The Mathematics of Infectious Diseases. *SIAM Review*, *42*(4), 599–653.

Hofbauer, J., & Sigmund, K. (1998). *Evolutionary Games and Population Dynamics*. Cambridge: Cambridge University Press.

Hölzl, M. M., Rauschmayer, A., & Wirsing, M. (2008). Engineering of Software-Intensive Systems: State of the Art and Research Challenges. In M. Wirsing, J.-P. Banatre, M. Hölzl, & A. Rauschmayer (Eds.), *Software-Intensive Systems and New Computing Paradigms* (pp. 1–44). Berlin, Heidelberg: Springer-Verlag. https://doi.org/10.1007/978-3-540-89437-7_1.

Hoogendoorn, S. P., & Bovy, P. H. (2001). State-of-the-Art of Vehicular Traffic Flow Modelling. *Proceedings of the Institution of Mechanical Engineers, Part I: Journal of Systems and Control Engineering*, *215*(4), 283–303.

Jelinski, D. E., & Wu, J. (1996). The Modifiable Area Unit Problem and Implications for Landscape Ecology. *Landscape Ecology*, *11*, 129–140.

Kermack, W. O., & McKendrick, A. G. (1927). A Contribution to the Mathematical Theory of Epidemics. *Proceedings of the Royal Society of London Series A, 115*, 700–721.

Lee, E. (2008). Cyber Physical Systems: Design Challenges. University of California, *Berkeley Technical Report No. UCB/EECS-2008-8*.

Levine, W. S. (Ed.) (1996). *The Control Handbook*. New York: CRC Press. ISBN: 9780-8493-85704

Lighthill, M. J., & Whitham, G. B. (1955). On Kinematic Waves II. A Theory of Traffic Flow on Long Crowded Roads. *Proceedings of the Royal Society of London, A229*(1178), 317–345.

Maier, M. W. (1998). Architecting Principles for Systems-of-Systems. *Systems Engineering, 1*(4), 267–284.

Marceau, D. (1999). The Scale Issue in Social and Natural Sciences. *Canadian Journal of Remote Sensing, 25*, 347–356.

May, A. D. (1990). *Traffic Flow Fundamentals*. Englewood Cliffs, NJ: Prentice-Hall.

Miller, J. H., & Page, S. E. (2007). *Complex Adaptive Systems: An Introduction to Computational Models of Social Life*. Princeton University Press. ISBN:9781400835522. OCLC 760073369

Mittal, S., & Zeigler, B. P. (2017). Theory and Practice of M&S in Cyber Environments. In A. Tolk & T. Oren (Eds.), *The Profession of Modeling and Simulation: Discipline, Ethics, Education, Vocation, Societies and Economics* (pp. 223–263). Hoboken, NJ: Wiley.

Muzy, A., & Zeigler, B. P. (2014). Specification of Dynamic Structure Discrete Event Systems Using Single Point Encapsulated Functions. *International Journal of Modeling, Simulation and Scientific Computing, 5*(3), 1450012.

Nagel, K., & Schreckenberg, M. (1992). A Cellular Automaton Model for Freeway Traffic. *Journal de Physique I, 2*(12), 2221–2229.

O'Neill, R. V., Johnson, A. R., & King, A. W. (1989). A Hierarchical Framework for the Analysis of Scale. *Landscape Ecology, 3*, 193–205.

Petty, M. D., & Weisel, E. W. (2003). A Formal Basis for a Theory of Semantic Composability. In *Proceedings of the Spring Simulation Interoperability Workshop*, 03S-SIW-054, Orlando, FL.

Rainey, L. B., & Tolk, A. (2015). *Modeling and Simulation Support for System of Systems Engineering Applications*. Hoboken, NJ: Wiley.

Rajhans, A., Bhave, A., Ruchkin, I., Krogh, B. H., Garlan, D., Platzer, A., & Schmerl, B. (2014). Supporting Heterogeneity in Cyber-Physical Systems Architectures. *IEEE Transactions on Automatic Control, 59*(12), 3178–3193.

Ratzé, C., Gillet, F., Müller, J. P., & Stoffel, K. (2007). Simulation Modelling of Ecological Hierarchies in Constructive Dynamical Systems. *Ecological Complexity, 4*(1–2), 13–25.

Reynolds, P. F. J., Natrajan, A., & Srinivasan, S. (1997). Consistency Maintenance in Multi-resolution Simulations. *ACM Transactions on Computer Modeling and Simulation (TOMACS), 7*(3), 368–392.

Salthe, S. N. (1985). *Evolving Hierarchical Systems: Their Structure and Representation.* New York: Columbia University Press.

Seck, M. D., & Honig, H. J. (2012). Multi-perspective Modelling of Complex Phenomena. *Computational and Mathematical Organization Theory, 18,* 128–144.

Tatikonda, S., & Mitter, S. (2004). Control Under Communication Constraints. *IEEE Transactions on Automatic Control, 49*(7), 1056–1068.

Tolk, A., Barros, F., D'Ambrogio, A., Rajhans, A., Mosterman, P., Shetty, S. S., ... Yilmaz, L. (2018a). Hybrid Simulation for Cyber Physical Systems: A Panel on Where Are We Going Regarding Complexity, Intelligence, and Adaptability of CPS Using Simulation. In *Proceedings of the Spring Simulation Multi-Conference: Symposium on Modeling and Simulation of Complexity in Intelligent, Adaptive and Autonomous Systems (MCIAAS)*, Article No. 3, Baltimore, MD: SCS/ACM.

Tolk, A., Diallo, S., & Mittal, S. (2018b). The Challenge of Emergence in Complex Systems Engineering. In S. Mittal, S. Diallo, & A. Tolk (Eds.), *Emergent Behavior in Complex Systems Engineering: A Modeling and Simulation Approach.* Hoboken, NJ: Wiley.

Tolk, A., Diallo, S. Y., King, R. D., & Turnitsa, C. D. (2009). A Layered Approach to Composition and Interoperation in Complex Systems. In A. Tolk & L. C. Jain (Eds.), *Complex Systems in Knowledge-Based Environments: Theory, Models and Applications* (pp. 41–74). Berlin, Germany: Springer.

Tolk, A., Diallo, S. Y., & Turnitsa, C. D. (2013). Applying the Levels of Conceptual Interoperability Model in Support of Integratability, Interoperability, and Composability for System-of-Systems Engineering. *Systemics, Cybernetics and Informatics, 5*(5), 65–74.

Tolk, A., & Muguira, J. A. (2003). The Levels of Conceptual Interoperability Model (LCIM). In *Proceedings of the Fall Simulation Interoperability Workshop*, 03F-SIW-007, San Diego, CA: IEEE.

Traoré, M. K., & Muzy, A. (2006). Capturing The Dual Relationship Between Simulation Models and Their Context. *Simulation Modelling Practice and Theory, 14*(2), 126–142.

Traoré, M. K., Zacharewicz, G., Duboz, R., & Zeigler, B. P. (2018). Modeling and Simulation Framework for Value-Based Healthcare Systems. *SIMULATION: Transactions of the SCS.* https://doi.org/10.1177/0037549718776765, accessed 24 July 2018

Ulanowicz, R. E. (1997). *Ecology, the Ascendant Perspective.* New York: Columbia University Press.

Vangheluwe, H, De Lara, J., & Mosterman, P. J. (2002, April 7–10). An Introduction to Multi-Paradigm Modelling and Simulation. In *Proceedings of the AIS'2002 Conference* (pp. 9–20). Lisboa, Portugal.

Voltera, V. (1931). Variations and Fluctuations of the Number of Individuals in Animal Species Living Together. In R. N. Chapman (Ed.), *Animal Ecology* (pp. 31–113). New York: McGraw–Hill.

Willekens, F. (2005). Biographic Forecasting: Bridging the Micro-macro Gap in Population Forecasting. *New Zealand Population Review*, *31*(1), 77–124.

Wu, J., & David, J. L. (2002). A Spatially Explicit Hierarchical Approach to Modeling Complex Ecological Systems: Theory and Applications. *Ecological Modeling*, *153*, 7–26.

Yilmaz, L., & Oren, T. I. (2004). Dynamic Model Updating in Simulation with Multimodels: A Taxonomy and a Generic Agent-Based Architecture. *Simulation Series*, *36*(4), 3.

Zeigler, B. P. (1976). *Theory of Modeling and Simulation*. New York: Wiley.

Zeigler, B. P. (1984). *Multifacetted Modelling and Discrete Event Simulation*. London: Academic Press.

Zeigler, B. P., Carter, E., Seo, C., Russell, C. K., & Leath, B. A. (2012, October 28–31). Methodology and Modeling Environment for Simulating National Health Care. In *Proceedings of the Autumn Simulation Multi-Conference* (pp. 30–46)., San Diego, CA.

Zeigler, B. P., & Hammonds, P. E. (2007). *Modeling and Simulation-Based Data Engineering: Introducing Pragmatics into Ontologies for Net-Centric Information Exchange*. Amsterdam, the Netherlands: Elsevier Academic Press.

Zeigler, B. P., Mittal, S., & Traore, M. K. (2018a). MBSE with/out Simulation: State of the Art and Way Forward. *Systems*, *6*(4), 40. https://doi.org/10.3390/systems6040040

Zeigler, B. P., Mittal, S., & Traoré, M. K. (2018b, April 15–18). Fundamental Requirements and DEVS Approach for Modeling and Simulation of Complex Adaptive System of Systems: Healthcare Reform. In *Proceedings of the Spring Simulation Multi-Conference (Spring-Sim'18)*, Baltimore, MD: SCS/ACM.

Zeigler, B. P., Praehofer, H., & Kim, T. G. (2000). *Theory of Modeling and Simulation: Integrating Discrete Event and Continuous Complex Dynamic Systems* (2nd ed.). New York: Academic Press.

Zeigler, B. P., & Sarjoughian, H. S. (2017). *Guide to Modeling and Simulation of Systems of Systems* (2nd ed.). Berlin, Germany: Springer.

Zeigler, B. P., Traoré, M. K., Zacharewicz, G., & Duboz, R. (2019). *Value-Based Learning Healthcare Systems: Integrative Modeling and Simulation Architecture*. London: The Institution of Engineering and Technology.

5

A Unifying Framework for the Hierarchical Co-Simulation of Cyber-Physical Systems

Fernando J. Barros

Department of Informatics Engineering, University of Coimbra, Coimbra, Portugal

5.1 Introduction

The representation of complex cyber-physical systems usually requires the combination of models expressed in different formalisms/paradigms. The interoperability of heterogeneous models becomes a major requirement for the simulation of complex systems. In this section, we propose the Hybrid Flow System Specification Formalism (HYFLOW) as a unifying framework for M&S of hybrid systems.

The advantages of a unifying formalism include the homogeneous treatment of all types of models. A unifying formalism makes all models interoperable by design, requiring no special *glue* operators traditionally necessary to combine heterogeneous models (Praehofer 1991). Unifying formalisms have been created in other areas, like physics, where, for example, the work of Newton provides the foundations of classical mechanics.

HYFLOW provides a description of modular hybrid models that combine sampling (Barros 2002) with discrete events (Zeigler 1984). HYFLOW models can be independently simulated since their interaction is exclusively based on message passing that guarantees model encapsulation since this type of communication requires no access to model state. Formalism co-simulation is supported by two basic operators: the ability to exactly represent dense outputs on a digital computer and the concept of generalized sampling (Barros 2002). These constructs have shown to provide the basis for describing a large variety of numerical methods, including first-order, second-order (geometric), and exponential integrators (Barros 2018). Some models of hybrid systems often exhibit chattering behavior, imposing a very large number of events during a short interval. This feature is

Complexity Challenges in Cyber Physical Systems: Using Modeling and Simulation (M&S) to Support Intelligence, Adaptation and Autonomy, First Edition. Edited by Saurabh Mittal and Andreas Tolk.

common, for example, in relay systems that require models to be switched (Bonilla et al. 2012). In same cases, model switching can slowdown simulation or even make it virtually impossible. HYFLOW can represent PID digital controllers and sliding-mode controllers, a solution for achieving chattering-free representations (Barros 2017). The formalism can also describe digital filters, zero-detectors, and Fluid Stochastic Petri Nets (Barros 2005, 2015).

HYFLOW provides a new approach to the representation of models described in multiple paradigms. Instead of treating models as heterogeneous, the formalism promotes a unifying view where all models are regarded as a particular realization of the basic HYFLOW model. The interoperability is guaranteed by construction, since models share the same underlying description. The semantics of Sampling-based Systems was established through the definition of the equivalent I/O System in Barros (2002). HYFLOW co-simulation algorithms for both atomic and network models are described in Barros (2016).

The HYFLOW approach to hierarchical co-simulation is depicted in Figure 5.1. HYFLOW model libraries can be developed to address domain specific requirements. They can include, for example, digital controllers, digital filters, or numerical integrators. Since model composition and interoperability is guaranteed by HYFLOW, hierarchical complex simulation models can be easily developed. Domain specific modeling formalisms (DSMFs) can be mapped into HYFLOW through the choice of an appropriate set of models. For example, Ordinary Differential Equations (ODEs) can be mapped into models that perform numerical the integration. A complex system can use different kinds of integrators, including geometric (Section 5.4.2), for energy conservative components, and exponential (Section 5.4.1), to represent stiff ODEs. Given that all HYFLOW models are interoperable by design, the simulation of complex systems can be achieved by components traditionally considered belonging to different formalism. For

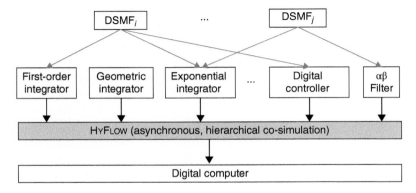

Figure 5.1 HYFLOW modular approach to M&S of cyber-physical systems.

example, a geometric integrator can be combined with a digital controller or a Fluid-Stochastic Petri Net to model a system. This approach enables the creation of new domain specific modeling languages and paradigms that choose a particular set of models as the basic blocks.

In our diagram ODEs do not play a first class role, used in more traditional views like System Dynamics (SDs) (Sterman 2001). Actually, SDs assume an analog computer paradigm that can solve ODEs exactly. However, this view is misleading, since nowadays ODE integration is likely to be performed by digital computers that use sampled-based algorithms that are prone to numerical errors. Several families of numerical integrators have been developed to improve the accuracy and efficiency of solutions to different types of ODEs. In Figure 5.1 we consider first-order, geometric (second-order) and exponential integrators, that have a HYFLOW representation. In our perspective, the key feature is not to simply offer support for ODE representation, but rather to expose the semantics of ODEs numerical integrators, enabling their composition and interoperability. These features become very important when using models not exclusively based on ODEs but including also other entities like digital controllers, for example. Failure to develop a unifying formalism, leads generally to *ad-hoc* implementations with loose defined semantics for *gluing* heterogeneous components.

Given HYFLOW unifying nature, concepts like heterogeneous and multi-paradigm models loose importance since all models become an instance of a common HYFLOW representation. Models become homogeneous, and belonging to the same paradigm.

5.2 Related Work

Co-simulation has its origins in the work on hierarchical and modular models developed by Zeigler (1984) where the concept of abstract simulator was introduced to enable the independent simulation of discrete event models. The High Level Architecture (HLA) provides a standard for the co-simulation of discrete event systems (Kuhl et al. 1999). Unfortunately, the representation of continuous systems did not accomplish the sound description that was achieved for discrete event models. Currently, modeling formalisms for continuous systems still rely on the nowadays extinct analogous machines (Burns and Kopp 1961; Henzinger 1996), and they were not updated to reflect the ubiquitousness of digital computers. The representation of continuous systems on a digital computer, while enable their co-simulation, is currently a major research challenge (Sztipanovits 2007; Neema et al. 2014; Tripakis 2015).

A standard for the co-simulation of hybrid systems has been proposed using the concept of Function Mock-up Interface (FMI) (Bastian et al. 2011). However, this approach achieves the interaction of models using piecewise constant signals, making it difficult to obtain accurate and efficient results. This limited representation of signals has also been used in other modeling approaches (Tripakis et al. 2013).

Modeling formalisms can have a discrete event representation at the atomic level (Vangheluwe et al. 2002). However, this approach has several shortcomings in the description of complex models. The representation, for example, of FSPNs by a composition of basic components is of utmost importance if our aim is to simplify the automatic translation of FSPNs into executable simulation models. As shown in Section 5.5, our modular representation based on HYFLOW makes it simple to map a FSPN into a HYFLOW network, requiring no code generation since it is based on the reuse of the predefined components. On the contrary, a solution based on the transformation of FSPNs into one (complex) atomic model will mostly require a compiler to perform the task. Moreover, this equivalent atomic model will become very difficult to be understood by a (human) modeler. Additionally, our modular approach enables FSPNs to be easily combined with components from other domain specific libraries to achieve complex representations based on multiple formalisms. The modular approach taken by HYFLOW enables also model co-simulation that is not promoted by monolithic transformation solutions.

5.3 The HYFLOW Formalism

The Hybrid Flow System Specification (HYFLOW) is a formalism for representing hybrid systems with a time-variant topology (Barros 2003). HYFLOW achieves the representation of continuous variables using the concept of multi-sampling and dense output (Barros 2000 2002), while the representation of discrete events is based on the Discrete Event System Specification (DEVS) (Zeigler 1984). HYFLOW has two types of models: basic and network. Basic models provide state representation and transition functions. Network models are a composition of basic models and other network models. Given its definition, a network provides an abstraction for representing hierarchical systems.

5.3.1 The Basic HYFLOW Model

We consider $\hat{B}$ as the set of names corresponding to basic HYFLOW models. A HYFLOW basic model associated with name $B \in \hat{B}$ is defined by

$$M_B = \left(X, Y, P, P_0, \rho, \omega, \delta, \Lambda_c, \lambda_d \right)$$

where

$X = X_c \times X_d$ is the set of input flow values

X_c is the set of continuous input flow values
X_d is the set of discrete input flow values

$Y = Y_c \times Y_d$ is the set of output flow values

Y_c is the set of continuous output flow values
Y_d is the set of discrete output flow values

P is the set of partial states (p-states)
$P_0 \subseteq P$ is the set of (valid) initial p-states
$\rho : P \to \mathbb{H}_0^+$ is the time-to-input function
$\omega : P \to \mathbb{H}_0^+$ is the time-to-output function
$S = \{(p, e) | p \in P, 0 \le e \le \nu(p)\}$ is the state set with $\tau(p) = \min\{\rho(p), \omega(p)\}$, the time-to-transition function
$\delta : S \times X^\phi \to P$ is the transition function
    where $X^\phi = X_c \times (X_d \cup \{\phi\})$ and ϕ is the null value (absence of value)
$\Lambda_c : S \to Y_c$ is the continuous output function
$\lambda_d : P \to Y_d$ is the partial discrete output function

The HYFLOW time base is the set of hyperreal numbers $\mathbb{H} = \{x + z\varepsilon \,|\, x \in \mathbb{R}, z \in \mathbb{Z}\}$, where ε is an infinitesimal value, such that $\varepsilon > 0$ and $\varepsilon < \dfrac{1}{n}$ for $n = 1, 2, 3, \ldots$ (Goldblatt 1998). The set of positive hyperreals is defined by $\mathbb{H}_0^+ = \{h \in \mathbb{H} \,|\, h \ge 0\}$. The discrete output of a component described by a HYFLOW basic model is constrained to be null (ϕ) when in states (s, e) with $e \ne \omega(p)$.

Figure 5.2 depicts the typical trajectories of a HYFLOW component. At time t_1 the component in p-state p_0 samples its input since its elapsed time reaches $\rho(p_0) = e$. The component changes its p-state to $p_1 = \delta((p_0, \rho(p_0)), (x_1, \phi))$, where x_1 is the sampled value and no discrete flow is present.

At time t_2 the discrete flow x_d is received by the component that changes to p-state $p_2 = \delta((p_1, e_1), (x_2, x_d))$, where x_2 is the continuous flow at t_2. At time t_3 the component reaches the time-to-output time limit and it changes to p-state $p_3 = \delta((p_2, \omega(p_2)), (x_3, \phi))$. At this time the discrete flow $y_d = \lambda_d(p_2)$ is produced. Additionally, component continuous output flow is always present and given by $\Lambda_c(p, e)$. Semantics of HYFLOW basic models is detailed in (Barros 2008). In the next section we provide the HYFLOW description of a pulse integrator, a numerical solver that efficiently handles piecewise constant signals.

5.3.2 Example: Pulse Integrator

The integration of ODEs plays a key role in describing many classes of dynamical systems. In previous work, we have developed general purpose ODE integrators (Barros 2015). However, more efficient algorithms can be used for particular types of signals. Piecewise constant flows, like square waves, can be integrated with a simpler and more efficient solver. An integrator for piecewise constant signals can be represented by the HYFLOW model:

$$M_\pi = \left(X, Y, P, P_0, \rho, \omega, \delta, \Lambda_c, \lambda_d \right)$$

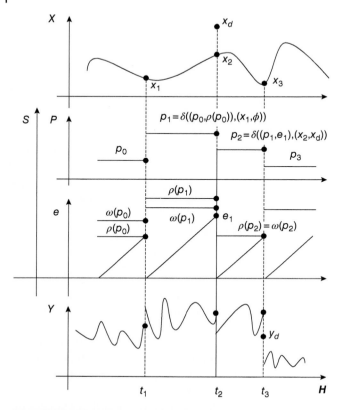

Figure 5.2 Basic HyFlow component trajectories.

where

$$X = Y = \mathbb{R} \times \mathbb{R}$$

$$P = \{(y, v, b) \,|\, y, v \in \mathbb{R}; \, b \in \{\top, \bot\}\}; \, P_0 = \{(y_0, 0, \top) \in P\}$$

$$\rho\left(y, v, b\right) = \infty$$

$$\omega(y, v, \top) = 0; \, \omega(y, v, \bot) = \infty$$

$$\delta(((y, v, \bot), e), (x_c, x_d)) = (y + v e_{std}, x_c, \top)$$

$$\delta(((y, v, \top), e), (x_c, x_d)) = (y, v, \bot)$$

$$\Lambda_c((y, v, b), e) = y + v e_{std}$$

$$\lambda_d(y, v, b) = y$$

The model obtains its efficiency by exploiting the features of the input signal. Contrarily to integrators that need to handle arbitrary input segments, this model avoids sampling, because all changes in the input value are signalled by a discrete flow. Model continuous flow output is piecewise linear and described by the

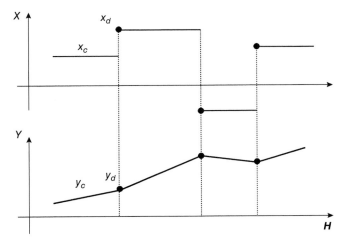

Figure 5.3 Pulse integrator input/output trajectories.

function Λ_c. This flow is available by other components that can sample it asynchronously at their own sampling rate. To facilitate model composition, all discrete flows received by the integrator are sent as discrete flow outputs. Typical pulse integrator input and output trajectories are depicted in Figure 5.3.

5.3.3 HyFlow Network Model

HyFLOW network models are compositions of HyFLOW models (basic or other HyFLOW network models). Let $\hat{N}$ be the set of names corresponding to HyFLOW network models, with $\hat{N} \cap \hat{B} = \{\}$. Formally, a HyFLOW network model associated with name N is defined by

$$M_N = (X, Y, \eta)$$

where

N is the network name
$X = X_c \times X_d$ is the set of network input flows

X_c is the set of network continuous input flows
X_d is the set of network discrete input flows

$Y = Y_c \times Y_d$ is the set of network output flows

Y_c is the set of network continuous output flows
Y_d is the set of network discrete output flows

η is the name of the dynamic topology network executive

The executive model is a modified HYFLOW basic model, defined by

$$M_\eta = \left(X_\eta, Y_\eta, P, P_0, \rho, \omega, \delta, \Lambda_c, \lambda_d, \hat{\Sigma}, \gamma\right)$$

where

$\hat{\Sigma}$ is the set of network topologies
$\gamma : P \rightarrow \hat{\Sigma}$ is the topology function

The network topology $\Sigma_\alpha \in \hat{\Sigma}$, corresponding to the p-state $p_\alpha \in P$, is given by

$$\Sigma_\alpha = \gamma\left(p_\alpha\right) = \left(C_\alpha, \{I_{i,\alpha}\} \cup \{I_{\eta,\alpha}, I_{N,\alpha}\}, \{F_{i,\alpha}\} \cup \{F_{\eta,\alpha}, F_{N,\alpha}\}\right)$$

where

C_α is the set of names associated with the executive state p_α
for all $i \in C_\alpha \cup \{\eta\}$
$I_{i,\alpha}$ is the sequence of influencers of i
$F_{i,\alpha}$ is the input function of i
$I_{N,\alpha}$ is the sequence of network influencers
$F_{N,\alpha}$ is the network output function

For all $i \in C_\alpha$

$$M_i = \left(X, Y, P, P_0, \rho, \omega, \delta, \Lambda_c, \lambda_d\right)_i \quad \text{if} \quad i \in \hat{B}$$

$$M_i = \left(X, Y, \eta\right)_i \quad \text{if} \quad i \in \hat{N}$$

The topology of a network is defined by its executive through the topology function γ, which maps the executive p-state into network composition and coupling. Thus, topology adaption can be achieved by changing the executive p-state. The use of HYFLOW dynamic topologies to represent mobile entities is described in Barros (2016).

HYFLOW network models are simulated by HYFLOW modular network simulators that perform the orchestration of basic or other network models. Network co-simulation is achieved by a general communication protocol that relies only on component interface. This protocol is independent from model internal details, enabling the composition of components that can be viewed as black boxes. A description of HYFLOW network master co-simulation algorithm is provided in Barros (2008).

5.4 Numerical Integration

Cyber-physical systems (CPSs) are a combination of software and psychical systems. While the former can be described using a discrete representation, physical elements usually require the use of continuous models. Ordinary Differential Equations (ODEs) play a key role in the description of continuous systems.

However, given requirements for efficiency, accuracy, and stiffness for example, there is no universal numerical integrator suitable to solve all families of ODEs. Since CPSs may have continuous models with different requirements, we consider that the ability to combine different numerical integrators is fundamental for representing complex CPSs.

Conventional ODEs numerical integrators require the transformation of arbitrary order ODEs into a large system of first-order equations that is commonly solved by a single integration algorithm. Thus, the prevalent approach does not promote model independence and it does not provide the basis for co-simulation. Co-simulation requires the development of new methods. Quantization (Zeigler and Lee 1998) achieved a representation of first-order integrators in a modular form, enabling the co-simulation of hybrid systems. Unfortunately, conventional first-order integrators do not provide the best approach to solve all types of ODEs. Other kinds of numerical methods have been developed to solve different families of ODEs. Geometric integrators, for example, are fundamental to represent second-order energy preserving systems that need to be simulated for long periods (Hairer et al. 2005). Other types of integrators can address stiffness using analytic approaches. This is the case of solvers that seek solutions in the form of exponential functions instead of using conventional polynomial interpolation, for integrating stiff ODEs (Hochbruck 2010).

Albeit important, ODEs do not have the exclusivity in systems representation. In particular, hybrid systems require the ability to combine ODEs with models that generate a discontinuous behavior. Examples include hybrid function generators, zero-detectors, digital controllers, and digital filters. Additionally, hybrid systems are affected by some serious problems that include chattering and Zeno- behavior (Johansson et al. 1999). To succeed, the co-simulation of CPSs need to tackle all these issues, while guaranteeing a sound semantics for achieving a deterministic simulation. Given the diversity of requirements, the problem of finding a unifying co-simulation representation for hybrid systems seems an overwhelming task. We consider, however, that an approach based on a reduced set of sound operators may provide a good solution for establishing a general co-simulation framework.

The efficient integration of ODEs also requires the use of adaptive step size numerical integrators that adjust the step in order to keep error within some bound. Conventional solvers adjust the sampling rate, performing an integration step when any of the equations exceeds the maximum allowed error (Epperson 2002). This type of integrator is synchronous, since all equations are integrated at the same (although variable) time step. We describe next two types of numerical integrators in the HYFLOW formalism.

5.4.1 Exponential Integrators

Common numerical integrators like Adams and BDF are based on polynomial interpolation. A different approach has been taken by exponential time differencing (ETD) methods that use the exact solution for the linear part of the ODE (Cox and Matthews 2002). ETD methods have several advantages over conventional integrators, including the possibility of using large step-size, even when integrating stiff ODEs. ETD methods consider ODEs composed by a linear part and a non-linear part in the form of:

$$y' = cy + F(x(t), y), y(0) = y_0$$

where F is a non-linear function. The *exact* solution for $t \in [0, T]$ is given in (Cox and Matthews 2002):

$$y(t) = y(0)e^{ct} + e^{ct} \int_0^t e^{-ct} F(x(\tau), y(\tau)) d\tau$$

For numerical integration, an approximation of F is required. Considering an integrator with a fixed step-size T, the solution for a zero-order approximation of F is given in (Cox and Matthews 2002):

$$y_{n+1} = y_n e^{cT} + \frac{e^{cT} - 1}{c} F(x_n, y_n)$$

The HYFLOW representation for a zero-order ETD that works at a fixed step-size T is given by:

$$M_x = (X, Y, P, P_0, \rho, \omega, \delta, \Lambda, \lambda_d)$$

where

$X = \mathbb{R} \times \mathbb{R}; Y = \mathbb{R} \times \{\phi\}$

$P = \{(\alpha, x, y) \mid \alpha, x, y \in \mathbb{R}\}$

$P_0 = \{(0, 0, y_0) \in P\}$

$\rho(\alpha, x, y) = \alpha; \omega(\alpha, x, y) = \infty$

$\delta(((0, x, y), h), (x_c, x_d)) = (T, x_c, y)$

$\delta(((\alpha, x, y), h), (x_c, x_d)) = \left(\alpha, x_c, y e^{ch} + \frac{e^{ch} - 1}{c} F(x, y) \right)$

$\Lambda c((\alpha, x, y), h) = y e^{ch} + \frac{e^{ch} - 1}{c} F(x, y)$

$\lambda_d(\alpha, x, y) = \phi$

with $\hbar = h_{std}$

The model specifies an initial value $\alpha = 0$, so it samples the input value at simulation start. After the beginning, the sampling period is set to T. The sampling interval can be modified by discrete flows that usually signal an input discontinuity.

ETD continuous flow output is an exponential function described by the continuous output function Λ_c. The ETD integrator can be sampled at an asynchronous rate, making composition with other HYFLOW models easy to achieve. An example showing the use of ETD integrators is given in Section 5.4.3. We provide next the description of Geometric integrators.

5.4.2 Geometric Integrators

The prevalent rule in most common modeling and simulation tools is to map the set of arbitrary order ODEs into a system of first-order ODEs (Fritzson 2015). This approach does not produce accurate results when simulating systems, like some energy-preserving systems, where solutions have properties that can only be observable after long intervals (Hairer et al. 2003). For example, in systems involving celestial mechanics, properties like the precession of planets, may require the simulation of hundreds of years to be characterized. The study of CPSs, like a space probe traveling in the solar system, also requires accurate models when simulated during long periods of time.

Since traditional methods based on decomposition are not acceptable for representing energy conserving systems, a direct representation of higher order ODEs have been proposed by Hairer et al. (2005). We describe next the HYFLOW representation of geometric ODEs. Given the second-order ODE

$$y'' = f\left(x(t), y\right) \quad \text{with} \quad y(0) = y_0, \ y'(0) = v_0$$

and using the variable $v = y'$, a fixed step-size T, second-order ODE, second-degree polynomial approximation, geometric integrator is described by the equations (Swope et al. 1982):

$$y_{n+1} = y_n + hv_n + \frac{1}{2}h^2 f_n$$

$$v_{n+1} = v_n + \frac{h}{2}\left(f_n + f_{n+1}\right)$$

Considering $y'' = f(x(t))$, the HYFLOW model of the geometric integrator is given by:

$$M_\Gamma = \left(X, Y, P, P_0, \rho, \omega, \delta, \Lambda_c, \lambda_d\right)$$

where

$$X = \mathbb{R} \times \phi$$
$$Y = \mathbb{R} \times \phi$$
$$P = \left\{\left(\alpha, y_n, v_n, f_n\right) \mid \alpha, y_n, v_n, f_n \in \mathbb{R}\right\}$$

$$P_0 = \{(0, y_n, v_n, f_n) \in P\}$$
$$\rho(\alpha, y_n, v_n, f_n) = \alpha$$
$$\omega\left(\alpha, y_n, v_n, f_n\right) = \infty$$
$$\delta\left(\left(\left(\alpha, y_n, v_n, f_n\right), e\right), \left(x_c, x_d\right)\right) = \left(T, y_{n+1}, v_n + \frac{1}{2}e_{std}\left(f_n + f_{n+1}\right), f_{n+1}\right) \text{ with}$$
$$y_{n+1} = y_n + e_{std}v_n + \frac{1}{2}e_{std}^2 f_n \text{ and } f_{n+1} = f(x_c)$$
$$\Lambda_c\left(\left(\alpha, y_n, v_n, f_n\right), e\right) = y_n + e_{std}v_n + \frac{1}{2}e_{std}^2 f_n$$
$$\lambda_d(\alpha, y_n, v_n, f_n) = \phi.$$

The recurrences are computed by the transition function δ and the sampling period T is specified by function ρ. The initial sampling period is set to 0 so the associated component can read the input and compute the initial value of f. The output flow is given by the second-degree polynomial provided by function Λ_c. Given that sampling rate can be adjusted by the ρ function, HYFLOW can also represent variable sampling geometric integrators (Hairer and Soderlind 2005). In the next section we provide an example of model composition that uses a geometric integrator.

5.4.3 Model Composability

Hierarchical and modular representation provide a powerful approach for describing complex CPSs. A divide-and-conquer strategy based on decomposition reduces complexity and promotes model reuse. A complex system can then be represented by a composition of simpler models that are easier to develop. Additionally, decomposition promotes library of models, enabling complex systems to be represented by the combination of pre-existing components. A modular representation promotes thus a faster design/test process of complex CPSs. Additionally, hierarchical models enable complex networks to be handled as basic models, further increasing our ability to tackle complexity. HYFLOW provides full support for hierarchical and modular models as described in Section 5.3. Given formalism semantics (Barros 2008), hierarchical co-simulation is also supported, enabling complex systems to be represented by networks composed by models developed independently, while guaranteeing models encapsulation, and intellectual property. As an example of HYFLOW model composability we consider the simple LC circuit with inductance L and capacity C connected to a voltage source $v(t)$, depicted in Figure 5.4. Capacitor voltage $u(t)$ is described by:

$$u'' = \frac{v - u}{LC}.$$

Circuit current $i(t)$ is given by:

$$i = Cu'.$$

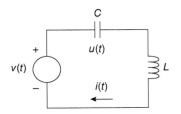

Figure 5.4 LC circuit.

Since the electric circuit has no dissipating elements (resistors), it becomes a second-order energy conservative system. A good solution is provided by the geometric integrator described in Section 5.4.2

For simulation, we consider a circuit with $C = 1\,\text{mF}$ and an inductance $L = 1\,\text{mH}$. The input voltage is defined as a triangular signal obtained by the integration of the square signal $\{-5, 5\}$ V with a half-period of 0.00628s. This circuit is described by Figure 5.4.

The HYFLOW network representation of the circuit is depicted in Figure 5.5, where the square signal s is integrated by the pulse integrator π to produce the triangular wave.

The HYFLOW network is defined by:

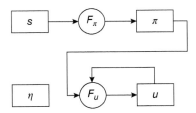

Figure 5.5 HYFLOW LC circuit network.

$$C = \{s, \pi, u\}, I_\eta = I_s = \{\ \}, I_\pi = \{s\}, I_u = \{u, \pi\}$$
$$F_\pi(s_c, s_d) = (x_c, x_d), F_u((u_c, u_d), (\pi_c, \pi_d)) = \left(\frac{\pi_c - u_c}{LC}, \pi_d\right).$$

The discrete flows produced by π are used by u as information on signal discontinuities. Simulation results for a period of 0.1 s are represented in Figure 5.6 that plots capacitor voltage and current.

While the geometric integrator is a sampled-based model, both the square wave generator and the pulse integrator are event-based models. HYFLOW, however, enables the seamless integration of these models without making any assumption about their internal behaviour, showing formalism composition and interface adaptation capabilities. HYFLOW can, thus, support the co-simulation of a large variety of models (Barros 2016).

A more complex example of HYFLOW model composability is provided in Barros (2016), where a defense scenario involving drones, airborne scanning radars, and pursuers is described. This example uses HYFLOW dynamic topologies for updating the communication links between entities that move in a 2D-space. A guidance digital controller is also employed for setting pursuers' trajectories.

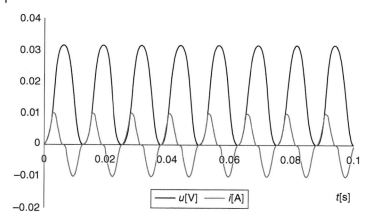

Figure 5.6 Capacitor voltage and current for a triangular input voltage.

ODEs has been considered as a homogeneous model paradigm and common approaches map a set of ODEs into an atomic model with a single numerical integrator (Fritzson 2015). However, as shown here, the ability to merge different integrators is fundamental to represent complex systems, where specific numerical methods need to be attached to models, for accuracy or stiffness reasons, for example. The multi-paradigm approaches fail to identify this problem. It addresses mainly the specification level, implicitly assuming an underlaying analog computer where numerical integrators are ideal/error-free. However, the actual challenge is a formalism, that like HYFLOW, enables the interoperability at the numerical level, since simulation is performed by digital computers and numerical integrators cannot be abstracted.

In the next section we provide a HYFLOW modular representation of Fluid Stochastic Petri Nets, a modeling formalism that extends Petri Nets, enabling the description of hybrid systems.

5.5 Fluid Stochastic Petri-Nets

HYFLOW can be used to represent other models/paradigms. Once a representation is developed, the new created elements can be composed with other HYFLOW models, effectively enabling multiparadigm M&S. As an example, we consider next the HYFLOW representation of Fluid Stochastic Petri-Nets (FSPNs) (Trivedi and Kulkarni 1993). Petri Nets have been widely used for modeling, simulation, verification, and analysis. Initially, created for studying discrete systems, many extensions have been developed, enabling Petri Nets to describe a large variety of systems (David and Alla 2010). In particular, FSPNs were introduced for describing hybrid systems that require a large number of tokens, since the explicit representation of each token would make a discrete Petri Net difficult to analyze and also time consuming to simulate. A fluid approximation is used, instead, becoming tokens represented by a real number whose value is governed by a piecewise constant rate.

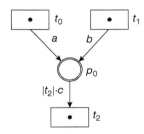

Figure 5.7 FSPN with variable rate $|t_2| \cdot c$.

For detailing the semantics of FSPNs we define $|t_k|$ as the number of instances of transition t_k currently active, and $|p_k|$ the quantity of tokens in place p_k.

In FSPNs, a transition t_k is considered active if $|t_k| > 0$. When a transition is active the corresponding flow is enabled and place content is influenced by that flow. On the contrary, when not active the corresponding flow is zero. Another constraint imposes that places can only contain positive values, i.e., $|p_k| \geq 0$.

Figure 5.7 represents a FSPN with transitions t_0, t_1, t_2, place p_0, constant flows a and b, and a variable flow controlled by the number of tokens in transition t_2.

When all transitions are enabled the content of place p_0 is described by:

$$\frac{d\,|p_0|}{dt} = a + b - |t_2| \cdot c.$$

When a transition is disabled, the corresponding flow is zero. For example, when $|t_0| = 0$, then $\dfrac{d\,|p_0|}{dt} = b - |t_2| \cdot c$.

FSPNs enable the representation of systems with both discrete and continuous semantics. The FSPN of Figure 5.8 models a manufacturing system with N machines that process at an (exponential) rate μ and breakdown at an (exponential) rate λ. Entities enter the system at rate a and are processed at rate $|t_1| \cdot d$. The initial number of entities to be produced is given by L and all machines are initially available, $|t_1| = N$. Given the semantics defined before, $|t_1|$ represents the number of machines available for production and $|t_2|$ is the number of machines being repaired (not working).

Figure 5.8 FSPN representation of machines subjected to breakdowns.

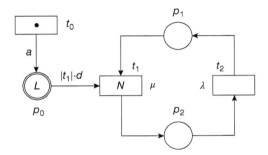

The HYFLOW description of FSPNs requires several types of models (Barros 2015). Discrete places are used to store (discrete) tokens. Reservoirs (holders of continuous places) are used to integrate the input flow. Given that FSPNs are constrained to piecewise constant flows, reservoirs exhibit a piecewise linear output and they can be described by the pulse integrators defined in Section 5.3.2. Given many transitions can be enabled at the same time, a conflict manager is necessary to decide what transition will actually take place, since when a transition is triggered it can disable other transitions that compete for the same resources (tokens). An additional component is required to represent transitions. In a Petri Net, a transition can have multiples instances scheduled to finish. A transition can be represented by the HYFLOW dynamic topology network of Figure 5.9, where each occurrence of a group of tokens scheduled to trigger is represented by a delay model. At the finish time, the delay is removed and the corresponding tokens are released to the Petri Net. These four models can be used to represent any FSPN at the benefit that the FSPN is not represented as a large monolithic model but rather is modeled as a modular network component with an explicit composition and coupling.

Although the mapping of FSPN into HYFLOW could be done with the help of a translating tool, given its modularity the mapping can also be made manually, without incurring in the expense of creating additional tools for modeling. Another benefit of the HYFLOW modular approach is that the four components developed to create a FSPN become available to define new formalisms where they can be used as building blocks. For example, if we relax the constraint of

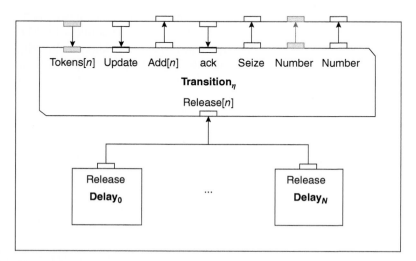

Figure 5.9 HYFLOW modular representation of a FSPN transition.

piecewise constant flows, another integrator, like the ETD described in Section 5.4.1, could be used to define the extend FSPN. Additionally, given that these four nodes are HyFLOW models they can be seamlessly combined with any other HyFLOW models to represent complex systems without any reference to any particular modeling paradigm. A modular approach was also developed to represent Stochastic Chemical Reaction Networks (SCRNs) (Barros 2012), where three HyFLOW models have been developed to enable a description of bio-molecular chemical reactions. If required, a combination of FSPNs and SCRNs HyFLOW models can be easily combined to create a new paradigm.

The HyFLOW model of the FSPN supports the conjecture expressed in Section 5.3 that the key aspect to support the composition of models described in different paradigms is the ability to develop a unifying formalism with well defined semantics for enabling model interoperability.

M&S of complex cyber-physical systems generally requires representations based on different modeling paradigms. This multi-paradigm approach enables experts to create models using the most appropriate formalism for each domain. A research challenge is to define a framework where the different formalism can communicate. HyFLOW provides a sound framework for supporting different numerical integrators as shown in Section 5.4 and also formalisms like FSPNs. Additional work is required for developing HyFLOW representations of other formalisms and numerical methods, providing support for a generalized M&S framework.

5.6 Conclusion

The HyFLOW formalism provides a unifying approach for representing sampled and discrete event based systems. HyFLOW networks keep model modularity enabling the co-simulation of hybrid systems. HyFLOW ability to represent continuous signals provides a framework for describing numerical integrators, the basic building blocks for computing the solution of differential equations. Since integrators are particular cases the HyFLOW basic model, they can be seamlessly combined for representing complex systems requiring the use of different numerical methods. HyFLOW enables the co-simulation of other paradigms, including Fluid Stochastic Petri Nets (FSPNs). Our results point to the generality of the HyFLOW formalism and its ability to describe a large variety of modeling paradigms. HyFLOW provides a new approach to the representation of multiple paradigms. Instead of treating models as heterogeneous, HyFLOW promotes a unifying view where all models are regarded as a particular realization of the same basic model. As future work we plan to develop HyFLOW representations of other formalisms, extending the sets of models that can be used to describe complex cyber-physical systems.

References

Barros, F. (2000, March 6–8). A framework for representing numerical multirate integration methods. In H. Sarjoughian, F. Cellier, M. Marefat, & J. Rozenblit (Eds.), *AI, Simulation and Planning in High Autonomy Systems*, Tucson, AZ, USA.

Barros, F. (2002). Towards a theory of continuous flow models. *International Journal of General Systems, 31*(1), 29–39.

Barros, F. (2003). Dynamic structure multiparadigm modeling and simulation. *ACM Transactions on Modeling and Computer Simulation, 13*(3), 259–275.

Barros, F. (2005, April 4–7). Simulating the data generated by a network of track-while-scan radars. In *Proceedings of the 12th Annual IEEE International Conference on Engineering Computer-Based Systems* (pp. 373–377). Greenbelt, MD, USA.

Barros, F. (2008). Semantics of discrete event systems. In R. Baldoni, A. Buchmann, & S. Piergiovanni (Eds.), *Distributed Event-Based Systems, Rome, Italy, 1–4 July*, ACM International Conference Proceeding Series (Vol. *332*, pp. 252–258). ACM.

Barros, F. (2012, December 9–12). A compositional approach for modeling and simulation of bio-molecular systems. In *Winter Simulation Conference* (pp. 2654–2665). Berlin, Germany.

Barros, F. (2015). A modular representation of fluid stochastic petri nets. In *Symposium on Theory of Modeling and Simulation*, San Diego: SCS Publishing.

Barros, F. (2015). Asynchronous, polynomial ode solvers based on error estimation. In F. Barros, M. Hwang, H. Prähofer, & X. Hu (Eds.), *Symposium on Theory of Modeling and Simulation, Alexandria, VA, USA, 12–15 April* (pp. 115–121). Red Hook, NY: Curran.

Barros, F. (2016). On the representation of time in modeling & simulation. In T. Roeder, P. Frazier, R. Szechtman, E. Zhou, T. Huschka, & S. Chick (Eds.), *Winter Simulation Conference. 11–14 December, Arlington, VA, USA* (pp. 1571–1582). Piscataway, NJ: IEEE.

Barros, F. (2016). Modeling mobility through dynamic topologies. *Simulation Modelling Practice and Theory, 69*, 113–135.

Barros, F. (2016). A modular representation of asynchronous, geometric solvers. In F. Barros, H. Prähofer, X. Hu, & J. Denil (Eds.), *Symposium on Theory of Modeling and Simulation, Pasadena, CA, USA, 3–5 April*. San Diego: SCS Publishing.

Barros, F. (2017). Chattering avoidance in hybrid simulation models: A modular approach based on the hyflow formalism. In *Symposium on Theory of Modeling and Simulation*. San Diego: SCS Publishing.

Barros, F. (2018, December 9–12). Composition of numerical integrators in the hyflow formalism. In *Winter Simulation Conference*, Gothenburg, Sweden.

Bastian, J., Clauß, C., Wolf, S., & Schneider, P. (2011, March 20–22). Master for co-simulation using FMI. In *Proceedings of the 8th Modelica Conference*, Dresden, Germany.

Bonilla, J., Yebra, L., & Dormido, S. (2012). Chattering in dynamic mathematical two-phase flow models. *Applied Mathematical Modelling, 36*, 2067–2081.

Burns, A. J., & Kopp, R. E. (1961). Combined analog-digital simulation. In *Proceedings of the December 12-14, 1961, Eastern Joint Computer Conference: Computers – Key to Total Systems Control, AFIPS '61 (Eastern)* (pp. 114–123). New York, NY, USA: ACM.

Cox, S. M., & Matthews, P. (2002). Exponential time differencing for stiff systems. *Journal of Computational Physics, 176*, 430–455.

David, R., & Alla, H. (2010). *Discrete, Continuous, and Hybrid Petri Nets*. Heidelberg: Springer.

Epperson, J. F. (2002). *An Introduction to Numerical Methods and Analysis*. New York: Wiley.

Fritzson, P. (2015). *Principles of Object-Oriented Modeling and Simulation with Modelica 3.3: A Cyber-Physical Approach* (2nd ed.). Piscataway, NJ: IEEE Press, Wiley-Interscience.

Goldblatt, R. (1998). *Lectures on the Hyperreals: An Introduction to Nonstandard Analysis*.
Number 188 in Graduate Texts in Mathematics. New York: Springer-Verlag.

Hairer, E., Lubich, C., & Wanner, G. (2003). Geometric numerical integration illustrated by the störmerverlet method. *Acta Numerica, 26*(6), 399–450.

Hairer, E., Lubich, C., & Wanner, G. (2005). *Geometrical Numerical Integration: Structure-Preserving Algorithms for Ordinary Differential Equations*. Number 31 in Springer Series in Computational Mathematics (2nd ed.). Berlin: Springer-Verlag.

Hairer, E., & Soderlind, G. (2005). Explicit, time reversible, adaptive step size control. *SIAM Journal of Scientific Computation, 26*(6), 1838–1851.

Henzinger, T. (1996, July 27–30). The theory of hybrid automata. In *Proceedings of the 11th Annual IEEE Symposium on Logic in Computer Science* (pp. 278–292). New Brunswick, NJ, USA.

Hochbruck, M. (2010). Exponential integrators. *Acta Numerica, 19*, 209–286.

Johansson, K. H., Egerstedt, M., Lygeros, J., & Sastry, S. (1999). On the regularization of zeno hybrid automata. *Systems and Control Letters, 38*, 141–150.

Kuhl, F., Weatherly, R., & Dahmann, J. (1999). *Creating Computer Simulation Systems: An Introduction to the High Level Architecture*. Upper Saddle River, NJ: Prentice Hall.

Neema, H., Gohl, J., Lattmann, Z., Sztipanovits, J., Karsai, G., Neema, S., ... Sureshkumar, C. (2014, March 10–12). Model-based integration platform for fmi co-simulation and heterogeneous simulations of cyber-physical systems. In *Proceedings of the 10th International Modelica Conference* (pp. 235–245). Lund, Sweden.

Praehofer, H. (1991). *System Theoretic Foundations for Combined Discrete-Continuous System Simulation* (PhD thesis). University of Linz.

Sterman, J. (2001). System dynamics modeling: Tools for learning in a complex world. *California Management Review, 4*, 8–25.

Swope, W., Andersen, H., Berens, P., & Wilson, K. (1982). A computer simulation method for the calculation of equilibrium constants for the formation of physical clusters of molecules: Application to small water clusters. *Journal of Chemical Physics, 76*(1), 637–649.

Sztipanovits, J. (2007, March 26–29). Composition of cyberphysical systems. In *Proceeding of the 14th Annual IEEE International Conference and Workshops on the Engineering of Computer-Based Systems* (pp. 3–6). Tucson, AZ, USA.

Tripakis, S. (2015, July 19–23). Bridging the semantic gap between heterogeneous modeling formalisms and FMI. In *Embedded Computer Systems: Architectures, Modeling, and Simulation* (pp. 60–69). Samos, Greece.

Tripakis, S., Stergiou, C., Shaver, C., & Lee, E. (2013). A modular formal semantics for ptolemy. *Mathematical Structures in Computer Science, 23*, 834–881.

Trivedi, K., & Kulkarni, V. (1993). FSPNs: Fluid stochastic Petri Net. In *Application and Theory of Petri Nets*, Lecture Notes in Computer Science (Vol. *691*, pp. 24–31). Springer Verlag.

Vangheluwe, H., de Lara, J., & Mosterman, P. J. (2002). An introduction to multi-paradigm modeling and simulation. In F. Barros & N. Giambiasi (Eds.), *AI, Simulation and Planning in High Autonomy Systems, 13–17 April, Lisbon, Portugal* (pp. 9–20). IEEE.

Zeigler, B. (1984). *Multifaceted Modelling and Discrete Event Simulation*. San Diego: Academic Press.

Zeigler, B., & Lee, J. S. (1998). Theory of quantized systems: Formal basis for DEVS/ HLA distributed simulation environment. In A. Sisti (Ed.), *Enabling Technology for Simulation Science II, volume 3369 of SPIE, Orlando, FL, USA, 13–17 April* (pp. 49–58). SPIE.

6

Model-Based Systems of Systems Engineering Trade-off Analytics

Aleksandra Markina-Khusid, Ryan Jacobs, and Judith Dahmann

The MITRE Corporation, McLean, VA, USA

6.1 Introduction

Increasingly today's societal capabilities are based on compositions of systems and system elements to provide the aggregate functionality needed to meet challenges across multiple domains. As these composite systems expand and adapt, their inherent complexity poses challenges for application of systems engineering methods which have been the hallmark of the discipline. Core to systems engineering is trade-off analysis – use of objective methods to assess alternatives to system architectures and to system components with respect to the system objectives. The systems engineering community has recognized that success of many of today's enterprises depends on our ability to effectively reason about and harness Systems of Systems (SoS) capabilities increasing the importance of addressing these challenges. This paper begins with a review of the relationships between systems of systems (SoS) and other composed systems, cyber-physical systems (CPS) and Internet of Things (IoT). Recognizing that CPS represent a class of SoS and as such can utilize the rich body of knowledge provided by SoSE, it explores the challenges in conducting trade-off analysis for SoS. The chapter then presents a set of approaches that can help to address these challenges. In particular, the chapter addresses the application of model-based tools for representation and analysis of SoS architectures, approaches to developing systems of systems objectives to ground SoS trade space analyses and the use of lightweight analysis

Complexity Challenges in Cyber Physical Systems: Using Modeling and Simulation (M&S) to Support Intelligence, Adaptation and Autonomy, First Edition. Edited by Saurabh Mittal and Andreas Tolk.
© 2020 John Wiley & Sons, Inc. Published 2020 by John Wiley & Sons, Inc.

approaches to manage the complexity of large SoS, and to identify promising options for more detailed trade-off analysis approaches.[1]

6.2 Systems of Systems (SoS), Cyber Physical Systems (CPS), and Internet of Things (IoT)

Henshaw in his article (de C Henshaw 2016) states:

> This paper ... compare[s] and contrast[s] the concepts of SoS, CPS, and IoT ... The purpose in doing so is to understand the implications for systems engineers; the starting point is recognition that they are not simply fashionable phrases for the same thing, but valid perspectives and constructs for types of industrially relevant complex systems.

To appreciate the relationship among these types of compositional systems, the characteristic features of each are described. The relationships among these are shown in Figure 6.1.

The International Standards Organization (ISO) defines systems of systems (ISO 2018):

> A **System of Systems (SoS)** is a set of systems that interact to provide a unique capability that none of the constituent systems can accomplish on its own.

In today's highly connected world, SoS surround us in a variety of industries and domains. Although a SoS within each domain may have its unique variations, five key characteristics have been identified that generally indicate a system is really a SoS. These include Operational independence of constituent systems (CSs), Managerial independence of constituent systems, Geographical distribution, Evolutionary development processes, and Emergent behavior (Maier 1998). The last indicator, emergent behavior, is particularly important as it's these cross-system capabilities that allow a SoS to achieve its "unique capability."

---

1 This chapter brings together work conducted by the MITRE Systems Engineering Technical Center Systems of Systems (SoS) and Model Based Engineering (MBE) Capability Areas draws directly from material which has been presented at the INCOSE International Symposium and in INCOSE INSIGHT, IEEE Systems and Aerospace Conferences and at the National Defense Industry Association Annual Systems Engineering Conference. The authors acknowledge the contributions of the MITRE SoS research team who contributed to this research including Tom Wheeler, Janna Kamenetsky, and the technical staff of the Systems Engineering Technical Center.

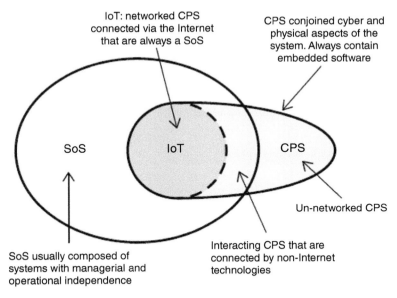

Figure 6.1 Relationship between SoS, CPS, and IoT (de C Henshaw 2016).

The US National Institute of Standards, describes CPS as

> **Cyber-physical systems (CPS)** can be described as smart systems that encompass computational (i.e., hardware and software) and physical components, seamlessly integrated and closely interacting to sense the changing state of the real world. These systems involve a high degree of complexity at numerous spatial and temporal scales and highly networked communications integrating computational and physical components. (NIST 2013)

Like SoS, CPS are composite systems where components may be either embedded or networked in which case CPS are an example of an SoS. Henshaw goes further to point out that in a position paper by Acatech (2011), their definition of CPS makes this relationship clear:

> Cyber-physical systems are systems with embedded software (as part of devices, buildings, means of transport, transport routes, production systems, medical processes, coordination processes, and management processes), which:
> - directly record physical data using sensors and affect physical processes using actuators;
> - evaluate and save recorded data, and actively interact with the physical and digital world;

- **are connected with one another and in global networks via digital communication facilities (wireless and/or wired, local and/or global);** [(emphasis added]
- have a series of dedicated, multi-modal human machine interfaces.

The third point, highlighting networking among the components of a CPS, builds on the Brook broad definition of SoS as "a system which results from the coupling of constituent systems at some point in their life cycles" (Brook 2016), making CPS as a type of SoS.

Finally, the IoT (TechTarget 2016) is described as

> The **Internet-of-things (IoT)** is a system of interrelated computing devices, mechanical and digital machines, objects, animals or people that are provided with unique identifiers and the ability to transfer data over a network without requiring human-to-human or human-to-computer interaction.

This makes IoT a subset of networked CPS and, as such, a type of SoS again as shown in Figure 6.1.

In his paper Henshaw (de C Henshaw 2016) concludes:

> What are the differences between the three concepts: SoS, CPS, and IoT? It seems that they represent different parts of the complex system problem, but that the IoT represents a convergence of SoS and CPS. Can they be treated with the same approaches, or are there fundamental differences that demand different techniques for their management? This turns out not to be the most useful question: it is not so much a question of different techniques between the three concepts, but rather the **need for SoS techniques matched to the design and operation of increasingly complex systems**. [*Emphasis Added*]

This chapter then describes the challenges faced by SoS in trade-off analyses and addresses supporting techniques which can be more generally applied for networked and interconnected CPS.

6.3 Systems of Systems Challenges for Trade-off Analysis

The characteristics of SoS and networked CPS contribute to systems engineering challenges that are not typically encountered when working at the system level. As a result, the System of Systems Engineering practice (SoSE) has emerged to help address challenges specific to SoS. Formally, SoSE is defined as "the process of planning, analyzing, organizing, and integrating the capabilities of a mix of existing and new systems into a system-of-systems capability that is greater than the sum of the capabilities of the constituent parts" (USAF Scientific Advisory Board 2005). Application of SoSE, like SE more generally, is based on analysis of options and trades for both composing an SoS and evolution of SoS capability to address changing needs. As will be discussed in this section, the characteristics of SoS, including networked CPS, pose challenges for the SoS trade-off analytics.

There are key challenges that differentiate the SoSE process from typical SE approaches. The INCOSE Systems of Systems Working Group has identified seven key challenges relating specifically to systems engineering of a SoS, which have become known as "SoS Pain Points" (INCOSE 2018). These pain points were identified in collaboration with the community of SoS practitioners and are described in detail in the INCOSE Systems Engineering Handbook (2015). These are shown in Figure 6.2.

Authority Relationships The managerial independence of the systems comprising an SoS and the limited authority at the SoS level over the constituent systems puts constraints on SoS trade-off analyses. The range of authority relationship in SoS have been described in terms of SoS types (Table 6.1). All of these types recognize autonomy of the constituents particularly since in most cases they independently provide value to a user independent of the SoS. Without top level authority changes in systems need to be negotiated, which may mean that the best technical solution for the SoS may not be acceptable to the constituent systems for a variety of reasons, and it effectively constrains the options which can be realistic

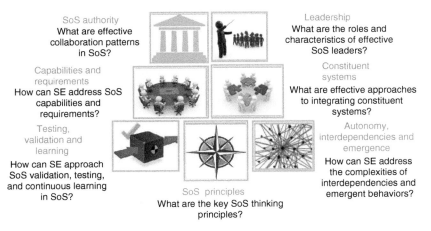

Figure 6.2 Systems of Systems pain points (Dahmann 2014).

Table 6.1 SoS types.

Directed	The SoS is created and managed to fulfill specific purposes and the constituent systems are subordinated to the SoS. The component systems maintain an ability to operate independently; however, their normal operational mode is subordinated to the central managed purpose
Acknowledged	The SoS has recognized objectives, a designated manager, and resources for the SoS; however, the constituent systems retain their independent ownership, objectives, funding, and development and sustainment approaches. Changes in the systems are based on cooperative agreements between the SoS and the system
Collaborative	The component systems interact more or less voluntarily to fulfill agreed upon central purposes. The central players collectively decide how to provide or deny
Virtual	The SoS lacks a central management authority and a centrally agreed upon purpose for the SoS. Large-scale behavior emerges – and may be desirable – but this type of SoS must rely on relatively invisible mechanisms to maintain it

Source: From SEBoK (2018).

considered for the SoS. This puts pressure on trade space analytics to consider a wider range of options and conduct trades which satisfy both constituents and SoS needs.

Leadership To address this requires the systems engineer to be able to step out of a purely technical role and advocate for a broader view of options and provide compelling evidence to system owners to consider systems changes beyond what would be expected for their system's needs. As will be discussed further below, model-based engineering tools can often provide visibility into the large SoS technical and operational or business context, which can provide the basis for a shared understanding of the broader context and provide a vehicle for developing a shared understanding across the stakeholders of the systems and SoS.

Constituent Systems' Perspectives SoS are composed of constituent systems which are typically developed independently of the SoS. These systems have their own objectives, which may or may not align with the objectives of the SoS. Trade-off analytics must consider options and trades from both a CS and a SoS point of view since, particularly when there is limited authority oat the SoS level, and changes in constituents. Analysis can be done from the CS perspective to understand how taking on additional responsibilities in the SoS can affect the ability of the CS to support their original tasks. The resulting insight can help the SoS authorities appreciate what they are asking of the CS, to better inform their own decisions when requesting changes in the way a CS supports the SoS.

Capabilities and Requirements Systems typically have a well-defined set of requirements. A typical CS has a well-defined set of requirements, whereas SoS typically have a desired set of capability objectives rather than requirements. A challenge for SoSE is characterizing these capability objectives and understanding how the systems and their requirements relate to and in fact, support these objectives. SoS trade-off analyses build on this understanding to identify options for adding or changing requirements on a constituent system to improve the SoS ability to address the capability objectives.

Autonomy, Interdependencies, and Emergence Composing systems into support a broader capability opens opportunities for new functionality generated by the interactions among the systems but it also leads to the risk of additional unexpected, and undesirable behaviors especially when the constituent systems are complex systems in their own right. This becomes an important consideration in assessing options for making changes in systems to improve SoS capabilities, to ensure that potential impacts of these changes on other systems or the SoS are considered in the trade-off analysis.

Testing, Validation, and Learning Because most SoS are composed of existing systems, the traditional approach of conducting end-to-end test and evaluation of an SoS faces practical challenges. The lack of authority over independent CS, time and financial constraints, and a continually evolving baseline call for other approaches to assessing risk in making changes to an SoS need to be employed. A more appropriate approach is to address the areas of greatest risk regarding the interactions and contributions of the CS to the SoS at key points in the SoS lifecycle (Dahmann 2012; Dahmann et al. 2010). By structuring an SoS architecture to partition functionality, impacts of changes in one system on others and the SoS, risks can be localized and allow for effective approaches to risk management. Using trade-off analytics to identify the most crucial and comprehensive testing activities can further enable more effective testing under SoS constraints.

Because SoS, including many networked CPS, are typically comprised of systems which were developed and fielded independently of the SoS and which continue to evolve to meet changing needs of their particular users, SoS development differs from that of a typical system and this affects SoSE implementation. The SoS Wave Model[2] is one of several approaches to systems engineering for a SoS (Cook 2016) which reflects the evolutionary nature of SoSE. At its core, the wave model illustrates the incremental process of achieving improved SoS capabilities based on changes made in the constituent systems. The approach is particularly well-suited to acknowledged SoS where some amount of cross-cutting authority exists to advance the SoS capabilities rather than solely advancing the CS.

---

2 See (Dahmann et al. 2011) for a discussion of the Wave Model.

In the wave model four key steps are repeated as the SoS evolution progresses (Figure 6.3). These steps are *Conduct and Continue SoS Analysis, Develop and Evolve SoS Architecture, Plan SoS Update,* and *Implement SoS Update.* Because SoS typically evolve over time, the steps are implemented repeatedly as new SoS needs are identified and changes in the SoS to address these needs are sought. In this process, SoS trade-off analysis supports the development and evolution of the SoS architecture and the collaborative planning for changes in constituent systems to support SoS evolution.

Conduct and Continue SoS Analysis focuses on characterizing the performance of the SoS. Here, the capability objectives of the SoS are established and data from SoS performance is assessed. Gaps or risks in the SoS meeting capability objectives are identified here along with an assessment of the root cause of any shortfalls or risks. This step sets the conditions for tradeoffs for making improvements in the SoS.

Develop and Evolve SoS Architecture identifies and assess options for changes in the SoS to address gaps or risks discovered during the first step. This step identifies architectural options and trades that can be made by some combination of elements in the SoS, including adding or updating CSs, changing the way CSs are employed, or other changes in the operation of the SoS. Assessing the options includes both technically focused considerations, such as system performance, as well as operationally focused trades, such as tactics, techniques and procedures (TTPs) and is constrained by the limits of the constituent systems. The activities in these first two steps form an iterative process analyzing architectural changes in conjunction with the existing SoS baseline. This includes explicitly addressing the perspectives of the CS to ensure the SoS-level trades including viable options for the CS. These trade-off analyses result in a set of recommended changes in the architecture of the SoS.

Plan SoS Update takes these recommended changes and, factoring in considerations of the constituent systems, sets priorities and plans for implementation.

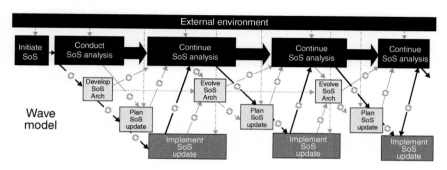

Figure 6.3 The SoS wave model (Dahmann et al. 2011).

Because SoS changes are the product of changes in constituent systems, implementation plans are driven by constraints of those systems. These include scheduling constraints of the various CS, since each CS typically has its own development plan and schedule with a focus on the primary users or customers of the system which also affects plans for testing and evaluation to maintain confidence in the CS and SoS capabilities. Because SoS are typically operational, plans for changes need to consider the risks associated with making changes to current operations. Many of these concerns can be addressed by trade-off analytics. For instance, trades can play important role in scheduling which CS undergo an update at a given point in time. If the update requires the CS to come offline, it is essential that the effects of those performance trades are understood in the context of the SoS ability to maintain its capabilities. Trades can also be employed to better assess which testing and evaluation options satisfy the most critical questions regarding CS and SoS performances.

Implement SoS Update implements the changes which feed back into *Continue SoS Analysis* phase, where the SoS baseline is continuously updated to facilitate a review and assessment of the available capabilities in light of changes in SoS capability needs and drivers, some of which come from independent changes in the CS.

6.4 Model-Based Architectures as Framework for SoS Trade-off Analytics

Use of model-based approaches to systems engineering is increasingly accepted by the SE community. The International Council of Systems Engineering (INCOSE) (2014) views model-based approaches as the heart of transforming SE to meet today's and tomorrow's complex system challenges. Organizations like the US Department of Defense have established strategies to promote this transformation across their domains. Model-based strategies, like the DoD Digital Engineering Strategy (DASD SE 2018) have focused mainly on use of modeling for the engineering of systems, showing great benefits related to cost savings and error reductions (Hause 2014). Antul et al. (2018) argue that these approaches are even more relevant to addressing the challenges of SoSE and are key to SoS trade-off analytics:

> As with system-level problems, MBE offers an approach to develop an unambiguous representation of an SoS architecture via a computer-based model. This model acts as a central repository for definition and description of SoS elements, enables generation of common architecture views, and can even support execution of integrated analysis tools.

Several modeling languages have been proposed for SoS architecture modeling, including the Systems Modeling Language (SysML), the Unified Profile for DoDAF/MoDAF (UPDM), the Unified Architecture Framework (UAF), and the Object Process Methodology (OPM). Several researchers advocate SoS modeling methodologies leveraging SysML (Huynh and Osmundson 2007; Rao et al. 2008; Gezgin et al. 2012; Lane and Bohn 2013). Mori et al. (2016) proposed extending SysML to describe[3] seven viewpoints or characteristics of SoS (Ceccarelli et al. 2015). Others have proposed languages specifically tailored to SoS, such as the Comprehensive Modelling for Advanced Systems of Systems (COMPASS) (Woodcock et al. 2012). Another example of a specialized architecture language is the SoS Architectural Description Language (SosADL) (Oquendo 2016) created to support formal descriptions of the architecture of software-intensive SoS (Antul et al. 2018). While practitioners have not converged on a standard for modeling of CPSs, attempts have been made at creating domain specific languages for this purpose leveraging existing standard languages such as Unified Modeling Language (UML) and Object Constraint Language (OCL) (Aziz and Rashid 2016).

For SoSE purposes, an architecture model of the SoS in an industry-standard language such as SysML offers an unambiguous, structured, executable, digital representation of the SoS architecture (Dahmann et al. 2017). The model serves as a unified repository of internally consistent SoS architecture information including CSs, SoS- and CS-level requirements, CS behavior, logical and physical interfaces, information and control flows between CSs that enable the SoS to perform end-to-end missions (Object Management Group 2015). An executable representation of the architecture facilitates simulation for purposes such as verifying coherent end-to-end operations and evaluating effectiveness. An overview of simulation methods applicable to systems engineering including executable architecture concepts is provided in Tolk et al. (2017).

The systems composed into an SoS architecture to support a mission are typically drawn from a variety of specialty areas such as in Defense SoS, sensors, weapons, platforms, communications. The diverse organizations responsible for these systems bring various perspectives to the SoS from their communities. Use of a common SoS architecture model provides the venue for effecting a common understanding of the SoS and the role of the systems in achieving SoS capability objectives. Monahan et al. (2018) argues that the initiative to stand up an SoS model should be viewed as an expression of Leadership in an SoS.

Standards-based modeling approaches provide both the SoS engineer and the constituent systems engineers a domain independent cross cutting integrated view across the systems and how they are expected to be employed in a SoS context. A model allows for representation of the complexity of the interrelations among systems in the SoS and addresses the needs of a wide range of stakeholders through built-for-purpose yet consistent custom views. Specificity provided by a model can help avoid misunderstandings about system behavior, system interactions, and interfaces among the owners of the constituent systems and SoS (Cotter et al. 2017).

As multiple stakeholders collaborate in building the model, it becomes imperative to establish model governance. SoS Authorities Pain Point is evident in the need to adjudicate what information is considered authoritative and how it is verified. As Monahan et al. points out,

> Modern SysML modeling tools allow for model configuration management with flexible package-level control of viewing and editing permissions. CSs owners can provide their respective systems' specifications to the level they are comfortable sharing, hiding sensitive and proprietary information while exposing interfaces and relevant functionality.
>
> Reymondet et al. (2016) advocate for establishing a formal model curator role for complex engineering enterprises. An effective model curator facilitates a shared understanding of model purpose, development practices and composition guidelines across the stakeholder community (Reymondet et al. 2016). Model curation can also benefit practitioners by providing model credibility, resulting in more buy-in from stakeholders.

Despite the advantages of a model-based architecture for SoSE and trade-space analytics, there are recognized challenges to adopting this approach which are tied to the characteristics of SoS. As summarized by Monahan et al., SoS stakeholders including both SoSE authorities and owners of complex CS may be reluctant to embark on Model Based Systems Engineering (MBSE) efforts due to the daunting volume of design information and number of candidate architecture options. There is often a concern that the required level of effort appears impractical and unaffordable. Another issue is accessibility of the needed or desired CS-level information. These concerns can be at least partially allayed through thoughtful and deliberate scoping of the SoS modeling efforts. Douglass (2015) notes that architecture modeling is most likely to succeed in providing value when it initiates with a clear goal in mind. Thus, it is crucial to ensure that SoS modeling begins to deliver return on investment within a modest time from the initiation, preferably several months. Scope of the modeling effort can be controlled through the choice of abstraction level, the number of end-to-end threads, and key measured of effectiveness. It is advisable to structure the modeling effort iteratively, starting with a small set of threads at a high level of abstraction, exploring trades that involve several key measures of effectiveness (MOEs). Subsequent investment into higher fidelity modeling should be guided by the initial analysis of risks and gaps in end-to-end SoS performance.

So why is this hard? What techniques can be used to address the challenges? A key enabler of model-based SoSE is the ability to efficiently develop large complex SoS architecture model. SoS can be large and comprise many systems, and the time required to develop a model framework for each mission architecture can raise the cost of entry for use of models to support mission engineering. Gathering the needed data to understand the current state of a large SoS can be difficult given the diversity of knowledgeable mission stakeholders. Providing intuitive tools to allow stakeholders to share knowledge in a way familiar to them can build confidence and speed up knowledge gathering. Finally, automated transform directly into a model again lowers the cost of entry for large mission architecture and reduces likelihood of errors or misunderstandings. Two techniques, use of a "base model" and a "CSV importer," illustrate ways to address these challenges to facilitate model-based SoSE architectures as a starting point for SoS trade-off analysis.

Base Model The effort required to build SoS architecture models can be reduced by starting the modeling process with a reusable SysML "Base Model" template, independently of the architecture size. Base Model captures key functional SoS architecture information and allows a modeler to represent domain-specific behavior in derivative models. In general, all effects chains involve a sequence of steps that must be conducted to achieve a desired result. Base Model also provides a generic state machine for SoS structural elements represented as SysML blocks. The state machine captures, for instance, Operational, Paused and Inoperable states and transitions between these states. Operational state is further decomposed into Normal and Degraded states. These states and transitions are broadly applicable to almost any SoS elements, and the use of the Base Model reduces modeling effort by leveraging the commonality. Specifics of each system's behavior can be captured by adding relevant behavioral detail to each of the generic states.

As an example, Antul et al. (2018) have created a base model as a SysML model implemented using the IBM Rational Rhapsody for Developers software. Because the intended user domain for this base model is DoD mission analysis, the effects-chain content of this base model represents a DoD mission thread consisting of six operational activities: Find, Fix, Track, Target, Engage, and Assess (F2T2EA), aligning with the US Air Forces conception of a typical "kill chain." Differences among effects chains will manifest in the types of systems, captured as SysML blocks, that each contribute to one or more effects-chain activities. To determine which system types should be included in the base model, Antul et al. reviewed defense effects-chain MBE models previously developed for several modeling efforts. It was determined that common elements across DoD effects-chains include Commanding Officer (CO), Command and Control (C2), Sensor, Weapon, Platform, and Operator. Any effects-chain scenario will require one of more instance of these elements, with each element contributing to one or more of the

F2T2EA operational activities. All system types representing devices rather than human actors inherit from an abstract Device block which possesses high-level behaviors common to all systems. These behaviors are captured in a state machine and include inoperable, operable, initializing, and paused states, with normal and degraded sub-states within the operable state. Higher-fidelity behavior can be further defined for each system type within the generic states. For example, abstract element Sensor may be extended to specify characteristics of a radar, a camera, or an acoustic sensor.

The base model is intended to serve as a template for modeling SoS architectures in MBE tools, reducing the level of effort required to stand up or extend an SoS architecture model (Figure 6.4). Use of a template across several modeling efforts ensures consistency in SoS modeling approach, enabling reuse of software leveraging APIs of the modeling tools, for instance, plug-ins that use information captured as SysML constructs as input to architecture analysis tools.

Importer Tool Tools can facilitate integration of SoS connectivity information into MBE tools, tightening the coupling between operational subject matter experts (SMEs), software engineers, and analysts – comma separated value (CSV) importer tool. Networked nature of SoS lends them well to representation as matrices where constituent systems appear in rows and columns and their connections in the cells, for example, DoDAF SV-3 (Systems-Systems Matrix). Such representation can be implemented in common spreadsheet tools such as Microsoft Excel.

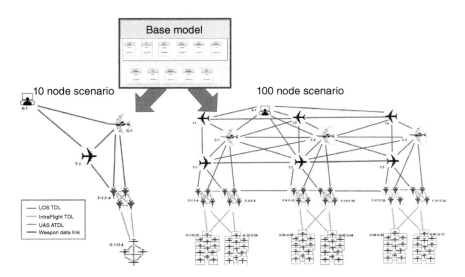

Figure 6.4 Base Model applied to models of different size (Albro et al. 2017).

The CSV importer tool allows to analysts convert a matrix representation of a SoS architecture into a SysML form within an MBE software tool. The benefits of this approach are twofold: (i) it facilitates transfer of knowledge from an operational SME who is not a user of MBE software and (ii) it automates the creation of links between SoS elements, which can reduce errors and modeling time.

The workflow for using the CSV importer utility is shown in Figure 6.5. Once the SoS architecture is conceptualized, analysts can then collaborate with SMEs to encode the architecture as an adjacency matrix. Matrix cells can be ones or zeros, where a one indicates existence of a connection between two systems and a zero indicates no connection. Alternatively, the cells can be topic titles in a publish–subscribe data exchange pattern. With this matrix populated, the importer tool automatically creates the SysML elements indicated in the matrix.

SoS MBE architecture modeling provides a foundation for addressing complexity and trade-off analysis in SoSE. Multiple stakeholders share the same data sets, promoting a common understanding of SoS architectures. This can help to uncover implicit assumptions made by CS owners about behaviors and capabilities of other CSs and provides a basis for conversations. It is the role of the SoS authority to facilitate this dialogue among CS owners. Another way architecture models help with managing complexity is by enabling executable verification of mission threads and analysis of corresponding requirements on CSs. MBE models also foster reasoning about architecture trades. SysML enables representation of trades both within the SoS and CS scopes with parametric diagrams.

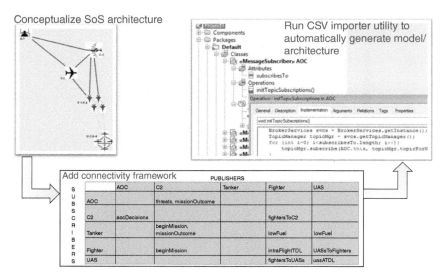

Figure 6.5 Illustrative workflow for using the CSV Importer (Albro et al. 2017).

Additionally, the computable nature of architecture data captured in SysML facilitates transfer of the data to lightweight analysis approaches to identify promising options as well as costlier M&S tools for more detail trade-off analysis approaches.

6.5 Establishing SoS Objectives and Evaluation Criteria

To assess a current SoS architecture and architecture alternatives in the Conduct/ Continue SoS Analysis steps of the Wave mode, evaluation criteria must first be identified. These criteria should reflect important consequences of decision makers' choices such as SoS performance, impacts on constituent systems, costs, risks, and expected effectiveness against capability objectives.

To define an appropriate set of evaluation criteria, the SoSE team should elicit SoS objectives from stakeholders. Poorly defined objectives may result in the SoSE team struggling to derive evaluation criteria and ultimately lead to the wrong solutions. A common example with SoS is when ambiguous goals are stated, such as a desire to increase resilience or other "ilities." Additionally, the tendency of analysts to select evaluation criteria based on readily available modeling and simulation tools for quantification of criteria, regardless of alignment with the SoS objectives, should be avoided.

To help guide the selection of objectives, Gibson et al. (2017) propose seven detailed steps applicable to SoS and other types of large-scale systems:

1) Generalize the question: Generalize the analysis problem and place it in context to ensure that it is not too narrowly focused.
2) Develop a descriptive scenario: Describe the current situation, including the good and the undesirable features of the problem, to broaden and deepen the team's understanding.
3) Develop a normative scenario: Describe the situation as it will be when the solution is operational, preserving the good features of the descriptive scenario and changing as many of the undesirable features as possible.
4) Develop the axiological component: Elicit decision makers' values, which may be incomplete and/or conflicting.
5) Prepare an objectives tree: Create a graphical representation of the objectives and values obtained in the previous steps.
6) Validate: Evaluate the products from the first five steps to identify omissions, inconsistencies, etc.
7) Iterate: Repeat these steps throughout the analysis process, gaining new perspective along the way.

This iterative process results in a validated objectives tree that helps clarify and organize the set of objectives. In an SoS, each constituent system has local stakeholders with their own objectives. Since some of these objectives may conflict with the needs of the SoS, needs and values of the CS owners need to be balanced with those of the SoS. This is a direct effect of the SoS authorities pain point, and it highlights the importance of identifying CS stakeholders in addition to the managing SoS authority when establishing objectives.

A tool that the authors have found useful for understanding the landscape of stakeholders and reasoning about their conflicting objectives is stakeholder value network (SVN) analysis (Cameron et al. 2011a, 2011b). SVN is an approach for modeling stakeholders, their relationships, and the importance of each relationship. The relationships considered are called value flows, which represent anything transferred between two stakeholders that contributes to their actions and interactions. The importance of these value flows can be elicited from stakeholders. With data on the stakeholders and relationships, a graph of all elements can be built. With a graph representation of the stakeholder network, centrality measures from graph theory can be used to assess the importance of each stakeholder from a primary stakeholder's perspective. In an SoS context, for example, the primary stakeholder may be the managing authority, and the CS owners and other influential organizations would be other stakeholders in the graph. Value flows between these stakeholders would include policy, capabilities, data, investments, etc. A notional example of a simple SVN model for an SoS is shown below in Figure 6.6.

The SVN model can be built in a SysML tool and linked with the SoS architecture. For example, Sease et al. (2018) proposed an approach for using SysML constructs to model SVNs. Doing so enables traceability from stakeholder importance to other aspects of the architecture, such as evaluation criteria and CS elements.

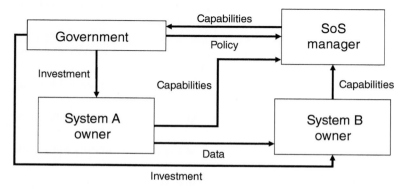

Figure 6.6 Notional SVN model for an SoS (Monahan et al. 2018).

Once the stakeholder landscape is understood, an objectives tree can be built. Evaluation criteria can be derived directly from this tree. As recommended by Gibson et al. (2017), the metrics should be measurable, objective, nonrelativistic, meaningful, and understandable. With a set of appropriate evaluation criteria, SVN analysis results can then inform prioritization of the criteria for multi-objective trade-off analyses.

6.6 Evaluating Alternatives

Establishing SoS objectives and metrics and representing SoS architecture in a model opens the options for analysis, however, the complexity of the SoS and the potential variety of options poses added challenges to the analysis option and tradeoffs in an SoS. The potential variety of options across potentially very different architecture alternatives and concepts of option can lead to a very large and complex trade space.

There are several approaches to address this. The first is to limit the number of options at the outset. This can be based on placing constraints on changes to components of the SoS. For example, by one approach is to basically accept a set of constraints on changes to constituent systems and focusing possible changes only in areas where there is a strong potential for agreement by the owners of the system themselves. The options can be further constrained by limiting alternative to those the system themselves have agreed to at the outset. This can be a practical strategy since it explicitly addresses the risk that CS will be unwilling to make changes that do not align with their local objectives and development plans. It can however, overly constrain options and discard which might be more effective options, which if the benefits were understood by the constituents they might be more willing to make changes than assumed at the outset.

For large complex SoS, another option is to use the end-to-end functional flow to partition the SoS into independent segments and address trades within each segment. This type of a partitioned SoS architecture has numerous advantages. Because the CS may make changes independently from the SoS, architecting an SoS with independent segments has the advantage of protection CS and the SoS as a whole from unanticipated impacts of changes in CS. It can also help manage testing and risk reduction over planned changes in the SoS but limiting the impacts of the changes and hence the scope for testing for the SoS effects of those changes. This strategy depends on the independence of the segments and does not consider trades across the segments which may offer benefits which if using this approach will remain unexplored.

An added consideration in SoS trade space analyses is the number of different dimension which need to be addressed in the in making trades, especially when the stakeholders in the SoS analyses value different considerations in their assessment of preferred alternatives. SoS improvements may be assessed in terms of in performance of the SoS, in the cost of making the changes for one or more of the CS, the extent to which the option requires changes to multiple CS which can lead to disruption or longer time to achieve the SoS improvements the disruption to the SoS. Unlike typical systems where there is a single owner and a relatively well-defined set of stakeholders, the SoS has owners and stakeholders at both the CS and SoS with multiple and potentially different and possibly conflicting objectives.

There are several approaches to addressing these issues. First, lightweight metrics can provide a practical approach to identify from a broad set of alternatives those which warrant fuller consideration, hence limiting the number of actual options to be considered while beginning with a broader more inclusive set of options. Second, the model-based SoS architecture can be leveraged to create a digital environment which employs data from the SoS model as inputs to a variety of modeling and simulation tools for the trade space analysis of the selected SoS options. This can facilitate richer concurrent analysis of options using a variety of tools while maintaining consistency across the analyses through shared "anchor" data from the SoS architecture model. Importantly, this can include operational simulations where changes in the SoS architecture can be explored in terms of the impacts on the user capabilities which are the fundamental driver for the SoS.

6.6.1 Lightweight Analytic Tools for SoS Trade Space Analyses

A workflow with lightweight analytic tools filtering the design space prior to more expensive M&S activities is shown in Figure 6.7. As is suggested above, a useful strategy to address complexity in the scope and scale of SoS trade space options is rapid trade space exploration using lightweight metrics to identify those options where investment in detailed analysis is deemed warranted. Lightweight analytic tools employing SoS architecture data enable fast exploration of a large trade space and identification of viable options for more detailed analysis with higher-fidelity modeling and simulation tools. For a comprehensive resource on modeling and simulation for SoS, the reader is referred to the text edited by Rainey and Tolk (2015).

Based on the networked nature of SoS, the use of structural network metrics drawn from network theory has been viewed as a promising approach to developing and applying lightweight measures to SoS alternatives.

> The networked nature of SoS suggests that using structural network metrics to evaluate SoS has promise. Network theory is a common approach for modeling and evaluating SoS (Han and Delaurentis 2006; Harrison 2016).

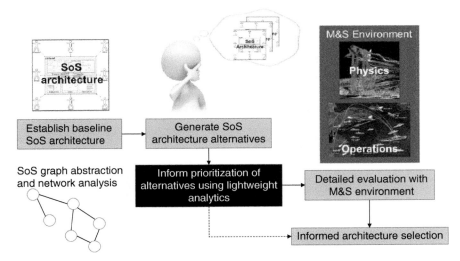

Figure 6.7 SoS trade-off analysis workflow with lightweight analytic tools and M&S tools (Antul et al. 2017).

A network model of an SoS typically represents constituent systems as nodes, and relationships between those systems (e.g., communications, contractual agreements) as links. This approach enables analysis of SoS with minimal computational cost, while maintaining relevant details of the architectural structure. (Monahan et al. 2018)

In particular, SoS resilience is viewed as an important consideration in comparison of SoS architecture alternatives.

Resilience is often represented as a combination of survivability and recoverability (Uday and Marais 2015). Thus, resilience is not only defined by the likelihood of failure but also the ability of systems to recover, or "bounce back", from unexpected disturbances in the environment. Many resilience metrics capture how system performance changes over time following a disruption event. However, acquiring the necessary performance data requires detailed SoS simulations or collection from actual operations, both of which are costly for large-scale SoS.

There is also a growing body of research investigating the resilience and robustness of networks. Network theory provides a rigorous, mathematical framework for assessing SoS architectures with limited computational cost. Rather than seeking to quantify resilience directly, graph theoretic metrics provide insights to robustness of SoS. Robustness is the ability of an SoS to maintain capabilities after a disturbance, which can be thought of as a contributing property to resilience. Algebraic connectivity, network diameter, average path length, and cluster size (or component size) are common network metrics for assessing network robustness. Though these metrics do not fully quantify resilience (since they do not explicitly account for recovery dynamics), they are still useful for assessing SoS architecture alternatives given the importance of robustness to resilience. (Monahan et al. 2018)

An example of this approach is presented in Antul et al. (2018) where the research

quantified robustness of three architecture variations of a notional military SoS and compared observed trends from this lightweight approach to one based on running detailed simulations to characterize performance of the same architectures. Virtual simulation experiments provided evidence of the predictive value of the graph-theoretic metrics for SoS analysis. The modeled scenario considered a notional joint engagement sequence, consisting of five radar systems, three fighters, and two air operations centers (AOCs) on the blue team. Radars and fighters were strategically placed to defend a single critical asset. 100 threat aircraft approach the critical asset from four surrounding locations, as shown in Figure 6.8. The three architecture alternatives (shown in Figure 6.9) included: the baseline (a), a robust alternative (b), and a vulnerable alternative (c). Ten threat cases were considered in the simulations, where each case disabled (i.e. removed) a single system from the architecture. Cases one through five each disabled a radar system, cases six and seven each disabled an AOC, and cases eight through ten each disabled a fighter. A reference case, case 0, in which no systems are disabled was also considered. Table 6.2 shows probability of engagement success P_{ES} from the simulation cases against a graph theoretic metric for robustness called algebraic connectivity. Comparing mean P_{ES} to algebraic connectivity, a similar trend is observed in that robustness increases for the second architecture and decreases for the third relative to the baseline.

Table 6.3 lists a set of network metrics which appear to be useful in assess SoS architecture robustness as a key dimension in SoS trade-space analytics. In each case, the higher the value of the metric, the more robust the architecture. Additional work is needed to apply these and other network graph theory-based approaches and other lightweight methods to enable SoS engineers to address as wide an array of options as possible through rapid assessment of numerous candidates to identify those selected options which offer the greatest potential and hence warrant detailed analysis.

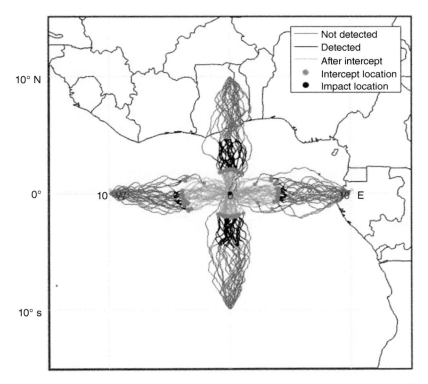

Figure 6.8 Notional scenario considered for network robustness analysis. The critical asset is shown in the center, with red lines indicating undetected threat aircraft and green dots representing successfully intercept threats (Antul et al. 2018).

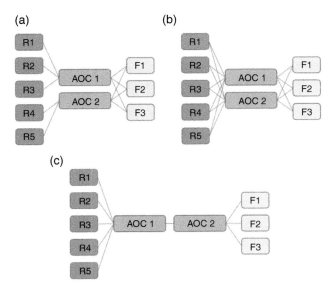

Figure 6.9 Considered SoS architectures, where (a) is the baseline, (b) the robust design, and (c) the vulnerable one. Radar systems are denoted by R, fighters by F, and AOCs by AOC (Antul et al. 2018).

Table 6.2 Robustness comparison results.

Case	System disabled	Baseline P_{ES}	Robust P_{ES}	Vulnerable P_{ES}
0	—	0.608	0.608	0.608
1	R1	0.448	0.448	0.448
2	R2	0.608	0.608	0.608
3	R3	0.454	0.454	0.454
4	R4	0.467	0.467	0.467
5	R5	0.454	0.454	0.454
6	**AOC 1**	**0.314**	**0.608**	**0.0**
7	**AOC 2**	**0.294**	**0.608**	**0.0**
8	F1	0.582	0.582	0.582
9	F2	0.461	0.461	0.461
10	F3	0.589	0.589	0.589
Mean P_{ES}		0.467	0.528	0.406
Algebraic connectivity		0.506	2.000	0.309

Source: From Antul et al. (2018).

6.6.2 Integrated Engineering Environment to Support SoS Trade Space Analysis

Once a tractable number of alternatives have been identified, the next challenge is to address these options using analysis tools to assess the relative merits of the selected alternatives. There exist a variety of models and simulations which can be employed to address dimensions of SoS alternatives. A major challenge is developing and applying cost effective approaches to address options form the multiple perspectives of diverse stakeholders, including the collection of needed data to support the analyses.

> Understanding trades in an SoS architecture requires results of analyses across multiple disciplines including operational simulations, physics-based models, and cost models. Typical approaches to SoS trade-off analysis rely on modeling and simulation (M&S) tools. One example is the work by Chattopadhyay (2009) and Ross and Rhodes (2015) who proposed a quantitative approach that can be applied with varying levels of model fidelity. In this methodology, alternative SoS architectures are evaluated in the dimensions of utility, cost, and participation risk. Utility is quantified by eliciting stakeholder utility functions over a set of SoS-level attributes (e.g., measures of effectiveness) and aggregating the single-attribute utility

Table 6.3 Network metrics for assessing SoS architectures.

Metric	Calculation
Algebraic connectivity	Algebraic connectivity represents the average difficulty of isolating a node within a connected network. A network is connected if there exists a path between every pair of nodes in the network.
Diameter	Network diameter is length of the longest shortest path that exists in a network.
Average degree	The average degree of a network is the average of node degrees within the network itself.
Natural connectivity	Natural connectivity represents the weighted number of closed walks that exist for all nodes in a network, where the walks are weighted by the factorial length of the walk. A closed walk is a path that traverses the network beginning and ending at the same node. As natural connectivity is based on the number of closed walks in the network, it is highly influenced by the number of redundant paths. It can also be proved that natural connectivity increases monotonically when edges are added to the network, which means a highly connected network will have a higher natural connectivity than a sparse network.
Degree diversity	Degree diversity represents how difficult it is to disintegrate a network, where an increase in degree diversity increases the number of nodes required to be removed to disintegrate the network.
Global clustering coefficient	The clustering coefficient of a given node in a network is based on the number of neighbors the given node has that are connected to each other, which could be visualized in a graph structure as a triangle where there is a connection between two neighbor nodes and those nodes both connect to the given node. The global clustering coefficient is the average clustering coefficient for the entire network. This value increases with the number of triangle relationships among nodes.
Global average distance	Global average distance is based on the sum of the shortest path lengths between each node pair in a network.
Effective graph conductance	Effective graph conductance is calculated from the sum of the effective resistance between all nodal pairs in a network. This metric is based on the concept of effective resistance within electrical engineering, which provides the total resistance of a circuit; in the context of graphs, each edge is seen as a resistor having an Ohm value of one with each pair of nodes being considered a circuit.

functions into a multi-attribute utility function. Others have used surrogate modeling techniques that approximate the responses of the M&S tools to alleviate the computational burden of detailed analysis (e.g., see Ender et al. 2010). (Monahan et al. 2018)

The value of treating data as well as models as critical engineering assets has been recognized by the US Defense Department Digital Engineering Strategy (DASD SE 2018). Goals of this strategy include formalizing the development, integration, and use of models to inform enterprise and program decision making, providing an enduring, authoritative source of truth and establishing the supporting infrastructure and environment to perform activities, collaborate, and communicate across stakeholders as shown in Figure 6.10.

These goals all speak to the challenges of SoS engineering and offer a basis for approaches to addressing these challenges, so of which have been discussed in this chapter. In particular, the use of model-based SoS architectures to provide a shared view of the SoS to the multiple stakeholders with their own interests, motivations, and perspectives, often with their own models and analysis tools. Beyond this, driving common, shared data through coordinated modeling and data management across an SoS can provide the backbone for coherency across the various models and analyses needed for a comprehensive analysis of trades across options.

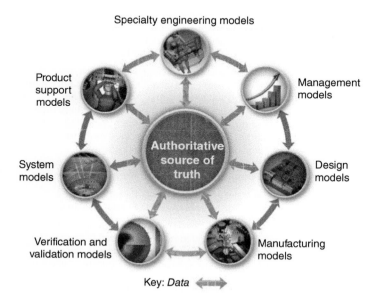

Figure 6.10 Shared data and models are core goals of the DoD Digital Engineering Strategy (DASD SE 2018).

Data is critical to effective engineering at any level and common data shared across models and analyses is key to successful SoS engineering. Typically, each organization invests considerable resource to develop data which is often not known or shared across an SoS. In SoSE, SoS architecture model data can serve as a core data resource, for both the light weight trade space exploration described in Section 6.6.1, but also as the core data to feed the analysis models and tools to implement more detailed analysis of the smaller number of selected alternatives. These concepts can be adapted and applied to SoS as shown in Figure 6.11.

Using the SoS architecture model as the source of core data and linking the model to the suite of tools and models which address key dimension of the SoS can provide the basis for an SoS digital engineering environment, which can provide a critical resource for SoS trade space analyses.

There are ongoing efforts to strengthen the digital link between MBSE architecture representations and M&S tools as well as other disciplinary analyses. Architecture options and parameterization captured in SoS architecture models represent the data required by other tools to assess performance, cost, and other aspects of the SoS. Automated information sharing between the central specification repository and analysis tools promotes traceability, speed, and accuracy of trade space evaluation. LaSorda et al. (2018) present an implementation of an architecture-centric approach to a trade study for a military satellite communications SoS. The study notably combines technical design considerations with acquisition strategy decisions for a holistic exploration of trades on performance, cost, and schedule (LaSorda et al. 2018). Dynamic dashboard displays can be utilized to aid decision makers in understanding of the trade space. Interactive graphical displays offer visual exploration of a wide range of architecture and operational options across multiple measures and design considerations (MacCalman et al. 2016). Although most existing literature focuses on using advanced trade space analysis techniques in the early conceptual design stage of complex systems, the SoS context makes such techniques relevant for continuous analysis of the evolving SoS architecture and operational concepts. (Monahan et al. 2018)

Interfacing a SoS model with other tools to assess performance, cost, and other aspects of the SoS, provides a shared representation of the architectures for analysis from different perspectives. Use of architecture model data as source for analysis of different trade-off dimensions can provide coherency to the suite of analyses conducted to assess options. It may be the case that some tools require added data over that provided in the architecture. Calibrating that added data to the shared architecture data helps to maintain coherence and integrity across the analyses.

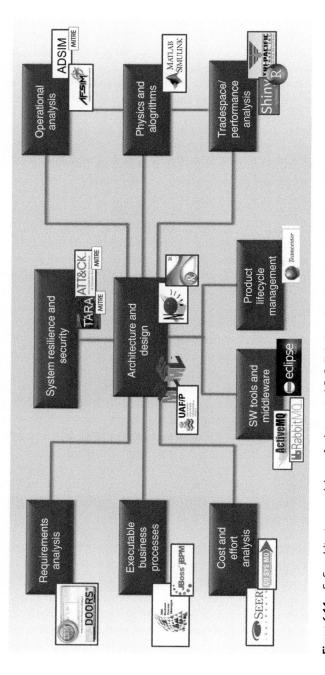

Figure 6.11 SoS architecture model core for integrated SoS digital engineering. Based on (Wheeler 2018).

Further, curating both the data and the models and tools, treating them as valued reusable assets, and implementing the integration within a reusable infrastructure can provide a cost-effective way to leverage a wide variety of models and tools for subsequent SoS trade space analyses. The combined effects of curated linked models, managed data, supporting infrastructure provide the capabilities needed to address SoS integrated analysis challenges in the near-term and they will continue to add value over time which is important given the evolutionary nature of most SoS.

For SoS in particular, it can be important to link the effects of SoS options to operational impacts since model systems of system capabilities are in the provision of some type of integrated operational capability which can only be evaluated in terms of the operational or business environment. The SoS technical trades are important, to ensure that the proposed changes provide needed technical capability. But it can be equally important that the user outcomes are realized when that SoS capability is fielded and operating in the expected user context. Which is to say that ensuring technical feasibility is an important prerequisite since it is key that systems work together as planned based on engineering across the systems supporting the mission, but it is also key that the SoS composition is fit for purpose in the user environment – physical, threat, etc. – and when executed leads to the expected outcomes under anticipated conditions, that the user outcomes are support by the technical SoS features.

SoS architectures can be complex, and it can be time consuming and error prone to have to manually instantiate these in today's operational simulations. Hence there is an advantage to investing in developing automated interfaces between architecture models and operational simulations, allowing for analysis of the effectiveness of the SoS in representative conditions and scenarios, following proposed concepts of employment, allowing for iterative analysis of technical changes and their impacts on user outcome. Automating this interface not only facilitates the conduct of the analysis of the mission effect or proposed or alternative SoS compositions, it also allows various users and stakeholders to view the proposed composition in their user environment, potentially overcoming CS inhibitions to making changes to meet SoS needs that extend beyond their own perspectives once they see the value to other users in the SoS context.

6.7 Summary and Conclusion

Increasingly systems engineers are facing the challenges of engineering composite systems which provide a suite of capabilities needed to meet user needs. These composite systems, which include CPS, share attributes and challenges of what have been addressed by the systems engineering community as "systems of systems" (SoS). As SoS become an increasingly dominant part of the overall system landscape, systems engineers are applying modeling and simulation approaches to address challenges posed by his class of complex system. This chapter has discussed the driving characteristics of SoS, which are shared by many CPS, and their implications for systems engineering, particularly in the application of trade space analysis which is at the heart of the systems engineering discipline. The challenges outlined in the SoS "pain points" all contribute to the complexities of representing and analyzing systems of systems, including CPS. Leveraging the knowledge developed for SoS can provide approaches for addressing complexity of CPS.

Model-based engineering practices provide a means to capture and share a common understanding of systems of systems elements and dependencies across key players in SoS engineering – including multiple independent and often diverse constituent systems owners and stakeholders. Structured techniques provide a means to define SoS objectives and metrics, SoS architecture models provide a core set of data to both apply lightweight metrics to assess key attributed of alternative SoS architectures and the anchor data for implementation of detailed analysis of select alternatives to evaluate options for evolution of the SoS thought its lifecycle. Continued development and integration of these model-based approaches is key to continued development of digitals engineering capabilities to address the challenges of this growing class of complex composite systems.

Trade-off analysis for SoS presents challenges in establishing evaluation criteria, modeling of candidate solutions, and evaluation of alternatives while addressing both CS and SoS-level engineering and investment decisions. Some techniques that help to alleviate these challenges have been described above. A structured approach to establishing goals and evaluation criteria is a critical early step in trade-off analysis, and SVN modeling shows potential for elucidating stakeholder priorities in an SoS context with conflicting objectives. Further, modeling of SoS architecture alternatives using MBSE tools and standards-based languages increases rigor of SoSE and holds promise for managing the complexity of SoSE analysis. The computable nature of architecture data captured in this way facilitates the use of lightweight analysis approaches to identify promising options for more detail trade-off analysis approaches. Lightweight metrics allow SoS architects to quickly explore large number of alternative solutions, recommending a computationally feasible subset for evaluation using higher-fidelity M&S tools. The high-fidelity analysis results in a multi-objective view of the solution space that the stakeholders and decision-makers navigate based on their preferences and constraints.

Initial work has been done to use lightweight metrics for evaluation of SoS robustness. Additional effort is needed to understand what network metrics are appropriate to support initial down-selection of SoS architectures for scalability, interoperability, and security. Ensuring traceability and consistency in models from problem formulation through making decisions is important due to the iterative nature of the SoSE trade-off analysis process. The practice of digitally linking the artifacts generated through the trade-off analysis process also represents opportunity for further development which offers promise for addressing complexity of CPS.

Disclaimer

The author's affiliation with The MITRE Corporation is provided for identification purposes only, and is not intended to convey or imply MITRE's concurrence with, or support for, the positions, opinions, or viewpoints expressed by the author. ©2019 The MITRE Corporation. Approved for public release. Distribution unlimited. Public Release case number: 18-2496-6.

References

Acatech (2011). Cyber-Physical Systems: Merging the Physical and Vitual World. In *Cyber-Physical Systems: Acatech Position Paper*, by National Academy of Science and Engineering (pp. 15–21). Berlin, Heidelberg: Springer.

Albro, S., Cotter, M., Kamenetsky, J., MacLeod, T., & Markina-Khusid, A. (2017). "Scaling Model-Based System Engineering Practices for System of Systems Applications: Software Tools." *NDIA 20th Systems Engineering Conference*.

Antul, L., Cho, L., Cotter, M., Dahmann, J., Tran, H., Jacobs, R., ... Ricks, S. (2018). "Toward Scaling Model-Based Engineering for Systems of Systems." *IEEE Aerospace Conference*.

Antul, L., Jacobs, R., Kamenetsky, J., & Markina-Khusid, A. (2017). Scaling Model-Based System Engineering Practices for System of Systems Applications: Analytic Methods. *NDIA 20th Systems Engineering Conference*.

Aziz, M., & Rashid, M. (2016). Domain Specific Modeling Language for Cyber Physical Systems. *International Conference on Information Systems Engineering* (pp. 29–33).

Brook, P. (2016). On the Nature of Systems of Systems. *INCOSE International Symposium* (pp. 18–21). Edinburgh, Scotland, GB.

Cameron, B. G., Crawley, E. F., Feng, W., & Lin, M. (2011a). Strategic Decisions in Complex Stakeholder Environments: A Theory of Generalized Exchange. *Engineering Management Journal, 23*, 37–45.

Cameron, B. G., Seher, T., & Crawley, E. F. (2011b). Goals for Space Exploration Based on Stakeholder Value Network Considerations. *Acta Astronautica, 66*, 2088–2097.

Ceccarelli, A., Mori, M., Lollini, P., & Bondavalli, A. (2015). Introducing Meta-Requirements for Describing System of Systems. *Proceedings of the IEEE International Symposium on High Assurance Systems Engineering*, 150–157.

Chattopadhyay, D. (2009). *A Method for Tradespace Exploration of Systems of Systems*. Cambridge, MA: Massachusetts Institute of Technology.

Cook, S. (2016, October). Some Approaches to Systems of Systems Engineering. *INCOSE INSIGHT* (pp. 17–22).

Cotter, M., Dahmann, J., Doren, A., Kelley, M., Markina-Khusid, A., & Wheeler, T. (2017). "SysML Executable Systems of Systems Architecture Definition: A Working Example." *IEEE Systems Conference*.

Dahmann, J. (2012). Integrating Systems Engineering and Test & Evalutaion in System of Systems Development. *IEEE Systems Conference*.

Dahmann, J., Markina-Khusid, A., Kamenetsky, J., Antul, L., & Jacobs, R. (2017). Systems of Systems Engineering Technical Approaches as Applied to Mission Engineering. *NDIA Systems Engineering Conference*.

Dahmann, J., Rebovich, G., Lane, J. A., & Lowry, R. (2010). "Systems of Systems Test and Evaluation Challenges." *5th International Conference on Systems of Systems Engineering*.

Dahmann, J., Rebovich, G., Lowry, R., Baldwin, K., & Lane, J. A. (2011). An Implementers' View of Systems Engineering for Systems of Systems. *2011 IEEE International Systems Conference*. Montreal, QC.

Dahmann, J. (2014). System of Systems Pain Points. *INCOSE International Symposium, 24*(1), 108–121.

Douglas, B. P. (2015). *Agile Systems Engineering*.

Ender, T., Leurck, R. F., Weaver, B., Miceli, P., Blair, W. D., West, P., & Mavris, D. (2010). Systems-of-Systems Analysis of Ballistic Missile Defense Architecture Effectiveness Through Surrogate Modeling and Simulation. *IEEE Systems Journal, 4*, 156–166.

Gezgin, T., Etzien, C., Henkler, S., & Rettberg, A. (2012). Towards a Rigorous Modeling Formalism for Systems of Systems. *2012 15th IEEE International Symposium* (pp. 204–211).

Gibson, J. E., Scherer, W. T., Gibson, W. F., & Smith, M. C. (2017). *How to do Systems Analysis: Primer and Casebook*. Hoboken, NJ: Wiley.

Han, E. P., & Delaurentis, D. (2006). A Network Theory-based Approach for Modeling a System-of-Systems. *11th AIAA/ISSMO Multidisciplinary Analysis and Optimization Conference* (pp. 1–16).

Harrison, W. K. (2016). The Role of Graph Theory in System of Systems Engineering. *IEEE Access*.

Hause, M. C. (2014). SOS for SoS: A New Paradigm for System of Systems Modeling. *2014 IEEE Aerospace Conference*. Big Sky, MT.

Henshaw, d. C., & Michael, J. (2016). Systems of Systems, Cyber-Physical Systems, the Internet-of-Things...Whatever Next? *Incose Insight, 19*, 51–54.

Huynh, T. V., & Osmundson, J. S. (2007). An Integrated Systems Engineering Methodology for Analyzing Systems of Systems Architectures. *Asia-Pacaific Systems Engineering Conference* (pp. 1–10).

INCOSE. (2014). Systems Engineering Vision 2025.

INCOSE. (2015). Systems Engineering Handbook: A Guide for System Life Cycle Processes and Activities. INCOSE-TP-2003-002-04.

INCOSE. (2018). INCOSE Systems of Systems Primer. INCOSE-TP-2018-003-01.0.

ISO. (2018). ISO/IEC/IEEE DIS 21839. *System of Systems (SoS) considerations in life cycle stages of a system.*

Lane, J. A., & Bohn, T. (2013). Using SysML Modeling to Understand and Evolve Systems of Systems. *Systems Engineering, 16*, 87–98.

LaSorda, M., Borky, J., & Sega, R. (2018). Model-Based Architecture and Programmatic Optimization for Satellite System-of-Systems Architectures. *Systems Engineering, 21*, 372–387.

MacCalman, A., Beery, P., & Paulo, E. (2016). A Systems Design Exploration Approach that Illuminates Tradespaces Using Statistical Experimental Designs. *Systems Engineering, 19*, 409–421.

Maier, M. (1998). Architecting Principles for System of Systems. *Systems Engineering, 1*, 267–284.

Monahan, W., Jacobs, R., Markina-Khusid, A., & Dahmann, J. (2018). Challenges and Opportunities in Trade-off Analytics for Systems of Systems. *Incose Insight, 4*, 22–28.

Mori, M., Ceccarelli, A., Lollini, P., Bondavalli, A., & Fromel, B. (2016). A Holistic Viewpoint-Based SysML Profile to Design Systems-of-Systems. *Proceedings of the IEEE International Symposium on High Assurance Systems Engineering* (pp. 276–283).

NIST. (2013). Foundations for Innovation in Cyber-Physical Systems Workshop Report. Columbia, MD.

Object Management Group. (2015). *OMG Systems Modeling Language (OMG SysML™) v1.4*, http://www.omg.org/spec/SysML/20150709/SysML.xmi

Office of the Deputy Assistant Secretary of Defense for Systems Engineering (DASD SE). (2018). Digital Engineering Strategy.

Oquendo, F. (2016). Formally describing the software architecture of Systems-of-Systems with SosADL. *11th IEEE Systems of Systems Conference* (pp. 1–6).

Rainey, L. B., & Tolk, A. (2015). *Modeling and Simulation Support for System of Systems Engineering Applications.* Wiley.

Rao, M., Ramakrishnan, S., & Dagli, C. (2008). Modeling and Simulation of Net Centric System of Systems Using Systems Modeling Language and Colored Petri-nets: A Demonstration Using the Global Earth Observation System of Systems. *Systems Engineering, 11*, 203–220.

Reymondet, L., Ross, A., & Rhodes, D. (2016). Considerations for Model Curation in Model-Centric Systems Engineering. *IEEE International Systems Conference.*

Ross, A., & Rhodes, D. (2015). An Approach for System of Systems Tradespace Exploration. In L. B. Rainey & A. Tolk (Eds.), *Modeling and Simulation Support for System of Systems Engineering Applications*. Hoboken: Wiley.

Sease, M., Smith, B., Selva, D., & Hummell, J. (2018). Setting Priorities: Demonstrating Stakeholder Value Networks in SysML. *28th Annual INCOSE International Symposium*, Washington, DC.

Systems Engineering Book of Knowledge (SEBoK). (2018). *Systems of Sytems (SoS)*. https://www.sebokwiki.org/wiki/Systems_of_Systems_(SoS).

TechTarget. (2016). *Internet of Things (IoT)*. https://internetofthingsagenda. techtarget.com/definition/Internet-of-Things-IoT.

Tolk, A., Glazner, C. G., & Pitsko, R. (2017). Simulation-Based Systems Engineering. In S. Mittal, U. Durak, & T. Oren (Eds.), *Guide to Simulation-Based Disciplines* (pp. 75–102). Springer.

Uday, P., & Marais, K. (2015). Designing Resilient Systems-of-Systems: A Survey of Metrics, Methods, and Challenges. *Systems Engineering, 18*, 491–510.

USAF Scientific Advisory Board. (2005). System-of-Systems Engineering for Air Force Capability Development.

Wheeler, T. (2018). MITRE's Integrated Engineering Environment. *Aerospace Digital Engineering Conference*.

Woodcock, J., Cavalcanti, A., Fitzgerald, J., Larsen, P., Miyazawa, A., & Perry, S. (2012). Features of CML: A Formal Modeling Language for Systems of Systems. *7th IEEE Systems of Systems Engineering Conference* (pp. 1–6).

7

Taming Complexity and Risk in Internet of Things (IoT) Ecosystem Using System Entity Structure (SES) Modeling

Saurabh Mittal[1], Sheila A. Cane[2], Charles Schmidt[1], Richard B. Harris[1], and John Tufarolo[3]

[1] *The Homeland Security Systems Engineering and Development Institute (HSSEDI)[TM], Operated by The MITRE Corporation, McLean, VA, USA*
[2] *Quinnipiac University, Hamden, CT, USA*
[3] *Research Innovations, Inc., Alexandria, VA, USA*

7.1 Introduction

Cyber Physical Systems (CPS) are systems that bring together cyber and physical worlds, i.e. computational and physical systems (through sensing and actuation). The computational and physical elements may be separated by a network channel that is local, or geographically separated, as in Internet. CPS have a specific use case, i.e. domain-specific, and correspond more to industrial sector (e.g. automobile, manufacturing, medicine, defense, etc.) that has complex machinery. CPS when connected through Internet is often referred to as Internet of Things (IoT), often marked as fourth Industrial Revolution: Industry 4.0 (Jazdi 2014). IoT has the same characteristic profile of a CPS, i.e. has sensing, computation infrastructure, actuation. However, IoT differs with CPS in usability, mobility, flexibility, scale, and the network protocols, that enable communication between the three fundamental parts of sensing, computation and actuation. While CPS has limited scalability and targeted user interaction, IoT has high scalability and wide user-profile across different sections of the society (Sehgal, Patrick, and Rajpoot 2014). Because of the presence of Internet in IoT, both the amount of data generated and the security and privacy concerns are high. Indeed, IoT is a more complicated version of CPS where complexities resulting from the scale abound. As IoT is interwoven with the fabric of society, the challenges to engineer IoT and study their effects before deployment as many of complexities are non-technical (e.g. policy, resources, etc.).

Complexity Challenges in Cyber Physical Systems: Using Modeling and Simulation (M&S) to Support Intelligence, Adaptation and Autonomy, First Edition. Edited by Saurabh Mittal and Andreas Tolk.
© 2020 John Wiley & Sons, Inc. Published 2020 by John Wiley & Sons, Inc.

There is growing recognition that the rapid deployment of new, connected IoT capabilities will change the risk profile within and across many existing technology-driven and technology-enhanced environments.

This chapter explores the nature of IoT as an advanced form of CPS and develops an IoT theoretical model. It describes how such a model could be used to perform risk assessments in IoT-centric systems. This chapter reflects the foundational research behind establishing this theoretical IoT model and related environment considerations, and is meant primarily to provide the community (both academic and government) with a foundation which can be built upon to analyze the risk implications of these systems as they become pervasive. The potential benefits of this research could be the development of automated risk assessment and mitigation tools to evaluate various IoT system instantiations during the design or implementation phases of an IoT adoption. These benefits would lead to a better understanding of the degree of risk a particular IoT adoption presents, and improve implementer decisions on risk mitigation strategies.

IoT bridges Information Technology and Operational Technology communities.[1] IoT, because of its interconnected IT and OT components, introduces the possibility of greater physical consequences from vulnerabilities within an IoT system or external threats to that system. Both IT and OT risk management are mature fields with established best practices. However, the body of these practices were developed back when IT and OT existed primarily in separate systems, and many of risk sources and effects are changed by the connection of these environments in a way that might not be readily apparent. For instance, the cost to the IT community for shutting down a portion of an IT network or modifying an IT network function may be minimal, while shutting down or modifying an OT system may result in substantial costs and the interruption of key services such as power. Because of this rapidly changing technological environment, there is a need for a model or framework that can help merge these risk perspectives and that can be used to explore the larger question of how systemic risks are increased given specific IoT deployments. While there is theoretical and practical modeling work being pursued from a technology perspective, there are no systematic models representing IoT systems along with their environment that could contribute to a comprehensive risk assessment leading to more complete risk mitigation approaches. A theoretical model describing the IoT ecosystem is needed to provide

---

1 For our purposes, IT is primarily concerned with data processes and OT is primarily concerned with physical processes and outcomes. IT and OT communities refer to respective practitioners in fields of technology where the purpose of technology is either to process information consumed by humans or other IT devices, or manage a machine or processes that effect multiple machines. The Internet of Things along with the expansion of connected technology is creating a convergence of these two fields.

methods to describe variations in IoT implementations, their relationships to the environment, and associated risk implications.

In this chapter, the authors consider risk as a function of "undesired functionality" in a system. Undesired functionality is a significant concern in highly complex systems. A device or system can exhibit undesired functionality either because such functionality was unintended (e.g. software vulnerabilities introduced in development causing unexpected behavior), or because it was unplanned (e.g. functionality intentionally designed into the device, but unexpected by the end-user). Unfortunately, as has been demonstrated in the IT world by security breaches and failures, such unexpected functionality is almost impossible to eliminate. Moreover, such unintended functionality in highly complex systems can have significant, cascading repercussions.

IoT devices can have a multiplicative effect on operational complexity. As noted earlier, IoT systems combine aspects of both IT and OT systems, which were previously largely independent of each other. The result is a manifold increase in the number of possible connections between devices and the potential consequences from undesired functionality in these systems. In an IoT system, these consequences include, but are not limited to, data consequences (data disclosure, data corruption, loss of data services, and unauthorized data access) and physical consequences (impacts on physical property, availability of critical services, availability of physical resources, potential impacts on the environment, and potential threats to human health and safety).

Taming system complexity is a difficult and ongoing challenge. To do that, it is critical to have a means to understand each system, its dependencies and interdependencies, and to understand how failures of one component impacts aspects of the whole. With the rise of IoT and its corresponding expanded impact pattern, an optimal mitigation strategy needs to consider both data and physical impacts.

The following section presents an approach to modeling systems with IoT devices. This modeling approach provides a way to apply rigor to the problem of tracing operational dependencies within such a system as a means to highlight potential vulnerability sources. By identifying the most likely cascading failure sources, and tracing the failure consequences to any eventual data or physical impact, one can determine how best to apply mitigations that avoid the most damaging cascades (Zimmerman and Restrepo 2009). While this model cannot predict the unpredictable (e.g., which systems will be attacked, or which will have a software vulnerability), it can help one understand the relationship between components and better clarify where risk mitigation strategies will be most effective.

IoT modeling is a relatively new concept. The National Institute of Standards and Technology (NIST) developed an IoT model to define the utility of devices constituting an IoT system using five elements (Voas 2016): sensors, aggregators, communication channel, external utility, and decision triggers. Other models

(Huang and Li 2010) have also tended to focus on a small number of IoT aspects (e.g. cause-effect chaining, focus on connectivity and network topology, etc.)

By contrast, the work presented here conceptualizes a unified model capable of capturing a broad array of factors impacting IoT systems. These include, but are not limited to, aspects of the devices themselves, connectivity, connected services, policies, users, physical structures, and many more. This chapter explores methods to capture all of these IoT system facets in a unified way to better support rigorous analysis methods. While the work presented here is limited to an initial IoT ecosystem theoretical model with the associated analytics yet to be developed, the authors believe the presented methods provide a promising path by which IoT systems can be understood with sufficient context and within their broader environment to support a more comprehensive framework for understanding IoT system risks. We demonstrate the usefulness of the approach by applying it to MIRAI botnet case study.

This paper describes the initial research results to develop a theoretical IoT model. Section 7.2 samples various IoT definitions and presents a device-centric world view. Section 7.3 presents an overview on SES theory that describes the pruning process as applicable to an SES model resulting in a Pruned Entity Structure (PES). Section 7.4 presents the IoT theoretical model. Section 7.5 applies it to the MIRA botnet case study. Section 7.6 discusses risks in IoT by describing the technical consequences required for risk evaluation, followed by the integrated risk evaluation framework. Section 7.7 closes the chapter with conclusions and some ideas on future work.

7.2 IoT Definition and Device-Centric World View

The IoT is a relatively recent and evolving phenomenon, sharing a concurrently evolving definition. When used as an acronym in itself, IoT signifies the entire IoT ecosystem. However, the usage of this acronym changes with every definition. Most IoT definitions arise from the fundamental perspective "anything connected to the Internet becomes a part of an IoT" (while neglecting to actually define "IoT") (Cisco n.d.). Most definitions acknowledge two semantic elements: "the Thing," or "Things" and "the Internet". *A Thing* is defined as a physical device which is supplemented with a computing capability (on-board or remote), such that it becomes "smart," and has both software and hardware interfaces allowing it to connect to the Internet (CASAGRAS n.d.; Mckinsey n.d.; SIG 2013). Some IoT definitions include business processes, services, applications, and infrastructure to support the various "things" (CERP-IoT n.d.; Haller n.d.), while other research perspectives include virtual entities, virtual personalities, and all the miscellany of commerce and culture (Berge 1973), including real people (NIC n.d.; Domingue, Fensel, and Traverso 2008; Sundmaeker, Guillemin, and Woelffle 2010; Group 2011; Lee et al. 2011).

A particularly compelling IoT definition is provided by UK Future Strategy (Group 2011):

> An evolving convergent Internet of things and services available anywhere, anytime as part of an all-pervasive omnipresent socio–economic fabric, made up of converged services, shared data and an advanced wireless and fixed infrastructure linking people and machines to provide advanced services to business and citizens.

Consequently, every IoT system has a purpose: i.e., it serves a community/a group of users/people. Furthermore, any device that gets connected to the Internet without a purpose introduces unnecessary risk. We define an IoT system as:

> A complex system consisting of a collection of interconnected smart devices that may span a single user, a community, a region, a nation or across nations; provides services within an emerging or an existing system; is distributed across the Internet; and incorporates inherent risk.

Together, all the devices deployed to serve a purpose/use-case, with all the underlying infrastructure, constitute an IoT ecosystem, largely called just an *IoT*. To fully model an IoT essentially requires modeling the *network* and the *things*, as well as the various *contexts* in which these two are used.

Figure 7.1 displays a device-centric view of this IoT definition. The IoT-device participates in multiple contexts, across multiple sectors/application-domains. Each ring represents an activity scope (e.g. local, near neighborhood, regional, state, national, and global reach). Each ring has a logical boundary representing the data exchange between different regions through the access points. As you move out towards the edges, the scope, scale, and possible effect of IoT implementation increases. Figure 7.1 also depicts two use-cases as applicable to two different sectors (for example, electricity and water sectors A and B respectively). It comprises of two use-cases/scenarios depicted as Walk 1 and Walk 2. Walk 1 is a user-oriented *regional* scenario using resources only in Sector A and operates up to a regional scale. Walk 2 utilizes the network and resources belonging to both Sector A and B and operates on a larger, *global* scale.

7.3 System Entity Structure (SES) Model

This research suggests that a SES ontological framework can be applied to modeling IoT systems, as we have defined above. The SES theory (Zeigler 1984; Zeigler and Zhang 1989; Zeigler, Praehofer, and Kim 2000) is a formal ontology framework to capture system aspects and their properties. SES is typically used to lay

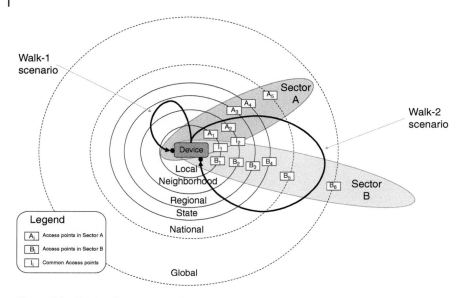

Figure 7.1 Device Centric View of an IoT system.

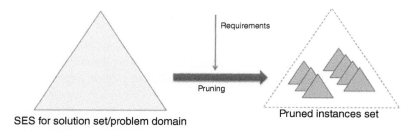

Figure 7.2 SES Application.

out the design space for complex information systems. Each configuration option in the design space may be applied to a specific use-case and thereby specify the needed architecture (Zeigler and Hammonds 2007; Mittal and Martin 2013). The underlying SES axioms provides the needed constraints to bring structure to the modeling process – these axioms can be found in (Zeigler 1984). A complete SES model formally describes a solution set containing all permutations and combinations available for modeling an actual system. Figure 7.2 conceptualizes the SES modeling process (Mittal and Martin 2013). It shows the complete solution set, represented by the large triangle, which represents the totality of a theoretical model. For each instantiation of the model, the theoretical model is pruned to only those elements required for the instance.

SES modeling semantics is constructed of the following elements:

- *Entity* (a physical entity or a concept represented as a label);
- *Aspects* (decomposition: *is made up of*): Denoted by a vertical bar (|);
- *Specializations* (can be of type: *is a type of*): Denoted by a double vertical bar (||);
- *Multi-aspect* (decomposition into similar type: *is made up of many such*): Denoted by a triple vertical bar (|||). It also has a variable *n*, that specifies the number of entities in the relationship; and
- *Variables* (each entity has variables that have a range and value): Denoted by ~.

As an example, consider the IoT-Inclusive System in Figure 7.3, represented as an SES. This diagram can be read as follows. An *IoT-Inclusive-System* has two aspects; *network aspect* and *physical aspect*, labeled *net-asp* and *phy-asp* respectively. These aspects are physically realized as the entities: *Network* and *Things*. The *Network* entity has a *connectivity-asp* aspect, which includes the entities *Connections* and *Resources*. *Connections* entity consists of many *Connection* entities. A *Connection* can be specialized using the *connect-mode-spec* into *Wired* or *Wireless* entities. A *Connection* entity through the *connect-protocol-aspect* has a *Communication Protocol*. A *Communication Protocol* entity can be specialized using the *protocol-spec* into *IP, Bluetooth, ZigBee,* or *802.x* entities. *Things* are comprised of many *LogicalDevice* entities. The multi-aspect *Things* and *Connections* may contain millions of *Logical Device* entities and millions of *Connection* entities.

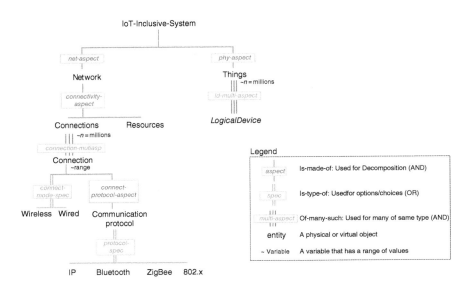

Figure 7.3 Notional IoT-Inclusive System.

Together with all the choices, the *IoT-Inclusive-System* SES represent all the possible architectures at a very high level of abstraction for a technology-only solution. From the permutation and combinations perspective, for example, assuming we have 10 Things and 50 connections between these 10 things (although a total of 100 connections are possible), there are about 4000 configuration options possible for the IoT-Inclusive System (2 Connection×4 Communication Protocol×10 Things×50 Connection). For brevity, *Resources* entity is not described in this example.

7.3.1 Pruned Entity Structure (PES) Model

The SES allows coverage of all the semantic permutation and combinations available for articulating any particular system in a formal manner. To systematically explore varying implementation options for specific use-cases (a.k.a. scenarios), the complete SES model is pruned into only the essential elements relating to the scenario. The resulting SES is called a Pruned Entity Structure (PES). The PES can be continuously pruned to reduce the available options to get closer to the problem-at-hand. Figure 7.4 shows the pruning process. The resulting PES acts as a Reference Architecture, as it provides enough constraints to represent the particular domain within a family of architectures. The pruning process ends when a PES results in a Component Entity Structure (CES) that requires no further pruning. The CES acts as a Solution Architecture. The methodology to use SES, PES, and CES for simulation models has been demonstrated in literature (Mittal and Martin 2013; Zeigler and Sarjoughian 2017).

The IoT-Inclusive System design space shown in Figure 7.3, when pruned for a specific use, such as 10 Things and 50 Connections, yields the PES in Figure 7.5. Once we have the PES, various use-cases can be constructed allowing entity navigation (*walks*) with the PES.

An execution of a use-case involving entities in SES is called a *walk*. Semantically, a walk represents the sequence of information flow between the entities. This is analogous to the walk as described in Section 7.2 for Figure 7.1. Figure 7.5 shows a walk with two *LogicalDevices* communicating over the *Network*. This walk is represented as a sequence of numbers on the PES entities. As a PES still contains a lot of permutations, a walk in PES is a higher level of abstraction. As the PES is continuously pruned (Figure 7.4), it eventually leads to a very specific implementation: an architecture of the IoT-Inclusive System with actual components (physical instances). This transforms PES into a CES (Mittal and Martin 2013). A simple two-device IoT from the PES is shown in Figure 7.5. Figure 7.6 shows the PES transformed into CES with multi-aspects converted into aspects linking actual instances of the entities.

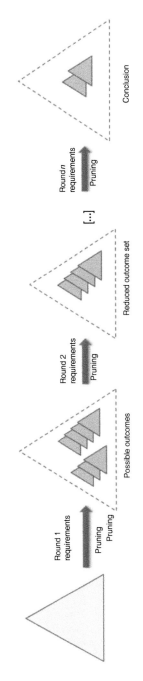

Figure 7.4 PES generation through an iterative process. *Source:* From Mittal and Martin 2013.

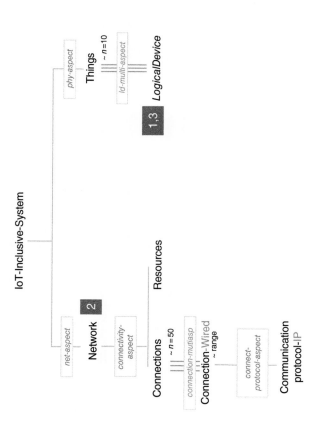

IoT-Inclusive-System

net-aspect

2 Network

connectivity-aspect

Resources

Connections

~ *n*=50

connection-mutiasp

Connection-Wired
~ range

connect-protocol-aspect

Communication
protocol-IP

phy-aspect

Things

~ *n*=10

ld-multi-aspect

1,3 *LogicalDevice*

Use-case: Two logical devices
communicating over network

Graph walk (in sequence):
1. Logical device
2. Network
3. Logical device

The communication from the leaf
nodes goes up the hierarchy until it
finds a common link to the parent of
the destination leaf

Figure 7.5 PES for IoT-Inclusive System.

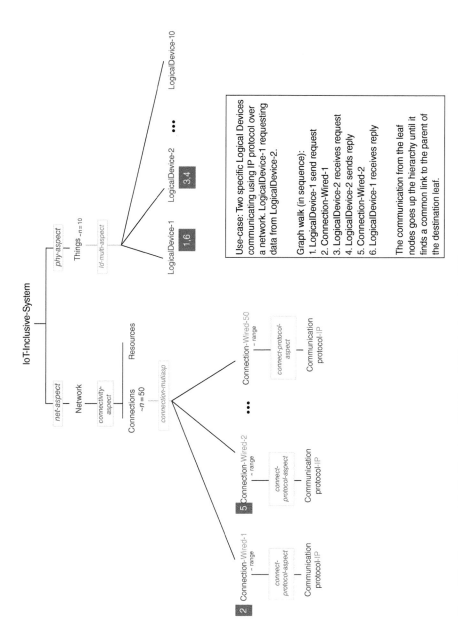

Figure 7.6 Component-Entity Structure (CES) from IoT-Inclusive system PES.

7.4 IoT Model

In order to model the IoT in its environment, we identify the following fourteen perspectives, modeled as SES aspects. Each perspective is necessary to understand the multi-dimensional nature of IoT. The perspectives are enumerated in Table 7.1.

The perspectives described in Table 7.1 portrays an abstract IoT-Inclusive system representation. Each perspective is modeled as an SES *aspect*. The SES model can help identify relationships between various IoT perspectives that need to be considered for an IoT deployed solution. These relationships are depicted by walks as described for Figures 7.5 and 7.6. Figure 7.7 provides a sample depiction of an IoT-Inclusive SES model. This model exhibits the following options (where the hypothetical numbers are used solely for illustration purposes):

> Ignoring the multi-aspect numbers, we have: $(2\ Behaviors) \times (4\ LogicalDevice$ types$) \times (3\ Services) \times (2\ Connection$ types$) \times (5\ CommunicationProtocol$ types$) \times (2\ Timeliness\ analytics$ types$) \times (6\ ComputingAnalytics$ types$) \times (4\ ResourceItems$ types), resulting in 12,000 IoT deployment options and the corresponding use-cases. This number does not even consider the combinations associated with entities: *Resource*, *DataElement*, *Domain*, *Capability*, *AttackEffects*, and *Orgs*. One can easily imagine the increase in permutations from thousands to millions of options.

As stated earlier, each structural option enumerated in the above permutation can be realized as a valid system architecture. This exploration also involves constraints and scores attributable to each of the nodes that may help in overall risk assessment. Conducting performance and risk analysis on each option shown in this high-level model would be resource intensive and cost-prohibitive. As more constraints are added by applying the pruning process, the IoT model moves closer to the specific IoT solution, and therefore, the risk quantification should become more tractable, both visually and mathematically.

The theoretical model brings forward 14 perspectives in a foundational IoT-Inclusive-System. It shows how various entities constitute cross-references in different. Associating various vulnerability types and risk scores to each of the entities in the abstract model yields the needed abstraction level to understand the risk impact at a macro level. Further specification of constraints helps us understand risk for a family of IoT-Inclusive-Systems.

7.5 Case Study: MIRAI Attack

This section presents how an IoT-Inclusive-System theoretical model can be used in a simple practical example.

Table 7.1 Fourteen IoT-Inclusive System perspectives.

ID	Perspective	Description
1	Behavior	Describes the overall macro IoT behavior as a whole. The behavior could be emergent (i.e. IoT resulted from devices just being connected to the Internet) or engineered (i.e. IoT resulted from following engineering practices).
2	Application Domain	Identifies various environmental variables and properties that monitor the usefulness of a particular IoT solution and its application to a single or multiple sectors. It also describes the effect a particular IoT solution has on a sectorial basis.
3	Physical structure	Describes the logical structure of a Thing: a sensor, an actuator, an aggregator assembling multiple inputs for a single output, or more specifically, a computing device having variable computational power levels.
4	Services	Describes various services a Thing consumes or provides to other components on the Network. Essentially, IoT solves a business problem or creates opportunities to develop new business cases, eventually leading to formulation of new business processes.
5	Resources	Describes resources, their management and supply-demand within an IoT system deployment
6	Information	Describes the knowledge structure as it is exchanged between resources, across network and various Things.
7	Network	Describes the network structure between two things. The geographical distance and relative compatibility between two things may ultimately decide what kind of resources are invoked along the pathways.
8	Vulnerability	Describes the vulnerability in the IoT deployment across multiple perspectives. The vulnerabilities need to be specified at the intersection of various perspectives define in this list. Various scoring systems such as Common Vulnerability Scoring System (CVSS) (First. org n.d.), Common Weakness Scoring System (CWSS) (Coley 2014), etc. can be used to assess vulnerability inherent in an IoT deployment.
9	Organizations	Describes various organizations having vested interest in a given IoT deployment. Multiple organizations may be involved in multiple IoT deployments requiring shared services, resources, information, and applications.
10	Security	Describes security mechanisms and technologies.
11	Privacy	Describe privacy mechanisms and technologies.
12	Analytics	Describes the analytics required to ascertain the correct functioning of an IoT deployment. These will involve both macro and micro levels so that the impact and influence of a Thing can be viewed in relation to a particular application domain or Sector as a whole. Eventually, such analytic frameworks will be used for instrumentation and control at a critical infrastructure level.
13	Policy	Describes the relationship between multiple perspective to specify and define the interactions and relationships at the intersection of various perspectives as applicable to various stakeholders.
14	Users	Describes the users (or stakeholders) involved in an IoT deployment. It may include the end-user, the transient user (i.e. a user who indirectly uses an IoT as in third-party services, e.g. marketing services), and the organizations that use the applications within a typical IoT deployment.

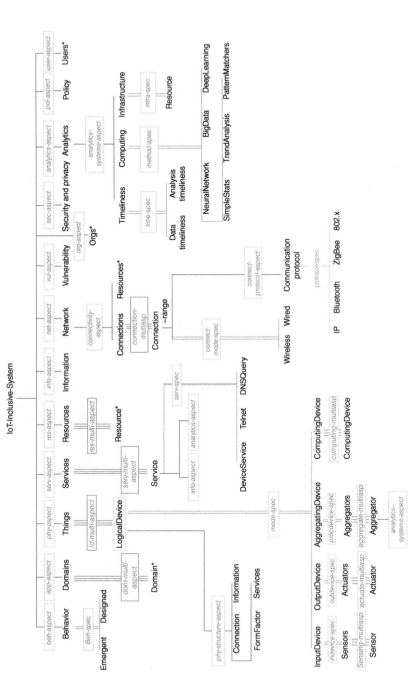

Figure 7.7 IoT-Inclusive System entity structure (sample depiction).

7.5.1   Description

In September of 2016, a massive Distributed Denial of Service (DDoS) was launched against select targets on the Internet. DDoS attacks are not new, but this attack was unprecedented in both its size (620 Gbps – almost double the volume of attack traffic as compared with the previous record attack reported by Akamai, a major provider of defenses against such attacks) and the significant number of IoT devices under control of the Mirai botnet, that contributed to the attack (Shugrue 2016). Previously, most members of botnets that were used for DDoS attacks were traditional endpoints, such as laptops and servers. The Mirai botnet, consisting of IoT devices, has been used in multiple other DDoS attacks, with some claims of traffic magnitudes up to 1.2 Tbps (Loshin 2016).

The Mirai virus and associated botnet represents a very simple example of an IoT attack scenario example. The only significant functionality of the targeted devices (beyond their vulnerability to attack) was their ability to send network traffic. In this regard, Mirai's objectives did not depend upon whether the compromised device was a toaster or a major industrial control system. The only dependency was on whether the virus was able to gain access to the device in question (Krebs 2016). The attack itself relied largely on brute-force, using fairly basic DDoS techniques to overwhelm targeted Internet services with network traffic (Dobbins 2016). Again, no device-specific capabilities were employed. Instead, the reason the Mirai botnet was so effective had more to do with reasons beyond the inherent function of the devices themselves. These devices were easily compromised since they lacked the security software used on PCs, such as anti-virus. A second important reason was because the targeted devices operated 24/7, unlike traditional PCs that are often shut down at the end of a business day (MalwareTech 2016). This meant the attackers were able to grow their botnet quickly (generally a vulnerable device was infected within 10 minutes of being connected to the network) (Dobbins 2016), and that, at any given point in time, a significant number of those compromised devices were available to participate in the attack.

Since that attack in September, the malware authors released the source code for the Mirai malware, and the number of devices compromised by this malware has more than doubled (Mimoso 2016). Future attacks using this malware and variations of it are inevitable, and may be harder to mitigate. Depending on the target selected, it is possible that more substantial impacts might be felt in the future.

7.5.2   Modeling the Mirai Use-case from the IoT SES Model

We now model the Mirai use case using the IoT SES model and demonstrate how the pruning process on the general IoT SES model can lead to the specific usage for Mirai case study. The main pruning process objective is to reduce the design choices and move towards an entity structure that can provide an architecture for a given problem.

As described in the pruning process, only one of the options from an SES-*specialization* is selected, which is then tagged with the parent entity. More details on the pruning process is available in Zeigler and Hammonds (2007). Using Figure 7.7 as the base SES, the following actions were taken as a part of the pruning process to yield an IoT-Inclusive-System SES relevant to the use-case of a normal IoT operation.

1) In the aspect, *beh-aspect*: *Behavior* is specialized to *Designed*, leading to the new entity label *Behavior-Designed*, replacing the original, *Behavior* entity label.
2) In the aspect, *app-aspect*: *Domain* is specialized to *InfoTech*, that has *Capabilities* entity specialized into *Capabilities-Desired*. There is no change in *AttackEffects* specification.
3) In the aspect, *phy-aspect*: *LogicalDevices* is specialized to *ComputingDevices*. There are many *ComputingDevice* of type *IPCameras* entities. This is new information that was not available in the parent SES. The original SES affords us the abstract structure that allows expansion at the any Entity level.
4) In the aspect, *serv-aspect*, there is no change.
5) In the aspect, *res-aspect*: Resource *Management* entity is specialized into *Configuration* as there may be configuration management present. There are no people involved in the current example people so *Persons entity is removed*. However, the Resource Item still display the choice of Information, System, and Roles entities. This shows that the pruning process can still retain choices and not have to be specialized at the lowest leaf level. *System* entity does not change. *Software* entity is specialized into *CameraSoftware* entity.
6) In the aspect, *info-aspect*: there is no change.
7) In the aspect, *net-aspect*: *Connection* specialized to *Wired*, *CommunicationProtocol* is specialized to *IP* and the *Resource* tree will use *System* entities, when it specifies the *ResoureItem* entity.
8) In the aspect, *vul-aspect*, there is no change.
9) In the aspect, *org-aspect*, there is no change.
10) In the aspect, *sec-aspect*, there is no change.
11) In the aspect, *analytics-system-aspect*: *Computing* is specialized into *SimpleStats* entity and *Infrastructure* is specialized into *Resources* entity.
12) In the aspect, *pol-aspect*, there is no change.
13) In the aspect, *user-aspect*, there is no change.

The above reduction yields in (without counting the multi-aspects):

(3 *Services*)×(2 *DataElement* types)×(3 *System* types)×(5 *Router* types)×(4 *AttackEffect* types)×(3 *Orgs*)×(2 *User* types), resulting in 2000 design and evaluation options, which is 3 orders of magnitude less than the 4.8 million options in the original abstract SES.

Figure 7.8 provides a summary view of the above tailing. Additional PES pruning/specialization is captured in Figures 7.9–7.13 to reflect the Mirai System. Note the red numbered labels in these figures are referenced in the "neutral walk" use-case described next. The neutral walk illustrates how the PES for Mirai can be navigated for a use-case for a normal operation of an IoT device: An End-user requests an image from his IP Camera over the Internet through a remote application.

Neutral Walk Sequence

1) *End User* begins procedure by sending a request
2) He opens an *Application*
3) *Application* is the *Camera Software*
4) *Application* invokes the *Communication Protocol – IP*
5) Request is received by *DeviceService*
6) *DeviceService* contacts the *IPCamera* that takes a picture
7) *IPCamera* sends the picture back to *DeviceService*
8) DeviceService invokes the *Communication Protocol-IP* to send the picture back to *Application*.
9) *Application* receives the picture
10) *EndUser* receives the picture

In the case of Mirai attack, once the IoT device (e.g. IP Camera) was loaded with Mirai malware, the IoT *Device Service* triggered DNS query (see Figure 7.8 *DNSQuery* service node) thereby contributing to the attack on DNS servers. Done at scale, the compromised IoT devices caused DDoS on DNS servers. This is shown in the highlighted step 5a that is inserted between the steps 5 and 6. This insertion turns the neutral walk sequence into an attack walk sequence:

Attack Walk Sequence

1) *End User* begins procedure by sending a request
2) He opens an *Application*
3) *Application* is the *Camera Software*
4) *Application* invokes the *Communication Protocol – IP*
5) Request is received by *DeviceService*
5a) *Device Service* contacts DNS Servers using *DNSQuery service*
6) *DeviceService* contacts the *IPCamera* that takes a picture
7) *IPCamera* sends the picture back to *DeviceService*
8) DeviceService invokes the *Communication Protocol-IP* to send the picture back to *Application*.
9) *Application* receives the picture
10) *EndUser* receives the picture

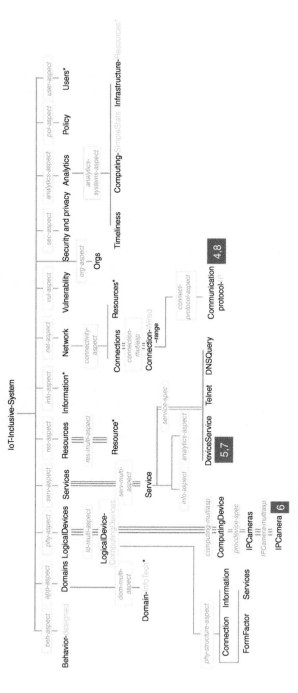

Figure 7.8 PES of IoT-Inclusive-System SES for Mirai System.

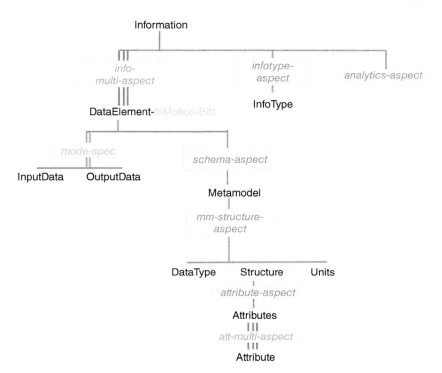

Figure 7.9 Information PES for Mirai System PES.

The attack walk sequence is only an illustration of how an existing normal walk sequence could be modified to model the attack that resulted in a new behavior impacting the overall system. While the shown attack sequence does not portray how the entire system behaves, it does indicate that the notional Mirai System PES does have the necessary detail, at the device level, that facilitated DDoS attack. More relevant walks that include the *DNSQuery* service can be constructed to model the entire Mirai case study. This is left as an exercise.

7.6 Risks in IoT

Both IT and OT have mature and robust risk assessment and management methodologies. However, now that these communities are connected by IoT, differences in each community's perspective on risk can lead to problems in developing full spectrum solutions that adequately address more complex issues in both IT and OT terms.

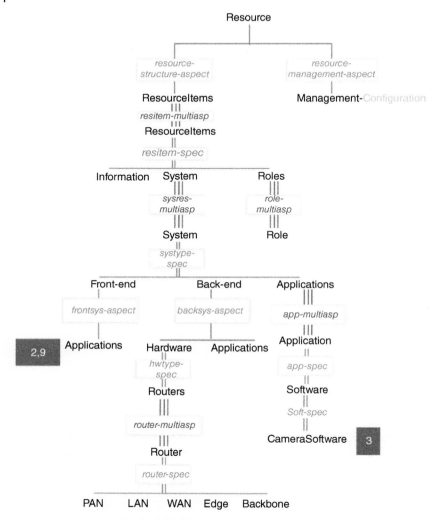

Figure 7.10 Resource PES for Mirai System PES.

7.6.1 IoT Technical Consequences

Many communities (including both IT and OT communities) express risk as a combination of the likelihood of an undesired event and the severity of that event's impact. More concisely, this concept is expressed as "Risk = Likelihood × Consequence." However, while there is broad consensus on the concepts that comprise risk, the risk characterization details can vary widely between communities. An outcome that might be acceptable to one community could be

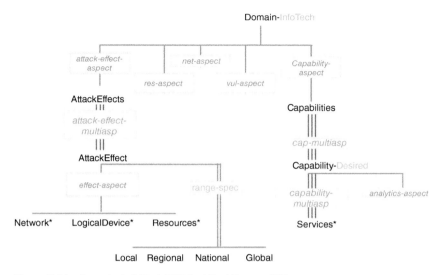

Figure 7.11 Domain-InfoTech PES for Mirai System PES.

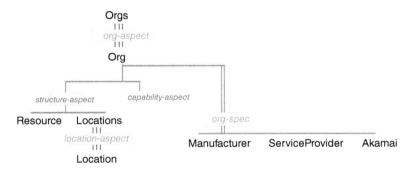

Figure 7.12 Orgs PES within the Mirai System PES.

Figure 7.13 Users PES within the Mirai System PES.

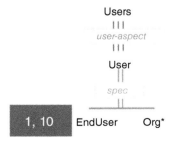

the worst-case-scenario for another. Since any quantitative severity expression necessarily codifies one community's risk perspective, such quantitative methods are not a good starting point when trying to create an integrated understanding that encompasses multiple communities. It is for this reason that the IoT technical consequence specification has to be expressed in qualitative terms, to begin with. It is believed, consequence characterization supported by a deterministic procedure for identifying such consequences, might be a useful tool in developing a risk understanding mutually acceptable to IT and OT communities.

Someone attempting to understand the risk posture of an IoT-Inclusive system may employ a deterministic procedure for the generation of technical consequences that might occur within the system. An undesired behavior of a technical capability results in a *technical consequence*. In order to generate technical consequences associated with a particular device or functional unit, one describes, with the greatest detail possible, the technical capabilities associated with that device or functional unit. These capabilities are then mapped to a short list of undesirable behaviors. These behaviors are:

- allowing undesired use;
- blocking desired use;
- slowing desired use; and
- changing desired use.

The combination of each capability with each of these undesired behaviors produces a set of technical consequence that is at least theoretically possible given the capabilities of the device or functional unit.

Understanding the potential technical consequences associated with a device or functional unit is only one part of a procedure for understanding the operational risk associated with that device. To understand operational risk, technical consequences need to be turned into operational consequences, which are expressed in terms of their impact on the IoT-Inclusive system as a whole. Doing this requires an understanding of the context in which a given technical consequence occurs. This sort of context could be provided by associating the technical consequence with elements in a model of the IoT system, such as is described by the SES model. By using such a model, one can place a technical consequence in the context needed to understand what, if any, operational consequence could result from that technical consequence.

7.6.2 Integrated Risk Evaluation Framework

The concepts provided in previous sections can be combined into a theoretical risk evaluation framework. The steps in this framework are as follows and shown in Figure 7.14.

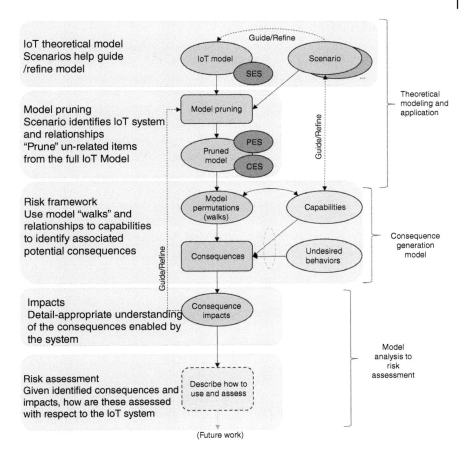

Figure 7.14 Combined risk evaluation framework.

1) **IoT Theoretical Model** – A set of scenarios describing nominal use of an IoT system in the SES framework.
2) **Model Pruning** – The SES producing PES and/or CES.
3) **Risk Framework** – The devices and/or functional units in the PES or CES are annotated with their attendant set of technical capabilities and nominal walks. The list of undesired behaviors are applied to the capabilities in these generated walks, leading to a set of potential technical consequences.
4) **Impacts** – By examining the generated technical consequences within the context of the walks in which they appear, one can derive a set of operational consequences associated with these walks to be used directly as part of a risk assessment.

7.7 Conclusions and Future Work

Both CPS and IoT contain three fundamental elements: sensing, computation, and actuation. These elements are separated through a network channel that has different characteristics based on the use-case of both the CPS and IoT. IoT differs from CPS in terms of scale at multiple levels: data, user, network, risk profile, etc. IoT is a more complicated version of CPS. Modeling both the CPS and IoT in all their richness in a lab-setting is a complex endeavor, simply because the environment that needs to be modeled is multi-faceted. CPS/IoT model requires description across multiple perspectives.

This chapter presented an IoT theoretical model in the form of a SES model as a potential way to identify risk in an IoT system. This model encompasses both IT and OT components of IoT as well as the environment surrounding these elements. It presented a device-centric world-view necessary to understand the expanded impact a device has on the larger technology-oriented ecosystem in different operational contexts. We believe that an SES-based methodology is a suitable means to identify and understand the dependencies and interdependencies between the model elements. The SES ontological framework provides a robust framework to construct an IoT model in an iterative manner both visually and structurally.

The pruning process facilitates the specialization of a generalized SES to identify the most relevant components of the IoT architecture. These components include the IT and the OT characteristics of IoT which were further identified by applying 14 different aspects of the conceptual IoT-Inclusive system. The resulting proposed model provides a starting point for developing a common set of terms describing the essential elements of the IoT and its environment, which are essential for creating a better understanding of the overall ecosystem and associated risks. The MIRAI Botnet case study showed how a model of the Botnet could be constructed from the SES description and potentially used to further our understanding of consequences and risk in an IoT ecosystem.

A significant amount of work needs to be done before the theoretical framework described in this paper could be practically employed as part of an actual risk assessment. Even an understanding of operational consequences is only one part of an overall risk assessment. One also needs to have a way to understand how likely a given consequence is within the system, lest one fixate upon the most disruptive consequences even if they are implausible. Eventually there is also a need to produce a quantitative understanding of an operational consequence by assigning it a severity. Such a severity is necessary to prioritize risks so that resources can be focused on the greatest sources of risk to the system. However, in an IoT-Inclusive system, such a severity assignment would need buy-in from both IT and OT practitioners. The hope is that by focusing on technical consequences as a first step, these communities will have a mutually agreed upon starting point for discussions leading to a mutually acceptable understanding of risk severity. Thus, technical consequences description a stepping stone to an understanding of risk that meets the combined needs of both IT and OT communities.

Future research could expand upon this work to develop a detailed risk calculation methodology, automated pruning algorithms, and eventually automated risk assessment and mitigation strategies adaptable to variations in IoT implementation from both static (design) and dynamic (implementation) perspectives. It should be cautioned, however, that the path from the general model to automated risk assessment and mitigation strategies is not short. Benefits of this future research include a way to enable improved stability and resilience for IoT enterprises, and an understanding of potential paths to improved situational awareness that can leverage industrial and commercial applications.

Acknowledgments

The Homeland Security Act of 2002 (Section 305 of PL 107-296, as codified in 6 U.S.C. 185), herein referred to as the "Act," authorizes the Secretary of the Department of Homeland Security (DHS), acting through the Under Secretary for Science and Technology, to establish one or more federally funded research and development centers (FFRDCs) to provide independent analysis of homeland security issues. MITRE Corp. operates the Homeland Security Systems Engineering and Development Institute (HSSEDI) as an FFRDC for DHS under contract HSHQDC-14-D-00006.

The HSSEDI FFRDC provides the government with the necessary systems engineering and development expertise to conduct complex acquisition planning and development; concept exploration, experimentation and evaluation; information technology, communications and cyber security processes, standards, methodologies and protocols; systems architecture and integration; quality and performance review, best practices and performance measures and metrics; and, independent test and evaluation activities. The HSSEDI FFRDC also works with and supports other federal, state, local, tribal, public, and private sector organizations that make up the homeland security enterprise. The HSSEDI FFRDC's research is undertaken by mutual consent with DHS and is organized as a set of discrete tasks. This report presents the results of research and analysis conducted under:

Task Order Number: 43161204,
Task Title: HSHQDC-16-J-00526:Core Research Program, Internet of Things (IoT) Modeling
Task Order Sponsor: Department of Homeland Security, National Protection and Programs Directorate

Purpose statement: The purpose of this research was to develop an IoT theoretical model, and describe how such a model could be used to perform risk assessments in IoT-centric systems.

The results presented in this report do not necessarily reflect official DHS opinion or policy.

Notice

This (software/technical data) was produced for the U. S. Government under Contract Number HSHQDC-14-D-00006, and is subject to Federal Acquisition Regulation Clause 52.227-14, Rights in Data – General. As prescribed in 27.409(b)(1), insert the following clause with any appropriate alternates:

Rights in Data – General (Deviation May 2014).

No other use other than that granted to the U. S. Government, or to those acting on behalf of the U. S. Government under that Clause is authorized without the express written permission of The MITRE Corporation.

For further information, please contact The MITRE Corporation, Contracts Management Office, 7515 Colshire Drive, McLean, VA 22102-7539, (703) 983-6000.

Approved for Public Release; Distribution Unlimited. Public Release Case Number 19-0575 / DHS reference number 16-J-00526-01

References

Berge, C. (1973). *Hypergraphs*. North-Holland, Amsterdam: American Elsevier Publishing Company.

CASAGRAS. (n.d.). RFID and the Inclusive Model for the Internet of Things. http://www.rfidglobal.eu/userfiles/documents/FinalReport.pdf (accessed 28 December 2016).

CERP-IoT. (n.d.). Internet of Things: Strategic Research Roadmap. http://www.grifs-project.eu/data/File/CERP-IoT%20SRA_IoT_v11.pdf (accessed 28 December 2016).

Cisco. (n.d.). The Internet of Things: How the Next Evolution of the Internet is Changing Everything. *CISCO IBSG*. http://www.cisco.com/web/about/ac79/docs/innov/IoT_IBSG_0411FINAL.pdf (accessed 28 December 2016).

Coley, S. (2014). *Common Weakness Scoring System (CWSS)*. McLean, VA: MITRE. https://cwe.mitre.org/cwss/cwss_v1.0.1.html (accessed 30 April 2019)

Dobbins, R. (2016, October 26). Mirai IoT Botnet Description and DDoS Attack Mitigation. *Arbor Networks.* https://www.arbornetworks.com/blog/asert/mirai-iot-botnet-description-ddos-attack-mitigation (accessed 8 January 2017).

Domingue, J., Fensel, D., & Traverso, P. (2008). *Future Internet – FIS 2008: First Future Internet Symposium.* Vienna, Austria: Springer.

First.org. (n.d.). *Common Vulnerability Scoring System v3.0: Specification Document.* https://www.first.org/cvss/specification-document (accessed 30 April 2019).

Group, UK Future Internet Strategy. (2011, May). Future Internet Report. https://connect.innovateuk.org/documents/3677566/3729595/Future+Internet+report.pdf (accessed 28 December 2016).

Haller, S. (n.d.). Internet of Things: An Integral Part of the Future Internet. http://services.future-internet.eu/images/1/16/A4_Things_Haller.pdf (accessed 28 December 2016).

Huang, Y., & Li, G. (2010). Descriptive Model for Internet of Things. *International Conference on Intelligent Control and Information Processing*, Dalian, China.

Jazdi, N. (2014). Cyber Physical Systems in the Context of Industry 4.0. *IEEE International Conference on Automation, Quality and Testing, Robotics,* Cluj-Napoca, Romania. IEEE. doi:https://doi.org/10.1109/AQTR.2014.6857843

Krebs, B. (2016, October 16). Hacked Cameras, DVRs Powered Today's Massive Internet Outage. *Krebs on Security.* https://krebsonsecurity.com/2016/10/hacked-cameras-dvrs-powered-todays-massive-internet-outage/ (accessed 8 January 2017).

Lee, G.M., Park, J., Kong, N., & Crespi, N. (2011). *IETF-The Internet of Things: Concepts and Problem Statement.* Internet Draft, IETF.

Loshin, P. (2016, October 28). Details emerging on Dyn DNS DDoS attack, Mirai IoT botnet. *TechTarget.* http://searchsecurity.techtarget.com/news/450401962/Details-emerging-on-Dyn-DNS-DDoS-attack-Mirai-IoT-botnet (accessed 8 January 2017).

MalwareTech. (2016, October 3). Mapping Mirai: A Botnet Case Study. *MalwareTech.* https://www.malwaretech.com/2016/10/mapping-mirai-a-botnet-case-study.html (accessed 8 January 2017).

Mckinsey. (n.d.). The Internet of Things. https://www.mckinseyquarterly.com/High_Tech/Hardware/The_Internet_of_Things_2538 (accessed 28 December 2016).

Mimoso, M. (2016, October 19) Mirai Bots More Than Double Since Source Code Release. *Threatpost.* https://threatpost.com/mirai-bots-more-than-double-since-source-code-release/121368 (accessed 8 January 2017).

Mittal, S., & Martin, J. L. R. (2013). *Netcentric System of Systems Engineering with DEVS Unified Process.* Boca Raton, FL: CRC Press.

National Intelligence Council (NIC). (n.d.). Disruptive Technologies Global Trends 2025. http://www.fas.org/irp/nic/disruptive.pdf (accessed 28 December 2016).

Sehgal, V. K., Patrick, A., & Rajpoot, L. (2014). A Comparative Study of Cyber Physical Cloud, Cloud of Sensors and Internet of Things: Their Ideology,

Similarities and Differences. *IEEE International Advance Computing Conference* (pp. 708–716). Gurgaon, India. IEEE.

Shugrue, D. (2016, October 5) 620+ Gbps Attack - Post Mortem. *Akamai*. https://blogs.akamai.com/2016/10/620-gbps-attack-post-mortem.html (accessed 8 January 2017).

SIG, IoT. (2013). *Internet of Things (IoT) and Machine to Machine Communications (M2M)Challenges and opportunities: Final paper May 2013*. Technology Strategy Board - IoT Special Interest Group.

Sundmaeker, H., Friess, P., Guillemin, P., & Woelffle, S. (2010). *Vision and Challenges for Realizing the Internet of Things*. European Research Project. CERP-IoT: Brussels.

Voas, J. (2016). *Network of "Things"*. NIST Special Publication. (pp. 800–183). https://doi.org/10.6028/NIST.SP.800-183

Zeigler, B. P. (1984). *Multifaceted Modeling and Discrete Event Simulation*. London, UK: Academic Press.

Zeigler, B. P., & Hammonds, P. E. (2007). *Modeling and Simulation-Based Data Engineering: Introducing Pragmatics into Ontologies for Net-centric Information Exchange*. Academic Press. https://www.elsevier.com/books/modeling-and-simulation-based-data-engineering/zeigler/978-0-12-372515-8 (accessed 10 August 2019)

Zeigler, B. P., Praehofer, H., & Kim, T. G. (2000). *Theory of Modeling and Simulation: Integrating discrete event and continuous complex dynamical systems*. Academic Press.

Zeigler, B. P., & Sarjoughian, H. (2017). *Guide to Modeling and Simulation of System of Systems (Simulation Foundations, Methods and Applications)*. Springer. https://www.springer.com/gp/book/9783319641331#aboutBook (accessed 10 August 2019)

Zeigler, B. P., & Zhang, G. (1989). The system entity structure: knowledge representation for simulation modeling and design. In L. Widman, N. Nielseen, & K. Loparo (Eds.), *Artificial Intelligence, Simulation and Modeling* (pp. 47–73). Hoboken, NJ: Wiley.

Zimmerman, R., & Restrepo, C. E. (2009). Analyzing Cascading Effects within Infrastructure Sectors for Consequence Reduction. *IEEE International Conference on Technologies for Homeland Security*. Waltham.

Part III

Simulation-Based CPS Engineering

8

Simulation Model Continuity for Efficient Development of Embedded Controllers in Cyber-Physical Systems

Rodrigo Castro[1], Ezequiel Pecker Marcosig[2], and Juan I. Giribet[3]

[1] *Departamento de Computación, FCEyN, Universidad de Buenos Aires and Instituto de Ciencias de la Computación, CONICET, Buenos Aires, Argentina*
[2] *Departamento de Ingeniería Electrónica, FIUBA, Universidad de Buenos Aires and Instituto de Ciencias de la Computación, CONICET, Buenos Aires, Argentina*
[3] *Departamento de Ingeniería Electrónica y Matemática, FIUBA, Universidad de Buenos Aires, and Instituto Argentino de Matemática Alberto Calderón, CONICET, Buenos Aires, Argentina*

8.1 Introduction and Motivation

The efficient development of embedded controllers for Cyber-Physical Systems (CPS) is under unprecedented pressure, driven by an upsurge of markets combining the Internet of Things (IoT) and flexible production systems (Marwedel 2018). Salient properties in this setting are the uncertainty and rapid evolution of requirements, pushed by a pervasive availability of cheap, yet powerful technology for sensors, actuators, and embedded computing platforms.

For decades the software engineering community has spent a tremendous effort in creating formal methods and tools to develop controllers for embedded systems, in particular for those of hybrid nature and with real-time constraints. Modern design methods for hybrid controllers tend to rely on unified modeling frameworks, which capture and combine together the expressive power of well-known modeling techniques such as hybrid automata (see, for instance, Branicky et al. (1998) and references therein).

Yet, most existing methods are still heavy, expensive, and hard to scale up for real applications. Model- and simulation-based techniques (Jensen et al. 2011) offer increasingly attractive capabilities to allow for rapid, yet robust prototyping and final delivery of embedded systems (Wainer and Castro 2011). Recently in (Tolk et al. 2018) it was recognized that Model-Driven Engineering is a strong

Complexity Challenges in Cyber Physical Systems: Using Modeling and Simulation (M&S) to Support Intelligence, Adaptation and Autonomy, First Edition. Edited by Saurabh Mittal and Andreas Tolk.
© 2020 John Wiley & Sons, Inc. Published 2020 by John Wiley & Sons, Inc.

candidate to facilitate the composability of hybrid models while relieving modelers from several underlying complexities.

Nevertheless, when adhering to a simulation-based control approach, we inevitably rely on a mathematical model that describes the behavior of the CPS to be controlled along with its (usually unpredictable) environment. This model should be as accurate as required by the desired quality of control to be obtained. What is the right level of detail for a given accuracy? Sacrificing details for simplicity implies the acknowledgement of uncertainties purposely introduced in the model through the unmodeled behaviors. How can this knowledge be brought into the design steps of robust controllers? When moving forward to stages of verification, validation, and accreditation of candidate models, errors between desired and observed behavior are inevitable, and therefore criteria must be defined to determine the acceptance or rejection of a model. Should a model require a rework, what is the right stage to go back to? Are analytical unmodeled dynamics to blame? Or we just made the model not robust enough?

We claim that these questions don't accept a universal nor straightforward answer, and that a methodology should help the designer of controllers for CPS to iterate freely through well defined stages to find the appropriate solution for each situation at hand.

In addition, a CPS can be seen as a hybrid *cybernetic and physical* plant to be controlled. Hybrid feedback loops are pervasive in CPS that integrate the algorithmic and physical domains, where computational *and* physical processes influence each other. This fact becomes a challenge when computing resources are limited, notably in scenarios where energy savings is a key requisite. In these settings (Marcosig et al. 2017), there is a need to provide safe trade-offs of scarce computing resources between physical stability issues and algorithmic capacity, while adapting to unpredictable environmental disturbances.

The model continuity-based methodology proposed in this chapter can be useful to accommodate smoothly an iterative development of hybrid controllers, including the consideration of trade-offs between the cybernetic and the physical aspects of a CPS.

8.1.1 Control of Cyber-Physical Systems

A typical closed-loop control system applied to a CPS (see Figure 8.1) is composed by the physical system to be controlled, the sensors used to acquire system information, the actuators used to modify the physical system behavior, and the controller, i.e., a processor running the control algorithm.

In the control engineering literature (Ozbay 1999), the conjunction of the physical system, sensors, and actuators is known as the *plant*. In many applications the interfaces with physical system (sensors and/or actuators) include themselves

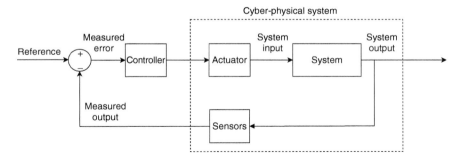

Figure 8.1 Closed-loop control system diagram. The system output is fed back to compensate system uncertainties and improve stability.

cyber and physical components, for instance when wireless sensor and actuator networks (WSAN) are used. The plant itself is a CPS with complex dynamics, and the objective of the control algorithm is to regulate the dynamical behaviour of the physical system in order to make its output follow a reference trajectory. The difference between the measured and desired outputs is fed back to the controller for computing the command signals for the actuators.

The problem of designing a control system (in particular for a CPS) usually involves three main classes of activities (Sánchez-Peña et al. 2007). The *experimental activities* deal directly with the plant. However, starting directly from the plant, it is not possible to carry out a rigorous and systematic analysis or design. Furthermore, in many situations experimentation with the system may be expensive or dangerous. Therefore, an activity prior to experimentation is necessary.

The *simulation-based activities* deal with a computational model of the plant. This computational model is a good representation of the system, but it inevitably demands making simplifying approximations, in many cases because the computing capacity is limited. However, working on a computational model (instead of experimenting directly with the physical system) reduces risks and costs.

The idea behind the simulation activity is to obtain the best approximation possible to the physical system that can be simulated considering the available computational capacity.

The underlying mathematical models may be too complicated for symbolic analysis aiming to design a strategy to control the system. Therefore, new simplifications must be imposed in order to be able to use mathematical tools to carry out the analysis and design of the controller, which leads to the *analytic activities*, where the problem is studied from a theoretical point of view.

The difference between the real system and the theoretical model can be quantified as uncertainties of the model, and need to be considered when the controller is designed. Although the controller design is based on the theoretical model, it has to provide a degree of robustness, in order to deal with the uncertainties. In many cases it is possible to provide theoretical results of the robustness of the system (Sánchez-Peña and Sznaier 1998), while in many other situations the robustness is validated, for instance, via Monte Carlo simulations.

The difference between the theoretical model and the real system is not the only uncertainty that must be considered when designing the controller. There are also disturbances such as sensor noise or actuator delays. Further, the hardware and/ or software running the control algorithm can also introduce uncertainties. This is particularly important in cyber-physical systems, where the controller and the physical may have a complex interaction, and many times is considered as uncertainty when the control algorithm is developed or, in some cases, ignored during the design of the controller and its effects are studied later, during the simulation or experimental validation.

The robustness property of the control algorithm is what makes it feasible to implement in real applications. However, in general, there is a trade-off between robustness and performance. A system capable of guaranteeing its operation in the presence of large uncertainties will generally do so at the expense of its performance, for instance making the control to react slower to changes in the plant. To improve the performance, a better knowledge of the system is required. For instance, it is possible to treat the actuators delays as uncertainty, and design a control algorithm capable of controlling the system even in the worst case scenario. In this case no previous knowledge of the delay is required. However, if a good characterization of delays is available, it can be leveraged during the analytic activities for designing the control law. Contrary to what might be often assumed, sometimes the existence of system's delays can in fact *improve* the performance of the controller (Moyne et al. 2008).

The three classes of activities explained for controller design are interlinked and often it is necessary to iterate between them in order to achieve the desired performance, or quality of control. This idea is depicted in Figure 8.2.

Ideally, the results obtained in analytic activities are validated through simulations and, then, experimentally, to ensure that the system has the expected behavior. If the proposed controller design is invalidated, a better tuning of the controller is necessary or, eventually, it is required to start over again from the analytic activity. This is the ideal situation. However, in practice, engineering decisions are made without a full support of a theoretical background, and are verified by numerical simulations or validated by experimental testing.

As Figure 8.2 shows, different stages can be performed during the simulation activity. Software in the loop simulation (SIL), allows validating the software in real conditions. To validate the software correctness, this strategy is useful to study how the controller performance may be degraded, for instance because of

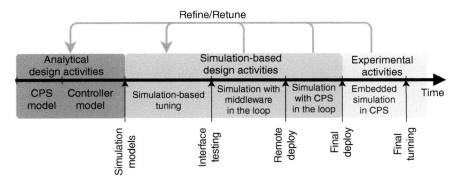

Figure 8.2 Typical activities and stages in a control design pipeline for a CPS.

software scheduling or latency. This is particularly useful when a middleware is a component of the control system, as will be shown later.

In several situations the simulation and experimental activities are combined to design the controller. Hardware in the loop simulation (HIL) is useful for testing system integration, interfaces and communications in a simulated environment. This simulation methodology allows to run experimental tests on some components of CPS (controller, sensors, or actuators), while the physical system is replaced by a computational model. This is particularly useful when experimental testing of the plant can not be carried out. To perform HIL simulations before the experimental validation in the control design workflow, allows identifying problems, and eventually redesign the controller, offering benefits in cost, time, and risks reduction.

The interaction between the different stages of the control design has encouraged the use of rapid prototyping strategies for the development of control systems. The basic idea of this methodology is to develop and validate the control algorithms, with models in simulated environments, and later transform the controller into a real-time control system prototype, created automatedly, and use this prototype to validate the algorithm with the plant under real operating conditions.

In the transition from a simulation environment to the experimental testbed, there are considerations that may be overlooked during the design of the control system such as, unmodeled sensors and/or actuators delays, sensors noise, actuators saturation, non-linearities, among others. But, there is another component that can affect the performance of the control system. In the transition from the simulation environment to the experimental hardware, there is a translation of the control algorithm, for instance, moving from a simulated environment to a C code that can be embedded in the corresponding hardware. Then, the algorithm evaluated in the simulation environment, in general is not the same that is evaluated during the experimental testing. New delays may be introduced which could affect the controller, degrading the performance of the system.

8.1.2 DEVS as a Formal Approach to Modeling and Simulation

The Discrete EVent Systems Specification (DEVS) (Zeigler et al. 2018) combines discrete event, discrete time, and continuous dynamics under a mathematically sound way. DEVS models are block-based units of self-contained behavior, which can be interconnected through input/output ports to create modular and hierarchical topologies of blocks. Details about formal aspects of DEVS are presented below in Section 8.2.1.

DEVS-based methodologies for M&S-driven engineering have been successfully implemented by integrating software development best practices to offer product life-cycle control. This becomes particularly relevant in scenarios where it is difficult to predict systems behavior as changes are introduced very frequently (see e.g. Bonaventura et al. 2016). Some of the main challenges faced are system complexity, tight delivery times, the quality and flexibility of the developed models and tools, communication of results in interdisciplinary teams, and big data-scale analysis.

Our approach aims at an end-to-end methodology for designing hybrid controllers based on *model continuity* for DEVS (Hu and Zeigler 2004). Our main idea is that a DEVS simulation model for a controller can evolve transparently from a desktop-based mocking up environment until its final embedded target without the need of intermediate recoding or reimplementation. In this context, the DEVS simulation engine plays the role of a "virtual simulation machine" for the DEVS model. Model continuity is a sound approach to mitigate the introduction of errors in the development process of controllers. Model continuity is becoming an increasingly recognized strategy (Cicirelli et al. 2018) where the DEVS framework is also seen as a candidate technology for CPSs (recently validated by Marcosig et al. 2017).

Previous works in the DEVS community (Henry and Wainer 2007; Castro et al. 2009, 2012; Moncada et al. 2013; Niyonkuru and Wainer 2015) dealing with embedded simulation adopted varied approaches to solve the real-time management or the coordination between the simulation executive and the underlying hardware (e.g. with or without an underlying operating system).

Yet, we believe that the current weakest link in these efforts lies in the connection between models with lower level sensors and actuators. This link represents a key interface layer between software abstractions and the physical platform. Typically, device drivers should cope with this requirement. Yet, being low-level platform-specific software artifacts, reprogramming or replacing device drivers can be a heavy process that demands specialized skills (often beyond those found with control designers).

8.1.3 The Simulation Model Continuity Approach

We propose a conceptual and practical framework to help reducing the obstacles described above. We combine simulation technology with a software middleware that abstracts away low level details of sensors and actuators.

In Figure 8.3 we provide a reference simulation-driven blueprint. The goal is to guide an end-to-end development process for embedded controllers relying on model continuity: the Control Model shall never leave its simulation executive, from the mocking up and tuning stages until the final deployment stage. No single line of code in the model is modified (nor translated/transformed) when landing into its target embedded platform.

Figure 8.3 presents four typical stages, or scenarios, encompassing those described in Figure 8.2 for simulation-driven control development.

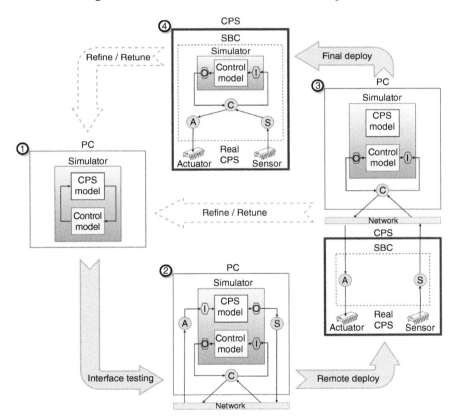

Figure 8.3 End-to-end model continuity. *Stage 1:* Standalone simulation-based controller design, *Stage 2:* Adding communication into the loop, *Stage 3:* Remote connection (CPS-in-the-loop simulation), *Stage 4:* Embedded simulation.

Stage 1: Simulation-based Controller Design. We assume that we count on a previously developed analytic model of the system to be controlled. It comes either from applying first principles from physics or from applying black-box identification techniques. The analytic models are then implemented as simulation models. The CPS Model represents the target real CPS. Based on this CPS Model, the Controller Model is in charge of controlling the plant's inputs so that it behaves in a desired way on its outputs. In Figure 8.3, *Stage 1*, we depict a closed-loop between a plant and controller models. This is the typical simulated scenario where engineers test different versions of their control strategies in a risk-free setting until simulated results are satisfactory.

Stage 2: Communication Middleware in the Loop. Stage 1 can be thought of as an idealized setting that considers perfect communication between the plant and its controller. Not only delays, but also congestion, information losses, data corruption, and many other non-idealities kick in when a real communication circuitry and/or network is introduced (Figure 8.3). The context is still a standard PC running some suitable OS. We incorporate two kinds of software entities, those that live within the simulator and those that run outside it. *Input* and *Output* simulation models (**I** and **O** hexagons) are in charge of the message exchanges with communication entities outside the simulator. The latter are considered to be the *Actuator, Sensor,* and *Controller* entities (**A, S** and **C** circles). These can be considered as a facade to a Communication Middleware that abstracts away the real CPS. In this new scenario the previously designed controller will probably need adjustments in order to take non-idealities into account and retain the required quality of control.

Stage 3: Remote Simulation-driven Control of the CPS. This next scenario gets rid of the CPS Model and replaces it by the real CPS itself. We assume that the CPS features some kind of Single Board Computer (SBC) capable of running communication software to exchange messages with their counterparts in the PC environment. Note that the CPS Model is simply unplugged from the loop. The Control Model must now deal with non-modeled behavior of the real CPS at the time of control design in *Stage 1*. Again, the new setting could affect noticeably the closed-loop CPS performance. This will likely be the case in contexts of rapid prototyping. At this stage, with new empirical evidence gathered, it becomes reasonable to refine or fine tune the simulation models, and return to Stage 1 for safer and faster experimentation.

Stage 4: Embedded Simulation at the Controlled CPS. Finally, the same Controller Model used throughout the previous stages is run on the target embedded platform. For the simulated controller nothing has changed, since it sees consistently the same **I** and **O** interfaces at all times, regardless of the platform it is running at. The simulator then can be thought of as a simulation virtual machine running on a particular SBC.

8.2   Background on Relevant Technologies

We briefly introduce some concepts that will be useful throughout the chapter.

Modeling and Simulation is key for understanding the behavior of hybrid systems, i.e. those that combine discrete and continuous dynamics interacting with each other. CPS fall certainly in this category, and in most cases of practical interest the resulting hybrid models don't accept closed-form analytical solutions.

The Discrete-Event System Specification (DEVS) is a mathematical formalism to describe hybrid systems. We will see that DEVS is able to also describe systems expressed both by discrete dynamics and by continuous differential equations.

Meanwhile, PowerDEVS is a simulation environment to work with DEVS models, featuring a graphical user interface for block-oriented design.

Finally, the Robotic Operating System (ROS) is a meta-operating system, or actually a software middleware running on top of an operating system, which provides a set of libraries and tools for robot development (in its broadest sense). We leverage this middleware for rapid prototyping.

These three technologies are key to build the simulation-based methodology presented along this work.

8.2.1   The Discrete EVent Systems Specification (DEVS) Framework

DEVS is a formal framework for model description, equipped with an abstract simulation algorithm that is independent of the nature of the described system. It has been shown that DEVS can describe exactly any discrete system and approximate continuous systems with any degree of desired accuracy, therefore being capable to simulate any kind of hybrid system that undergoes a finite number of state changes within finite time intervals (Zeigler et al. 2018).

A system modeled with DEVS is defined as a modular and hierarchical composite of submodels, which can be either behavioral (Atomic) or structural (Coupled). Submodels interact by means of messages sent through input/output ports, in a block-oriented fashion.

In terms of **behavior**, a DEVS **Atomic** model A_M is defined by the following tuple: $A_M = \{S, X, Y, \delta_{int}, \delta_{ext}, ta, \lambda\}$, where S is the set of states, X is the set of accepted input messages, and Y is the set of available output messages.

Four *dynamic functions* define behavior: ta, δ_{int}, δ_{ext}, and λ. The time advance $ta : S \rightarrow \mathbb{R}_0^+$ is the lifetime of each state $s \in S$. After $ta(s)$ units of time an internal transition given by the internal transition function $\delta_{int}: S \rightarrow S$ is triggered (assuming no external input events arrived). In case an external event arrives, $\delta_{ext} : S \times \mathbb{R}_0^+ \times X \rightarrow S$ (the external transition function) is triggered, with e being the elapsed time for a given state s and $0 \leq e < ta(s)$. Every time a new state s' is calculated (either by invoking δ_{int} or δ_{ext}), a new lifetime $ta(s')$ is calculated and the elapsed time e is reset to 0. Finally, $\lambda: S \to Y$ is the output function that can be invoked to emit output messages.

We will see in subsequent sections how the DEVS *dynamic functions* interact with messages to/from actuators/sensors at those DEVS models that play roles of Input or Output mappers (Figures 8.6 and 8.7).

In terms of **structure**, a DEVS **Coupled** model C_M interconnects Atomic and Coupled components together through their input/output ports. It can be described by the following tuple: $C_M = \{X, Y, D, EIC, EOC, IC, Select\}$, where: X and Y are sets of input and output messages respectively, D is the set of components' names, IC is the set of internal couplings among members of D, EIC is the external inputs coupling relation (links between external input ports and internal components), and EOC is the external output coupling relation. *Select* is a tie-breaking function to assign execution priorities when several internal or external transition functions are scheduled for the same simulation time.

The DEVS formalism is closed under coupling: any hierarchical composition of DEVS atomic and coupled models defines an equivalent atomic DEVS model. DEVS is by definition an asynchronous formalism, where each atomic model controls its own clock (time advance $ta(s)$). The composition of several atomic models into a larger coupled model preserves this timing independence.

This flexible coupling will permit a straightforward procedure for connecting or reconnecting pieces of the control-based models in our model continuity proposal.

For practical modeling and simulation, we shall adopt the PowerDEVS toolkit (Bergero and Kofman 2010) which will be described below.

8.2.2 PowerDEVS Simulator

There exist a vast spectrum of DEVS simulation tools, such as ADEVS, CD++, DEVS Java, PythonPDEVS, and PowerDEVS, just to name a few. We chose PowerDEVS for several reasons: it is easy-to-use for users not familiar with DEVS, and is very efficient in terms of performance for the simulation of hybrid systems (Van Tendeloo and Vangheluwe 2017) (and hence suitable for control applications).

PowerDEVS is an open-source DEVS-based simulator appropriate for hybrid system modeling and real-time simulation (Bergero and Kofman 2010). A salient feature is its graphical design interface, where atomic and coupled models are represented by blocks that can be wired and interchanged by simple drag-and-drop actions (see an example screenshot in Figure 8.10). Therefore, it is suitable for a non-DEVS expert, by hiding away the internals of the formalism, while the block-oriented visual metaphor matches closely the typical block diagrams widely used in control engineering.

PowerDEVS is also the flagship tool for the Quantized State System (QSS) family of numerical methods (Cellier and Kofman 2006) that can solve ordinary differential equations efficiently in the context of a discrete-event simulation. This way, by means of QSS methods PowerDEVS provides a unified framework to represent discrete and continuous dynamics in a hybrid system with a single coupled model, while featuring real-time simulation capabilities.

PowerDEVS is composed of various independent *programs*: Model Editor, Atomic Editor, Preprocessor, and Simulation. The first one is an IDE to edit and configure the simulation model, the second is a graphical editor for DEVS atomic models, the Preprocessor is in charge of translating the model editor files to a standalone executive, and finally a graphical simulation control interface allows for user interaction with the simulation execution. The simulator is programmed in C++ and each atomic model (built either with the graphical interface or a text editor) is mapped to a C++ file that is linked with the main simulator.

8.2.3 The Robot Operating System (ROS) Middleware

Despite its name, ROS is not an operating system, but rather a middleware library that provides an abstraction layer on top of an operating system (Quigley et al. 2009). ROS allows both experts and non-experts for developing software for (but not limited to) robotic applications in a way that makes its use agnostic of low-level hardware layers. ROS provides common interfaces fostering reusability and code sharing. Even when there exist other middleware options, ROS is a largely accepted library, with a vast and growing open-source community.

A system running ROS is composed of several processes called *nodes*. ROS nodes are software applications that perform different functions. Owing to its modular structure, a node can be started or stopped at any time facilitating debugging. ROS allows nodes for communicating over networks in a seamless way. It provides two communication mechanisms between nodes: *Topics* and *Services*. Topics follow a publisher/subscriber architecture, allowing many-to-many one-way communication. Publishers and subscribers are not aware of the existence of each other. Unlike topics, services are suitable for synchronous request/reply interaction between nodes. Both topics and services use ROS *messages* which are strictly typed data structures.

ROS relies on a master node that keeps track of all the nodes on the network, and available services and topics to which a node can subscribe. If a node is interested in particular data it must subscribe to the corresponding topic. Then, every time a message is published on that topic, the callback function of every listening node is executed. ROS has a varied and growing set of tools that facilitate data and network topology visualization, managing node parameters and message transformations, measure topic bandwidth, just to mention a few. ROS is provided with bags used for data storage, thus real data can be later played back. We adopted the *kinetic* flavor of ROS due to its massive adoption and the fact that it is a long-term support version.

8.3 DEVS over ROS (DoveR): An Implementation of the Model Continuity-Based Methodology

In this section, we show a specific implementation of the blueprint methodology introduced in Section 8.1.3 for model continuity.

In this particular exercise we make some key assumptions: (i) The target CPS system is a robotic platform which can be itself under construction, i.e., the model of the plant itself is a potential source of errors, (ii) the robot is controlled by means of a single-board computer (SBC) capable of running a standard embedded Linux OS, (iii) the model development platform is a regular PC system running standard desktop Linux, and (iv) there is a local network infrastructure for the communication between the PC and the SBC using standard technology (Ethernet/WiFi).

In this work the concept of real-time is taken as soft real-time in a relaxed sense. There are several alternatives to make this approach more stringent, using a more strict treatment of real-time. For instance, by adopting a DEVS simulator that considers explicitly the concept of message deadlines (Henry and Wainer 2007), or a DEVS simulator that relies on an RTOS (Bergero and Kofman 2010) providing worst case timing guarantees. They should only bring along better timing features to the final product. In this sense, we purposely stand on a "Commercial Off-The-Shelf" scenario, and study what can be achieved under this vanilla setting. In other words, we accept that the real-time simulation performs only a best-effort treatment of messages' timestamps, where overrun situations could arise (i.e., the simulation clock falls behind the wall clock, see Cellier and Kofman 2006), possibly affecting the quality of the controller.

In order to apply the model continuity-based methodology to embedded systems, we rely on the ROS middleware to abstract away subtle issues of hardware and subsystems communication. There exist a myriad of ROS packages freely available, developed by an active community for a variety of sensors and actuators. Typically, software systems depend on well-defined layers of abstraction (Figure 8.4), each one devoted to a particular duty and providing neighbor layers with a standard interface. ROS together with the Hardware Abstraction Layer (HAL) (provided by the OS kernel) can efficiently abstract lower-level layers from the user/application level. Meanwhile, hybrid controllers lie on the application layer. A DEVS-based simulation engine fills the gap in-between.

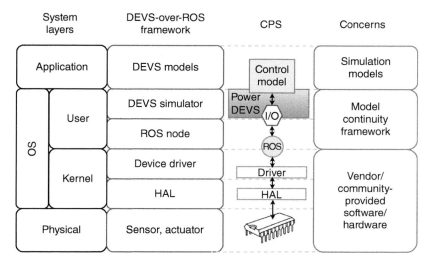

Figure 8.4 Abstraction layers to provide separation of concerns.

We shall refer to this composition of DEVS over ROS as the **DoveR** middleware. In our work, DoveR is implemented by extending the PowerDEVS toolkit.

The overall solution aims to provide the control automation community with a simulation platform adequate to develop hybrid systems and capable of linking transparently with the underlying hardware. Even when there exist other proposals that add an abstraction layer on top of ROS (Crick et al. 2017), we rely on PowerDEVS as it features modular interconnection of blocks for continuous and discrete models (an attractive and natural paradigm in the automation community) and because it offers excellent performance properties (Van Tendeloo and Vangheluwe 2017) making it particularly suitable for real-time scenarios. Also, in Monteriù (2016) the authors propose a similar approach, but lacking a step-by-step, incremental, and end-to-end design sequence to develop the controller.

Below are the stages that comprise the DoveR design methodology:

Stage 1: Simulation-based Controller Design. In this stage we simply adopt PowerDEVS as our modeling and simulation toolkit of choice.

Stage 2: DoveR in the Loop. In this step we adopt DoveR as the middleware that will allow for communication with the real system. Input and Output DEVS model I and O are particular DEVS atomic models with the ability to interact with information coming from outside the simulation engine (see Section 8.3.1).

Stage 3: DoveR for Remote Simulation-Driven Control with CPS in the Loop. The obvious next step is to get rid of the System Model and replace it by the robot itself (Figure 8.5). Then, we have to deal with non-modeled behavior

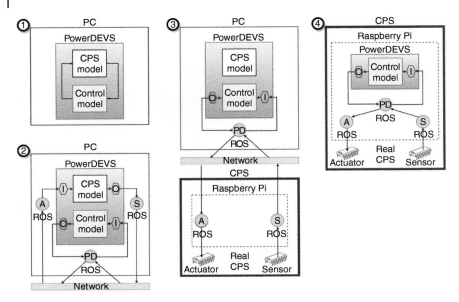

Figure 8.5 DoveR for End-to-end model continuity. *Stage 1:* Standalone simulation-based controller design (DEVS toolkit), *Stage 2:* Adding DoveR: DEVS over ROS and network-in-the-loop, *Stage 3:* Connecting to target: robot-in-the-loop simulation, *Stage 4:* Embedded simulation: embedded controller with DoveR framework.

that can strongly affect closed-loop performance. Even though we try to work with acceptable models, there will always exist some not-considered effects. As it was previously stated, we are providing a methodology for rapid prototyping. Therefore, it is not advisable to spend large amounts of time trying to develop a perfect model.

Stage 4: DoveR for Embedded Simulation at the Controlled CPS. Finally, the same Controller Model used throughout these steps, and implemented on PowerDEVS, is run on the final embedded platform (Figure 8.5). For the simulated controller nothing has changed, since it sees the same interfaces as in the previous stages, as if it were already running on the PC. PowerDEVS can be thought of as a virtual machine for models running on the SBC.

8.3.1 Communication Between the PowerDEVS Engine and the ROS Middleware

In order to establish a bridge between PowerDEVS and ROS we choose UDP network sockets. We avoid TCP because its connection-oriented nature imposes extra overhead of no value in robotic applications, where retransmission of lost messages is typically useless (new samples should preempt old ones).

When performing real-time simulation (Cellier and Kofman 2006), internal events stick to wall-clock time. The root-coordinator in a DEVS framework keeps track of the atomic model presenting the earliest time advance about to expire (also known as the **imminent** models). While waiting for said earliest imminent time stamp, the DEVS root-coordinator is in idle state and has the chance to process messages incoming from the ROS middleware.

When the simulation starts, a DoveR atomic model requests for listening to a particular UDP port, and it is registered as a *listener* of that port. Each UPD port is assigned to an individual thread that remains listening throughout the entire lifetime of the simulation. Each UDP port and its associated thread can be shared by several atomic models that, upon request, are added to the list of listeners for that port.

For interprocess communication between listener threads and the main PowerDEVS thread we rely on Linux IPC Message Queues. A single FIFO queue enqueues messages arriving to PowerDEVS, identifying the UDP port number they come from. Once a message is popped out from the queue, the associated port is read and each atomic model subscribed as a listener is notified of the arriving message by triggering its DEVS external transition function (see Figure 8.6).

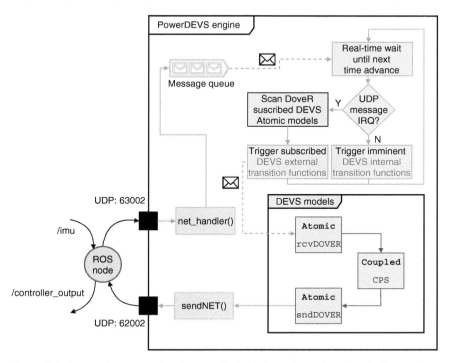

Figure 8.6 Network message handling at the ROS/PowerDEVS interface. Delivery of messages to subscribed atomic models.

Therefore, ROS messages play the role of external events in a DEVS framework, which consist of a value and a time stamp arriving at an atomic model.

In contrast, outgoing messages from PowerDEVS (Figure 8.7) depart directly to a corresponding UDP port (Figure 8.8). ROS then receives a UDP datagram, conforms a proper ROS message, and publishes it on a corresponding topic (Figure 8.9). Note that ta, dint, dext, and lambda functions in Figure 8.7 correspond to the time advance, internal and external transition, and output functions defined in Section 8.2.1.

In DEVS we approximate continuous systems by using its Quantized-State System (QSS) counterpart (Kofman and Junco 2001). Through this class of systems, we approximate a continuous trajectory by a piecewise polynomial trajectory. The update instants of these polynomials are asynchronous and get dynamically determined by the QSS accuracy control. However, we must serialize the messages sent out periodically to ROS, according to a given time period T. Once T expires, each polynomial corresponding to each signal to be sent to ROS is sampled, and then a UDP datagram is conformed by packing the samples together.

In Figure 8.10 we can see a screenshot of the PowerDEVS GUI with the DEVS special atomic models **rcvROS** and **sndROS** for input and output messaging (respectively) within the DoveR framework.

8.3.2 Embedded Simulation On a Raspberry Pi

We deployed PowerDEVS into a Raspberry Pi board running Ubuntu Xenial 16.04. We generate the simulation model files on a PC (our model development platform) and deploy them over the network into the SBC using ssh. Only the Preprocessor and the Simulation Interface modules of PowerDEVS are needed in the SBC. Likewise, we installed ROS-Base (without GUI tools). The PowerDEVS installation on the SBC proved quite easy after downloading it from the SourceForge repository. However, some precautions need to be taken: some libraries in the repository are precompiled for PC architecture (amd64 and i386) and must be re-compiled to run on an Armv7l architecture. Also the qmake package from the Qt library must already be installed. Finally, PowerDEVS uses the Scilab package as a numerical back end for complex mathematical operations (available for Debian and Ubuntu). After a successful installation, the BackDoor toolbox is required to open a communication channel between PowerDEVS and the Scilab workspace.

Even when it is not mandatory, we took advantage of the standard link between PowerDEVS and Scilab. The latter provides PowerDEVS with the capacity of performing more sophisticated numerical manipulations, and to register, plot,

```
void    sndROS::init(double t,...) {va_list parameters;
va_start(parameters,t);
port = atoi(va_arg(parameters, char*));...
sigma = INF;}
double sndROS::ta(double t) {return sigma;}
void    sndROS::dint(double t) {sigma = INF;}
void    sndROS::dext(Event x, double t) {
double *xv;
char msg[200]; xv = (double*) (x.value);
sprintf(msg,"%f",xv[0]);sndNET(port, ip, msg, strlen(msg));sigma = 0;}
Event sndROS::lambda(double t) {return;}
```

Figure 8.7   sndROS atomic model functions.

```
void sndNET(int port, char* ip, char* data,int size){...
/* Create new socket. */
int sockfd = socket(AF_INET, SOCK_DGRAM, 0);
bind(sockfd, (struct sockaddr *) & local_addr,sizeof(struct sockaddr_in));
int result = sendto(sockfd,data,size,0,(struct sockaddr*)&
    remote_addr,sizeof(remote_addr));
close(sockfd);}
```

Figure 8.8   sndNET method (PowerDEVS Engine).

```
def main():
...
rospy.init_node('PowerDEVS_Control') #ROS node declaration
udpport_read_yaw=62002 #Port to read the Control Action for the yaw angle
udp_server_socket_yaw=socket.socket(socket.AF_INET,socket.SOCK_DGRAM
    ) #UDP socket
udp_server_socket_yaw.bind(("", udpport_read_yaw))
#New ROS topic (where controller output will be published)
ROSpublisher = rospy.Publisher('controller_output',Twist,
    queue_size=10,latch=True) #ROS Publisher object
msg_read_list=[server_socket_yaw] # UDP ports to listen to
while True: read,_,_=select.select(read_list
    ,[],[],0.001) # timeout = 0.001 for s in read:
        yaw_ctrl_msg,addr = s.
    recvfrom(18) # UDP message with controller output from PowerDEVS
        rospy.loginfo("[PowerDEVS-Ctrl] Message received from
        PowerDEVS.")yaw_ctrl_msg = Twist() # create empty ROS message
            yaw_ctrl_msg.angular.z = float(yaw_ctrl) # fill in ROS message
            ROSpublisher.publish(yaw_ctrl_msg) # publish message to ROS topic
            rospy.loginfo("[PowerDEVS-Ctrl] Message succesfully
        published in ROS topic.")
continue
```

Figure 8.9   Excerpt of a ROS node for DoveR.

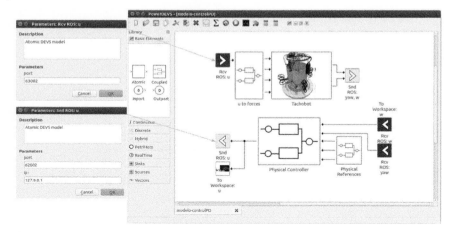

Figure 8.10 PowerDEVS screenshot. Dark gray (receive) and light gray (send) DoveR atomic models.

and analyze data recorded during experiments with the real CPS in the loop. This facilitates greatly the rapid prototyping approach.

8.4 An Experimental Robotic Platform: Hardware and Models

TachoBot (Marcosig et al. 2018), the *Test Assembly for the Control of Hybrid Objectives* (Figure 8.11a) is a custom-made experimental platform, crafted with off-the-shelf components. We will adopt TachoBot as the CPS of choice to test our methodology.

TachoBot is a 3-degree-of-freedom (3-DOF) vehicle designed for studying hybrid control strategies in cyber-physical systems. The robot features eight lateral propellers to manage the position on the plane and the rotation angle around its vertical axis. Said actuators are mounted in pairs, thus resulting in four independent forces acting on the robot (Figure 8.11b). Two central propellers (spinning clockwise and counterclockwise, respectively) are in charge of considerably reducing the friction with the underlying surface, allowing for displacements even with extremely small control efforts. This central actuator must lift a weight of approximately 3 kg.

A mathematical model is used to design the controller that will be connected to the *real* plant, following the model-based design (MBD) methodology (Jensen et al. 2011). As it was mentioned before, even simple mathematical models can provide well designed controllers. For systems being built concurrently with the design of its controller, the model might be improved iteratively.

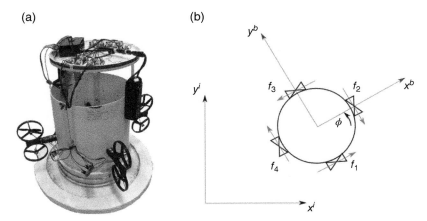

Figure 8.11 TachoBot: *Test Assembly for the Control of Hybrid Objectives* roBot. (a) Implementation. (b) Physical Model consisting of a rigid-body with 3-DOF subjected to four external forces.

8.4.1 Continuous Robot Model and Discrete Regulation Controller

The robot can be seen as a rigid body with 3-DOF subject to forces $\mathbf{F} = [f_1, f_2, f_3, f_4]^T$ in the vehicle body-fixed frame b (see Figure 8.11b). Therefore, dynamic equations describing its movement can be derived by applying the Newton–Euler equations (8.1). The state vector $\mathbf{x}(t) \in \mathbb{R}^6$ specifies the *pose* of the vehicle: $\mathbf{x} = [x, v_x, y, v_y, \phi, \omega]^T$, where x, y, and ϕ are the linear position (respect to a given inertial frame i, see Figure 8.11b) and the yaw angle, respectively, and v_x, v_y, ω the corresponding linear and angular velocities. Parameters M, R, and I_z are respectively the mass, radius, and moment of inertia of the TachoBot. $\dot{\mathbf{x}}$ is the time derivative of state vector $\mathbf{x}$, i.e. linear and angular velocities and accelerations.

$$
\dot{\mathbf{x}} = \begin{bmatrix} 0 & 1 & 0 & 0 & 0 & 0 \\ 0 & 0 & 0 & 0 & 0 & 0 \\ 0 & 0 & 0 & 1 & 0 & 0 \\ 0 & 0 & 0 & 0 & 0 & 0 \\ 0 & 0 & 0 & 0 & 0 & 1 \\ 0 & 0 & 0 & 0 & 0 & 0 \end{bmatrix} \mathbf{x} + \begin{bmatrix} 0 & 0 & 0 & 0 \\ \cos(\phi)/M & \sin(\phi)/M & -\cos(\phi)/M & -\sin(\phi)/M \\ 0 & 0 & 0 & 0 \\ \sin(\phi)/M & -\cos(\phi)/M & -\sin(\phi)/M & \cos(\phi)/M \\ 0 & 0 & 0 & 0 \\ R/I_z & -R/I_z & R/I_z & -R/I_z \end{bmatrix} \mathbf{F}
$$

$$(8.1)$$

This is a continuous-time non-linear system. Having this model at hand, we are able to study its transient (rise and settling times, overshoot, etc.) and steady-state (zero gain, steady-state error, etc.) responses, and the saturation limits. Moreover, we can identify the variables we can measure (system outputs) and the variables that allow us to modify the behaviour of the system (system inputs). In addition, we can identify the disturbances that possibly affect the system. From Eq. (8.1) we see that the inputs are the forces in $\mathbf{F}$, while the outputs $\mathbf{y} = [\mathbf{y}_x, \mathbf{y}_y, \mathbf{y}_\phi]^T$ are some components of the state vector $\mathbf{x}$ corrupted with sensor noises, more specifically $\mathbf{y}_x = x + \eta_x$, $\mathbf{y}_y = y + \eta_y$, and $\mathbf{y}_\phi = \phi + \eta_\phi$, where $\eta = (\eta_x, \eta_y, \eta_\phi)$ are the sensor noises. Along with the CPS, we have a specification of its desired response $\mathbf{y}^{des}$. We use a controller C to automatically generate and feedback the outputs of the system to make it follow the desired response (Figure 8.12). The design process of C is the so called *control problem design*.

Physical systems are frequently described by differential equations or DESS (Zeigler et al. 2018). Conversely, controllers are usually performed on a computer or embedded system, and thus run in discrete-time and are described as DTSS. Among the wide range of feedback control strategies we opted for a Proportional-Integral-Derivative (PID) control strategy (Ozbay 1999), which is by far the most commonly used class of controllers because its easy and intuitive functioning. Actually, we consider 3 PID controllers designed to generate $\mathbf{F}(t)$ in such a way that the output $\mathbf{y}(t)$ follows the desired response $\mathbf{y}^{des}(t)$. The PID outputs are given by the vector $\mathbf{u}^{pid} = [u_x, u_y, u_\phi]^T$. Each output is made up of a linear combination of the error $\mathbf{e}_i = \mathbf{y}_i^{des} - \mathbf{y}_i$, and its integrals and derivatives, where $i \in \{x, y, \phi\}$.

The three PID outputs $\mathbf{u}^{pid} = [u_x, u_y, u_\phi]^T$ relate to the system input $\mathbf{F}$ in Eq. (8.1) by means of transformation matrix T given in Eq. (8.2). In order to convert controller outputs to system inputs, we use the Moore–Penrose pseudoinverse $T^{\dagger}$, which gives the minimum-energy control action. Those forces are in turn translated to the eight propellers.

$$\mathbf{u}^{pid} = \frac{1}{2} \begin{bmatrix} 1 & 0 & -1 & 0 \\ 0 & -1 & 0 & 1 \\ 0.5 & -0.5 & 0.5 & -0.5 \end{bmatrix} \mathbf{F} \tag{8.2}$$

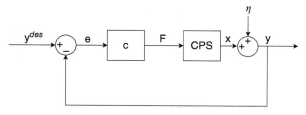

Figure 8.12 Ideal closed-loop, composed of: (a) plant (Robot or CPS), (b) controller C, (c) adder, (d) desired response $\mathbf{y}^{des}$, (e) error $\mathbf{e}$, (f) plant output $\mathbf{y}$, and (g) plant input $\mathbf{F}$.

8.4.2   Hardware Description

The control board is a Raspberry Pi 3 (Model B) computer which includes a quadcore 64 bit processor, 1 GB RAM, 32 GB external flash, and a built-in WiFi module. It runs an Ubuntu Xenial (16.04) operating system and communicates with sensors and actuators via an I^2C bus. Each force f_i in Eq. (8.2) is exerted by a pair of DC coreless motors attached with propellers. These motors are controlled by individual pulse width modulation (PWM) signals generated with a PWM controller PCA9685. Zero thrust corresponds to both motors on a pair working at 50%.

DC motor drivers consist of power MOSFETs transistors, while the central propeller uses a commercial electronic speed controller (ESC).

The angle and angular speed measurements are provided by an Inertial Measurement Unit (IMU) comprised of a three-axis accelerometer, gyroscope, and magnetometer. The MPU9250 sensor was selected due to the availability of its ROS node. Finally, two lithium polymer (LiPo) batteries are attached to the robot to supply energy.

When the TachoBot is in hovering mode (floating on the surface) all the 8 lateral propellers are working at 50% of PWM, aiming at taking advantage of the linear range of the motors. Any slight difference below or above this value can produce a rotation and/or displacement of the robot.

8.5   Experimental Case Study: Developing a Controller with a Model Continuity-Centered Methodology

Henceforth, we will follow the steps described in Section 8.3 to develop a PID controller for the TachoBot robot, running the PowerDEVS simulator in real-time mode embedded in a Raspberry Pi. The experiments were conducted with TachoBot mounted on a rotating platform in order to isolate it from axial displacements and deal only with the angular dynamics. The goal is to stabilize the angle around a reference value of 1 rad.

In Stage 1 we close the loop in a purely simulated setting. We start with the continuous time non-linear system model from Section 8.4. Since the working range of lateral propellers is around PWM at 50% (see Section 8.4.2), it is valid to assume a linear relation between the PWM signal and the force. We adjust the PID gains until we get a desired response and then store the resulting PID gains. In Figure 8.13 we can see that the angle (solid line) follows the reference of 1 rad, whereas the angular speed (dotted line) reaches zero.

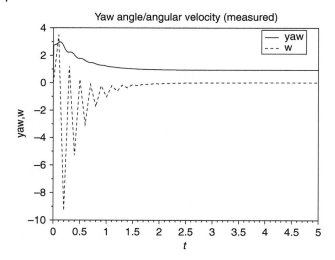

Figure 8.13 Closed-loop response with the simulation-based designed controller (Stage 1).

Introducing a wireless network and the ROS middleware into the loop might demand additional tuning for PID in Stage 2. A screenshot of the simulation model for this second step is shown in Figure 8.10. Resulting curves are in Figure 8.14 which differs from the ones in Figure 8.13 due to the delay introduced by adding two ROS nodes, one for PID controller and another connected to the robot model, running on the PC.

Still at Stage 2, with the simulator and ROS running on the PC, we need to assess the delay introduced by the ROS middleware. To this end, a ramp signal generated by a PowerDEVS source block was sent back to PowerDEVS through a ROS loop. In this case, two nodes were considered: one attached to PowerDEVS and another acting as a loopback, both running on the PC. The delay is the difference between the superimposition of the sent and received signals. From Figure 8.15a it can be appreciated that, for this test, the delay introduced by the ROS middleware is in the order of 7 ms.

Since this delay is almost constant it could be compensated with classical control techniques, such as a Smith predictor (Ozbay 1999). As the next step, we moved both ROS nodes to the Raspberry Pi, incorporating a WiFi network in between. Consequently a bigger and non-constant delay, in the order of 80 ms, can be seen in Figure 8.15b.

Now in Stage 3 we remove the System Model and introduce the TachoBot into the loop. First we assess the validity of the model we have been using up until now with the addition of network latencies, the ROS middleware and potential issues

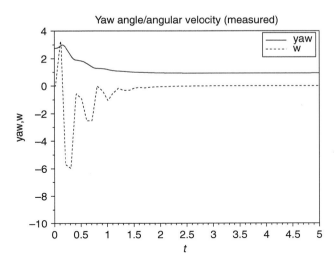

Figure 8.14 Closed-loop response after introducing ROS and network in the loop (Stage 2).

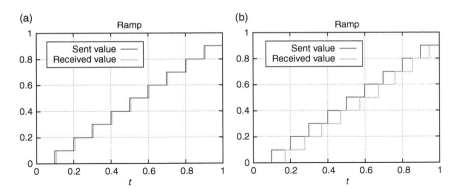

Figure 8.15 Delay introduced by ROS measured as the difference between sent and received signals: (a) local loop delay and (b) remote loop delay.

introduced by the hardware and the operating systems. With this in mind, we supplied the System Model and the TachoBot with a square wave and measured the resulting angle and angular speed. We performed adjustments on model parameters such that the measurements (Figure 8.16a) and the simulation (Figure 8.16b) looked quite similar. It can be thought of as a manual identification procedure. The model parameter adjustments led, in turn, to retune PID control gains.

Finally, in Stage 4 we embed the PID controller simulation model within the PowerDEVS simulator on the Raspberry Pi.

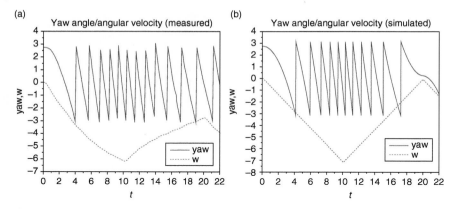

Figure 8.16 Model refinement by parameter tuning (Stage 3): (a) measurements and (b) simulation.

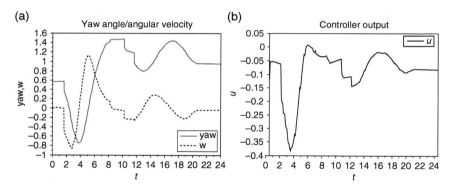

Figure 8.17 Embedded simulation (Stage 4): (a) closed-loop response, and (b) controller output.

What we see in Figure 8.17 are the resulting responses obtained after using the model continuity approach.

In Figure 8.17b we have the signal generated by the Control Model and used to feed the real propellers. In Figure 8.17a we have the response of the TachoBot: the solid line corresponds to the angle that we want to follow the reference value of 1 rad, and the dotted line refers to the angular speed which tends to zero. At the beginning a delay of about one second can be noticed due to initialization issues. Between one and two seconds, the angle remained stuck with zero angular speed due to the static friction on the rotating platform. Meanwhile, the controller output (right pane) started to increase with a negative slope because of the PIDs integral term. The robot released as soon as it was able to overcome the friction. It is followed by an interval, from two to four seconds, where the robot starts moving in the opposite direction due to mechanical issues not captured by the original model (later, a malfunctioning motor was detected). Between 8.3 and 10.4 seconds the angle remained stuck again because of the static friction, but in that case the controller output started to increase with a positive slope. Once again, the robot released as soon as it was able to overcome the friction. From then on, the angle response appears to behave underdamped. From 20.5 seconds onwards, the angle gets stuck once more with zero speed but closer to the reference. A small slope can be noticed in the controller output.

This process showed that we managed to successfully apply the methodology proposed based on the DoveR middleware, obtaining in a very straightforward way a controller of yaw angle embedded in the final platform. The process also helped to learn that the rotating platform affects robot behavior since it adds a static friction term not considered in the model in Section 8.4. However, the PID controller showed its robustness against unmodeled effects. From this point on, a second iteration of the full methodology can be applied from Stage 1, if desired, in order to enrich the system model for capturing more dynamics and develop incrementally a more robust controller.

8.6 Challenges of Implementing DoveR

We stated in Section 8.3 that our methodology assumes we count on an embedded system running a standard Linux OS to execute PowerDEVS and ROS. Even when it is not uncommon to have a microcontroller capable of running an OS it could be argued that this is a strong requirement. However, a Linux-based OS presents several advantages that simplifies the controller development process. Actually, this is not a limitation of the methodology but a simplification. We could adapt DoveR to produce code that can be run bare-metal on an ARM microcontroller with minimal modifications, without any OS whatsoever. This idea is supported by (Moncada et al. 2013) which presents a PowerDEVS extension that automatically generates code for ARM-based embedded systems. As it was claimed in Section 8.2, a C++ file is produced every time a simulation model is run in PowerDEVS.

Something similar could be argued about the lack of strong real-time guarantees. On the PowerDEVS side, it depends indeed on how it is run since the time advance function is tied to the OS clock. Yet, PowerDEVS can also be run over a real time OS, providing explicit time guarantees to the process, thus minimizing the chances of overrunning due to competing processes. Tests have been conducted under Linux RTAI reporting latency bounds (Bergero and Kofman 2010).

Meanwhile, several approaches are available in the literature for the development of embedded systems under varied real-time constraints in the context of DEVS (Hong et al. 1997; Xiaolin and Zeigler 2001; Henry and Wainer 2007; Song and Kim 2005; Furfaro and Nigro 2009; Moallemi and Wainer 2010; Wainer 2016). For instance, in Moallemi and Wainer (2011) Imprecise-DEVS (I-DEVS) was introduced to develop real-time and embedded systems based on the DEVS formalism, integrating imprecise computation to improve predictability under transient overloading. I-DEVS enables model designers to assign priorities to the model behavior and balance the execution based on the priorities assigned.

On the ROS side, there are no real-time guarantees. However, the ROS2 evolution, currently under development, is considering real-time as a key feature from day one (Maruyama et al. 2016).

8.7 Concluding Remarks and Future Work

We presented a framework that fosters simulation-model continuity to develop hybrid embedded controllers for Cyber-Physical Systems.

The framework is accompanied with a sample end-to-end methodology where a simulator plays the role of a real-time virtual machine for the simulation models. Thus, the desired controller model can evolve transparently from a desktop-based mocking up environment until its final embedded target without the need of intermediate recoding or reimplementation.

We adopted the DEVS formalism for modeling and simulation of hybrid systems, that has been suggested before as a viable technology for model continuity in embedded systems. Yet, the issue of dealing with low level intricacies at the hardware/software interface stands as a bottleneck for flexible simulation-driven development. We selected the ROS middleware to mitigate this problem, making the hardware/software interfaces as modular, reusable, and transparent as possible to developers of both DEVS simulation engines and DEVS controller models.

The control designer is then free to study the performance of a candidate controller by choosing its favorite workflow, without a need of being a DEVS expert or a ROS developer. A typical workflow can include to close the control loop in a purely simulated context (e.g. a non real-time PC), or retain the controller at the PC while interacting (now in real-time) over a network with the real robot, or to directly deploy the controller to the target embedded platform for full embedded control of the real system. The ROS nodes can be easily moved back and forth between the platforms to assess the effects of the introduced latency and data overhead.

We tested the DoveR strategy and tool in a case study where a custom made, crafted robotic system is being built concurrently with the design of its controller.

We successfully verified the usefulness and flexibility of our methodology. ROS nodes for sensors and actuators provided the required abstraction from the perspective of the controller model implemented in the PowerDEVS toolkit. The transparent portability of ROS and PowerDEVS between a PC platform and a Raspberry Pi embedded target was a key feature.

The modifications introduced into the PowerDEVS engine abide by the standards of a DEVS real-time abstract simulator, assimilating smoothly the message-based communication with ROS nodes via UDP sockets. This opens up the possibility of reusing previous efforts in DEVS-based automatic control studies and adapts them smoothly to realistic scenarios working with real robots (or physical plants in general).

Next steps include a thorough characterization of real-time performance limits imposed by latencies introduced by DoveR. We will also investigate alternative mechanisms of communication between PowerDEVS and ROS other than network sockets. Additionally, we will leverage the recording capabilities of ROS in order to provide immediate feedback from the robot to the PC (enabling for systematic optimization of the controller parameters). Finally, we will develop a position tracking control for the robot.

Acknowledgments

Ezequiel Pecker Marcosig thanks the Peruilh Foundation for their PhD Fellowship support.

References

Bergero, F., & Kofman, E. (2010). PowerDEVS: A Tool for Hybrid System Modeling and Real-Time Simulation. *Simulation, 87*(1–2), 113–132.

Bonaventura, M., Foguelman, D., & Castro, R. (2016). Discrete Event Modeling and Simulation-Driven Engineering for the ATLAS Data Acquisition Network. *Computing in Science and Engineering, 18*(3), 70–83.

Branicky, M. S., Borkar, V. S., & Mitter, S. K. (1998). A Unified Framework for Hybrid Control: Model and Optimal Control Theory. *IEEE Transactions on Automatic Control, 43*(1), 31–45.

Castro, R., Kofman, E., & Wainer, G. (2009). A DEVS-based End-to-end Methodology for Hybrid Control of Embedded Networking Systems. *IFAC Proceedings Volumes, 42*(17), 74–79.

Castro, R., Ramello, I., Bonaventura, M., & Wainer, G. A. (2012). M&S-Based Design of Embedded Controllers on Network Processors. In *Proceedings of the 2012*

Symposium on Theory of Modeling and Simulation DEVS Integrative M&S Symposium, 2012.

Cavanini, L., Cimini, G., Freddi, A., Ippoliti, G., & Monteriù, A. (2016). rapros: A ROS Package for Rapid Prototyping. In A. Koubaa (Ed.), *Robot Operating System (ROS): The Complete Reference (Volume 1)* (pp. 491–508). Cham, Switzerland: Springer International Publishing.

Cellier, F. E., & Kofman, E. (2006). *Continuous System Simulation* (1st ed.). New York, NY: Springer Science & Business Media.

Cicirelli, F., Nigro, L., & Sciammarella, P. F. (2018). Model Continuity in Cyber-Physical Systems: A Control-Centered Methodology Based on Agents. *Simulation Modelling Practice and Theory*, *83*, 93–107.

Crick, C., Jay, G., Osentoski, S., Pitzer, B., & Jenkins, O. C. (2017). Rosbridge: ROS for Non-ROS users. In H. I. Christensen & O. Khatib (Eds.), *Robotics Research* (pp. 493–504). Cham, Switzerland: Springer.

Furfaro, A., & Nigro, L. (2009). A development methodology for embedded systems based on RT-DEVS. *Innovations in Systems and Software Engineering*, *5*(2), 117–127.

Hong, J. S., Song, H.-S., Kim, T. G., & Park, K. H. (1997). A Real-Time Discrete Event System Specification Formalism for Seamless Real-Time Software Development. *Discrete Event Dynamic Systems*, *7*(4), 355–375.

Hu, X., & Zeigler, B. P. (2004). Model Continuity to Support Software Development for Distributed Robotic Systems: A Team Formation Example. *Journal of Intelligent and Robotic Systems*, *39*(1), 71–87.

Hu, X., Zeigler, B. P., & Couretas, J. (2001). DEVS-on-a-chip: implementing DEVS in real-time Java on a tiny internet interface for scalable factory automation. In *Proceedings of the 2001 IEEE International Conference on Systems, Man and Cybernetics* (vol. 5, pp. 3051–3056). IEEE.

Jensen, J. C., Chang, D. H., & Lee, E. A. (2011). A model-based design methodology for cyber-physical systems. In *Proceedings of the 2011 7th International Wireless Communications and Mobile Computing Conference* (pp. 1666–1671). Istanbul, Turkey.

Kofman, E., & Junco, S. (2001). Quantized-State Systems: A DEVS Approach for Continuous System Simulation. *Transactions of the Society for Modeling and Simulation International*, *18*(3), 123–132.

Marcosig, E. P., Giribet, J. I., & Castro, R. (2017). Hybrid Adaptive Control for UAV Data Collection: A Simulation-Based Design to Trade-off Resources Between Stability and Communication. In W. K. V. Chan & others (Eds.), *Proceedings of the 2017 Winter Simulation Conference, Las Vegas, USA* (pp. 1704–1715). Piscataway, NJ: IEEE.

Marcosig, E. P., Giribet, J. I., & Castro, R. (2018). DEVS-over-ROS (DOVER): A Framework for Simulation-Driven Embedded Control of Robotic Systems Based on Model Continuity. In M. Rabe (Ed.), and others*Proceedings of the 2018 Winter Simulation Conference* (pp. 1250–1261). Piscataway, NJ: IEEE.

Maruyama, Y., Kato, S., & Azumi, T. (2016). Exploring the performance of ROS2. In *Proceedings of the 13th International Conference on Embedded Software* (pp. 5:1–5:10). Pittsburgh, PA: ACM.

Marwedel, P. (2018). *Embedded System Design: Embedded Systems Foundations of Cyber-Physical Systems, and the Internet of Things* (3rd ed.). Cham, Switzerland: Springer.

Moallemi, M., & Wainer, G. (2010). Designing an interface for real-time and embedded DEVS. In *Proceedings of the 2010 Spring Simulation Conference* (pp. 154–161). Orlando, FL, USA: SCS.

Moallemi, M., & Wainer, G. (2011). I-DEVS: imprecise real-time and embedded DEVS modeling. In *Proceedings of the 2011 Spring Simulation Conference* (pp. 95–102). Boston, MA, USA: SCS.

Moncada, M., Kofman, E., Bergero, F., & Gentili, L. (2013). Generación Automática de Código para Sistemas Embebidos con PowerDEVS. In *Proceedings of the XV Workshop on Information Processing and Control (RPIC)*.

Moyne, J. R., Khan, A. A., & Tilbury, D. M. (2008). Favorable Effect of Time Delays on Tracking Performance of Type-I Control Systems. *IET Control Theory and Applications*, *2*(3), 210–218.

Niyonkuru, D., & Wainer, G. A. (2015). Discrete-Event Modeling and Simulation for Embedded Systems. *Computing in Science and Engineering*, *17*(5), 52–63.

Ozbay, H. (1999). *Introduction to Feedback Control Theory* (1st ed.). Boca Raton, FL: CRC Press.

Quigley, M., Conley, K., Gerkey, B., Faust, J., Foote, T., Leibs, J., ... Ng, A. Y. (2009). ROS: An Open-Source Robot Operating System. In *Proceedings of the Open-Source Software Workshop of the International Conference on Robotics and Automation (ICRA)*.

Sánchez-Peña, R., Quevedo-Casín, J., & Puig-Cayuela, V. (Eds.) (2007). *Identification and Control. The Gap between Theory and Practice* (1st ed.). London, UK: Springer.

Sánchez-Peña, R., & Sznaier, M. (1998). *Robust Systems: Theory and Applications* (1st ed.). New York, NY: Wiley.

Song, H. S., & Kim, T. G. (2005). Application of Real-Time DEVS to Analysis of Safety-Critical Embedded Control Systems: Railroad Crossing Control Example. *Simulation*, *81*(2), 119–136.

Tolk, A., Barros, F., D'Ambrogio, A., Rajhans, A., Mosterman, P. J., Shetty, S. S., ... Yilmaz, L. (2018). Hybrid Simulation for Cyber Physical Systems: a panel on where are we going regarding complexity, intelligence, and adaptability of CPS using simulation. In *MSCIAAS: Spring Simulation Multi-Conference* (pp. 681–698). Baltimore, MD: SCS.

Van Tendeloo, Y., & Vangheluwe, H. (2017). An Evaluation of DEVS Simulation Tools. *Simulation*, *93*(2), 103–121.

Wainer, G., & Castro, R. (2011). DEMES: A Discrete-Event Methodology for Modeling and Simulation of Embedded Systems. *Modeling and Simulation Magazine, 2,* 65–73.

Wainer, G. A. (2016). Real-Time Simulation of DEVS Models in CD++. *International Journal of Simulation and Process Modelling, 11*(2), 138–153.

Yu, H., & Wainer, G. A. (2007). eCD++: An Engine for Executing DEVS Models in Embedded Platforms. In *Proceedings of the 2007 Summer Computer Simulation Conference*, Piscataway, NJ, USA: IEEE.

Zeigler, B. P., Muzy, A., & Kofman, E. (2018). *Theory of Modeling and Simulation: Discrete Event and Iterative System Computational Foundations* (3rd ed.). San Diego, CA: Academic Press.

9

Cyber-Physical Systems Design Methodology for the Prediction of Symptomatic Events in Chronic Diseases

Kevin Henares[1], Josué Pagán[2], José L. Ayala[1], Marina Zapater[3], and José L. Risco-Martín[1]

[1] Complutense University of Madrid, Madrid, Spain
[2] Technical University of Madrid, Madrid, Spain
[3] Swiss Federal Institute of Technology Lausanne, Lausanne, Switzerland

9.1 Introduction

In this chapter we propose a robust methodology for predictive modeling and optimization applied to complex systems addressing symptomatic crises. The proposed methodology is not constrained by the data availability. This system consists of a framework to generate knowledge from multi-source data. The data can be collected from multiple and heterogeneous sources with questionable reliability. From the knowledge generation, we can predict and actuate a complex system (e.g. neurological diseases) without an analytical description. In the following pages, we describe a real case study: the migraine disease.

9.1.1 Predictive Modeling in Mobile Cloud Computing and Health

Over the last two decades, there has been an explosion of data generation using the Information and Communication Technologies (ICT) which has led to a new industrial era in many fields, such as agriculture, communications, or health. This exponential growth of unprecedented knowledge generation requires Big Data analytic solutions.

The Internet of Things (IoT) embraces heterogeneous architectures, methodologies, and elements of many different scenarios in vertical way, encompassing always-connected devices that acquire process and transmit data. When these

Complexity Challenges in Cyber Physical Systems: Using Modeling and Simulation (M&S) to Support Intelligence, Adaptation and Autonomy, First Edition. Edited by Saurabh Mittal and Andreas Tolk.
© 2020 John Wiley & Sons, Inc. Published 2020 by John Wiley & Sons, Inc.

always connected devices meet Cloud computing, we call it Mobile Cloud Computing (MCC), and when applied to healthcare within the MCC framework, it is called eHealth. It is also known as mobile health (mHealth) when eHealth uses mobile devices such as smartphones, wearable devices, or tablets for the healthcare practice.

In an eHealth application three major elements of an MCC network can be distinguished: (i) a Wireless Body Sensor Network (WBSN), (ii) an intermediate element or gateway (such as a smartphone), and (iii) a big computing facility such as Data Centers. In WBSN, sensors are placed on the body surface and register physiological and environmental human conditions in an unobtrusive way. In the paradigm of mHealth, the patients are named as digital patients. Digital patients do telemonitoring, self tracking, and self-diagnosis. The impact of eHealh applications is high, as it is projected to reach 30 billion Wearable Medical Devices by 2020, with a compounded annual growth rate of 42.9% between 2014 and 2019.[1]

There are three major challenges in mHealth to make this growth happen smoothly: (i) perform an unobtrusive ambulatory data acquisition, (ii) provide an accurate prediction, detection or diagnosis in real time, and (iii) increase the autonomy of monitoring devices (Figure 9.1).

In this chapter, we present a real case study gathering data from multiple sources and using real ambulatory devices. On the contrary, most of the current approaches in the literature use few variables, or data gathered from a database, or not monitored in an ambulatory fashion. In addition, we propose the use of a subjective and personalized pain scale, instead of traditional pain scales.

With this, we propose a prediction methodology of a complex system without an analytical description, for the migraine disease. In the literature, most prediction and detection problems focus on diseases with well described symptomatic responses, such as diabetes, arrhythmia or epilepsy. We present this predictive modeling methodology using a Model Based System Engineering (MBSE) framework. MBSE allows flexibility, modularity and fast prototyping. MBSE, combined with the use of the DEVS (Discrete Event System Specification) formalism allows us to study the behavior of a monitoring node to make predictions in real time prior to a final industrial implementation.

9.1.2 Energy Efficiency in IoT

The form-factor of the technology of current batteries limits the autonomy of WBSNs (Vallejo, Recas, Del Valle, & Ayala, 2013). Despite the use of low power microcontrollers and more efficient wireless communication interfaces, there is

---

1 https://rockhealth.com/reports/the-future-of-biosensing-wearables

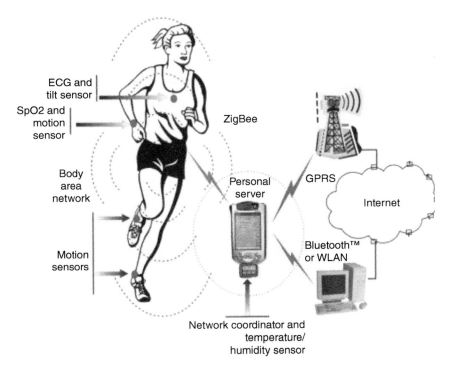

Figure 9.1 Example of a Mobile Cloud Computing network. Original Image: OFSRC. UL (2018).

still room for improvement of the energy efficiency in the monitoring devices and other elements of the network. However, the continous availability of the monitoring devices is still a challenge.

The computing capabilities of current smartphones have increased considerably in the recent years. These devices can perform complex computing tasks, and they neeed not be always connected to the backend – to the Data Center – offloading tasks to these facilities. However, this workload balancing has consequences, and this problem must be addressed from a holistic perspective.

Some efforts to address this issue are in practice through the implementation of new technologies such as Fog computing and Edge computing. These are the latest technologies that implement the workload off-loading using small computing facilities closer to the data-source than to the Data Center. However, in this book chapter we are going to deal with the problems of a real case of impact of energy utlization in an MCC scenario. The main challenge of energy effciency in MCC is the optimization of current infrastructures and high scale deployments. There are many approaches handling this problem in the state of the art. Major efforts have focused in (i) the data processing, (ii) the radio interface, and (iii) the workload balancing. Our contribution to these topics and presented in the following sections are:

- Data processing: Consumption aware proactive processing policies in the monitoring nodes, combined with optimized sensing and predictive model generation.
- Radio interface: trade-off between radio and processing power, and prediction accuracy of predictive models running in the monitoring devices.
- Workload balancing: holistic energy optimization in a real scenario with real nodes, solving a real problem and managing workload balancing.

9.1.3 The Migraine Disease

The migraine is one of the most disabling neurological diseases. It affects around 15% of the Europe population and 10% of the population worldwide (Stovner & Andree, 2010) leading to high economic costs for private and public health systems. In 2012, each migraine patient led to costs of €1,222 per year in Europe (according to the study in Linde et al. 2012), but more recent studies report the average cost of migraine per patient per year to €12,970 for patients with chronic migraine (more than 15 days with pain per month) and €5,041 for patients with episodic migraine, where around 60% correspond to the loss of labor productivity (Research & de Sevilla, 2018).

Migraine is mostly hereditary. It is a social disease that affects more women than men. Currently there is no cure for the migraine and patients take the pills when they feel the pain and it is too late. Migraines not only are composed of pain phases, but are associated with a cascade of neurological processes. Some migraine sufferers experience symptoms that may occur from three days to hours before the pain starts (Giffin et al., 2003). These symptoms are called premonitory symptoms and they are subjective and unspecific: nausea, yawns, tearing, etc. Some patients also suffer from auras. Auras are objective and specific perceptual disturbances such as losing vision that occurs commonly within 15–30 minutes before the onset of pain. The most efficient way to stop this process and avoid the pain is to take of specific drugs in advance. Therefore, the action mechanism of the medicine is able to block the symptoms before they appear. Even after the episodes, the migraine patients report hangovers, phases composed of the so-called postdromic symptoms. Some of them are fatigue, nausea, or dizziness.

Because of the pharmacokinetics of current drugs for treatment of migraine in the acute phase, premonitory symptoms and auras are not helpful to stop the pain sometimes, as it is difficult to estimate the onset of pain. Goadsby et al. (2008) demonstrate that the earlier the intake, the more effective the treatment. In addition, Hu, Raskin, Cowan, Markson, and Berger (2002) demonstrate that specific migraine treatments, such as rizatriptan can abort the migraine within 30 minutes. Other specific treatments, such as sumatriptan, reduce this time to 10 minutes before the crisis starts.

It is known that there are changes in hemodynamic variables when a migraine occurs. Hemodynamic variables are regulated by the Autonomous Nervous System like body temperature, electrodermal activity (EDA), heart rate (HR), or oxygen saturation. We have already shown that it exists an evidence that these changes occur before the pain starts (Pagán et al., 2015), going further than the state of the art (Ordás et al., 2013; Porta-Etessam, Cuadrado, Rodríguez-Gómez, Valencia, & García-Ptacek, 2010). Using an ambulatory and unobtrusive WBSN: (i) the condition of the human body before, during, and after the pain have been measured, (ii) a new method for pain objectification is proposed, and (iii) a study on the predictability of migraine attacks has been performed.

Since there was no mechanism that allows knowing the onset of pain, a prediction system becomes necessary. Predicting the onset of the attack will allow patients to act in advance in order to avoid the pain or reduce its intensity considerably. A case study analyzed the consequences of applying this predictive mechanisms to 2% of European migraine sufferers, concluding that would incur in savings of more than €1272 million (taking into account 76% prediction accuracy) (Pagán, Zapater, & Ayala, 2018).

9.1.4 Modeling and Simulation in the Design of CPS

We are now in the era of integrating previous well-studied embedded, real-time, and control systems in a large scale complex cyber-physical systems. One of the major challenges is how to integrate technologies developed by different communities into a coherent, robust, energy-aware CPS. This chapter tackles the design and implementation of current complex CPSs using different model-based methods, tools, and methodologies, addressing objectives like performance, smooth integration of cyber and physical subsystems, robustness, and energy consumption.

We describe the whole process of conceptualization, design, synthesis, and analysis through the use of Model-Based System Engineering (MBSE) principles. This allows us a straightforward shared communication between all the stakeholders involved in the design of the final product. Furthermore, the use of a M&S standard like Discrete Event Systems (DEVS) formalism (Zeigler, Praehofer, & Kim, 2000) facilitates the composition of models and the interaction with the physical system. Additionally, since DEVS separates the model from the simulator and there exist several simulation engines around the world, there is no need to build simulators, i.e. we must be focused only on the models.

The whole methodology, architecture, and some results are detailed in the following sections.

9.2 General Architecture

In this section, the architecture and conception of different modules that compose the system is presented. First, the architecture is described as a conceptual design, followed by Section 9.3, describing the actual implementation. Overall, the proposed methodology is aware of on-board energy efficiency and workload-balancing energy efficiency in the whole MCC network.

The proposed methodology is designed and implemented through the named Critical-Events Robust Prediction System (CERPS). CERPS is composed of three subsystems to compute predictions: (i) Data Acquisition Systems (DASs), (ii) Robust Prediction Systems (RPSs), and (iii) Expert Decision System (EDS). These systems perform predictions of critical events based on the processing of heterogeneous data collected from different kind of sources. These predictions are centralized in a single EDS that will operate with the data collected by one or multiple DASs. In addition, depending on the nature of these data, an RPS can be included between a DAS and an EDS. Figure 9.2 shows a high-level distribution of models in the CERPS architecture.

In the following sections, each one of these subsystems is explained in detail.

9.2.1 Data Acquisition Systems (DAS)

The architecture and implementation of each DAS depends on the data source, and they are usually distributed in independent units. It should be noted that we are dealing with heterogeneous data sources, and because of this, some DASs are implemented as physical monitoring nodes, and others as remote web services. Moreover, the system can provide information from several sensors or services at the same time, combining its data, if necessary. Depending on the use that will be given to that data, they will be used for prediction purposes by one or multiple RPSs (or even none, driving directly the EDS).

However, the distributed nature of DASs imposes certain limits in the systems design work-flow, due to the need to correlate the data. This problem will be solved later, when necessary, in the Data Drivers section ahead.

9.2.2 Robust Prediction Systems (RPS)

RPSs are the most important component of the architecture and are composed of four subsystems: (i) a Data Driver (DD), (ii) a Sensor Status Detector (SSD), (iii) a Prediction System (PS), and (iv) a Decider. Among them, the Prediction System (PS) represents the core of CERPS (Figure 9.3).

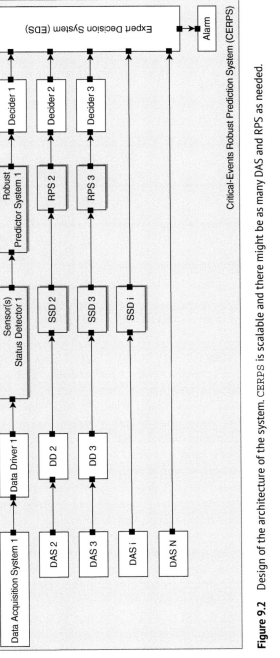

Figure 9.2 Design of the architecture of the system. CERPS is scalable and there might be as many DAS and RPS as needed.

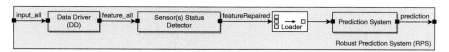

Figure 9.3 Elements that compose the `Robust Prediction System` architecture. RPSs are the most important subsystems, and they can be independent entities by themselves, out of the system.

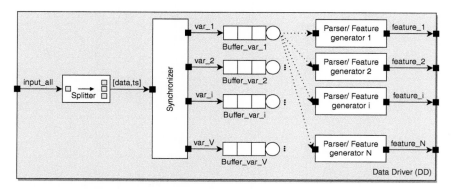

Figure 9.4 Example of `Data Driver` architecture. It is shown how one measured variable can lead to one or several features to be used in the subsequent modules.

9.2.2.1 Data Drivers (DD)

`Data Drivers` (DDs) pre-process the data obtained by DASs. They are composed of different modules, as shown in Figure 9.4. The `Splitter` generates a pair of *(variable, timestamp)* each sample gathered from any data source used by the DASs. Each pair of values goes through a `Synchronizer` that validates the value of the timestamp and stores the value of the variable in a FIFOs buffer (one per variable). Then, some `Parser/Feature Generator` module computes and correlates these data and generate features. It is important to notice that the number of generated features does not have to match the number of input variables: several variables can be computed into a single feature and several features can be obtained using a single variable.

These modules will vary depending on the type of the received data (qualitative or quantitative data). Quantitative data usually correspond to the measurements of physical magnitudes varying along time and sampled at a certain rate. Qualitative variables represent mostly non time-dependent information such as events or actions happening occasionally.

Moreover, depending on the particular use case, DDs can be placed before or after SSDs. If they are placed before them, the input is a matrix of data and timestamps and the output an array of synchronized features (that may contain errors).

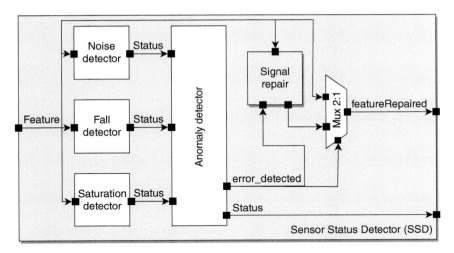

Figure 9.5 `Sensor Status Detector` architecture. It detects data errors and warns other modules to maintain the accuracy of the prediction.

Conversely, if they are placed after SSDs, the input is an error-free matrix of data and timestamps and the output a reliable array of synchronized features.

9.2.2.2 Sensor Status Detector (SSD)

Measurements in real scenarios are usually unreliable. They are susceptible to data loss and they can be exposed to multiple error sources. SSDs are responsible for providing robustness to the system. They check whether exist errors in the measured signals and generate alarms when something is wrong. Temporally, they repair the signals based on the statistics of buffered past data. Their components are shown in Figure 9.5.

While no error is detected, the output data correspond to the input signal and no alarms are triggered. Besides, this input signal, coming from the Data Driver or directly from a DAS, is analyzed by three error detectors (saturation, fall, and noise detectors). All these components generate independent alarms. Their activation depends on simple thresholds or they can be the result of more complex procedures, like fuzzy logic algorithms. These three alarms are grouped in the Anomaly Detector component, which will raise a definitive alarm that will notify to the following components of the presence of errors in the signal. When this happens, the Signal Repair component is also activated.

Generating predictions when an error is detected would lead to low reliability, erroneous solutions, and alarm failures. To avoid this situation, the Signal Repair component generates estimations of the data until the error is solved or enough time has passed for the estimations to stop being reliable. When this component is activated, the multiplexer drives its output to the main output of SSD.

Some solutions to the aforementioned errors would be the replacement of the sensors if they are damaged, or avoiding noisy environments or excessive movements. The operation of the Signal Repair module is based on the use of past samples to generate values that follow the trend that the signal carried before being erroneous. The algorithm applied will depend on the circumstances of the use case. Considering computing capabilities and accuracy of the repairing process, for those scenarios where the signal repairing process runs on devices with constrained computing capabilities, time-series algorithms are suitable to perform the recovery. Conversely, if it runs on a remote server with high processing capabilities a more complex algorithm can be used, such as Gaussian Process Machine Learning (GPML) algorithm. It is important to note that, depending on the algorithm used, it will not only use past samples of the signal managed by the current SSD, but external ones could be used – as an example, time-series models with exogenous inputs use past information from other features, whereas GPML only needs own past data.

9.2.2.3 Prediction System (PS)

The Predictor is the component where predictions are computed. It is composed of three different subsystems: (i) the Sensor Dependent Model Selection System (SDMS2), (ii) a set of Predictors, and (iii) the Linear Combiner.

As aforementioned, when some error is detected in the SSDs, the Signal Repair system is activated. Nevertheless, as it is based on previous samples, this procedure is only reliable during certain time. After a while, the past data is not reliable enough to generate predictions and the damaged data being recovered must be discarded. In these situations, with a missing input the prediction models do not work. However, the presented methodology takes this into account and it defines different prediction models for each subset of inputs. In this way, when one input data is discarded, other predictive model is activated using only the remaining variables to generate the predictions. Thus, this procedure, as the previous repairing of the signals, increases the overall robustness of the system. With this, the system does not stop but keeps computing predictions maintaining some level of accuracy in the prediction (Figure 9.6).

Consequently, there will be as many Predictors components as the number of supported combinations of active features. The one that is to be used is elected by the SDMS2 that activates the corresponding Predictor component based on the alarm signals triggered by the SSDs. In the worst case, when all V variables

(or features) but one are damaged there would be $VC = \sum_{i=1}^{V} \binom{V}{i}$ possible

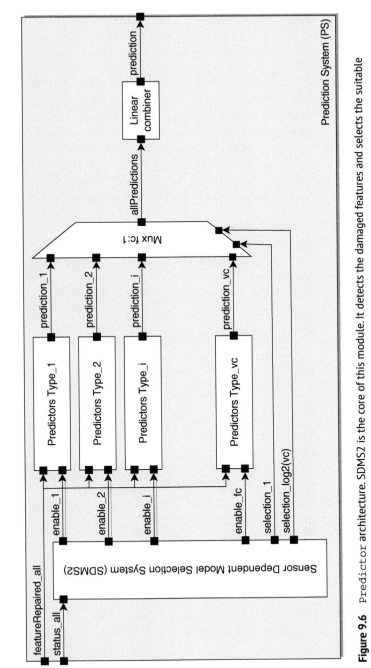

Figure 9.6 Predictor architecture. SDMS2 is the core of this module. It detects the damaged features and selects the suitable Predictor block.

variable combinations to create predictive models. The generation of accurate predictions with one or two features is neither common nor accurate in these cases, and they are not considered as well. Thus, usually a subset $vc \subset VC$ is selected, generating vc `Predictor` modules.

Each `Predictors` component includes one or more predictive models. These predictive models calculate their output h steps ahead (seconds, minutes,...), being h the prediction horizon (i.e. the time between declaration of a hypothetical event and the event itself). This allows to include models that consider different prediction horizons and combine them. Moreover, several types of predictive algorithms can be included in a `Predictors` component. There are different suitable algorithms that operate over time series data, such as state-space algorithms, time series analysis, artificial neural networks, or grammatical evolution.

In previous studies, it has been demonstrated that a combination of predictions (generated by different predictive models) provide more accurate results and larger prediction horizons (Pagán et al., 2015). This combination is done in the `Linear Combiner` that weighs the predictions of the active `Predictors` component and generates a single result that will be passed to the `Decider`.

9.2.2.4 Decider

The `Decider` gets the unified predictions generated by `RPS` and uses them to activate a *local_alarm* signal that will be received by the `Expert Decision System`. Depending on the case, it will use either the last prediction or the last N predictions (accumulated in a buffer). Figure 9.7 shows the architecture of the `Decider`.

Several mathematical functions can be used to raise the alarm (see Figure 9.8). Some of them are (i) binary threshold decider, where the local alarm raises when the current prediction or a weighted average value of buffered predictions exceeds a threshold (Figure 9.8a), (ii) a general case of sigmoid function where the alarm is a softer version of the binary one (Figure 9.8b), and (iii) a fuzzy logic function that represents the alarm as a result of the fuzzification of the current prediction or individual past predictions (see Figure 9.8c).

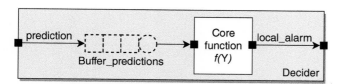

Figure 9.7 `Decider` architecture. This model generates local alarms based on a core function.

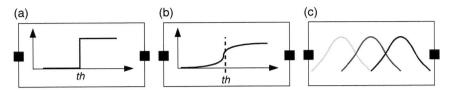

Figure 9.8 Three examples of core functions for the Decider module. (a) Binary function (b) sigmoid function (c) fuzzy logic.

In systems with only one DAS and RPS, the output generated by the Decider corresponds to the final alarm signal that notifies the critical event. In more complex systems with several Deciders their signals go to EDS, the last module of the system that generates the final decision.

9.2.3 Expert Decision System (EDS)

The EDS module is the one that triggers the final alarms in CERPS. It is a computer-based system and is fed with both the alarm signals of the Deciders and outputs of previous modules (as shown in Figure 9.2). As these inputs can be affected by problems as data loss or unlabeled data, the automatic generation of decision algorithms is not suitable. Instead, *Active Learning* (AL) algorithms are used. It is a semi-supervised machine learning that interacts with the user (or other information sources) by providing the output when special cases occur (as the arrival of new unlabeled data). It can operate with a greater variety of algorithms, such as Support Vector Machine algorithms, decision-tree based algorithms and Adaptive Neuro-Fuzzy Inference System. In the case of the CERPS, this decision model determines the occurrence of new critical events.

9.3 Software Model and Physical Implementation

As mentioned in previous sections, the CERPS architecture is applied to the migraine disease. Specifically, the objective of the presented system is to predict migraine episodes with enough lead time. To do that, migraine patients are monitored in an ambulatory way, measuring four hemodynamic variables: skin temperature (TEMP), electrodermal activity (EDA), heart rate (HR), and oxygen saturation (SpO2). An example of a six hours monitoring phase is shown in Figure 9.9.

To predict the symptomatic crisis, the first step is to generate a model of the migraine pain. To do this, an adjustment process of the registered subjective pain curve was carried out during the experiments. It is known that the pain rises faster

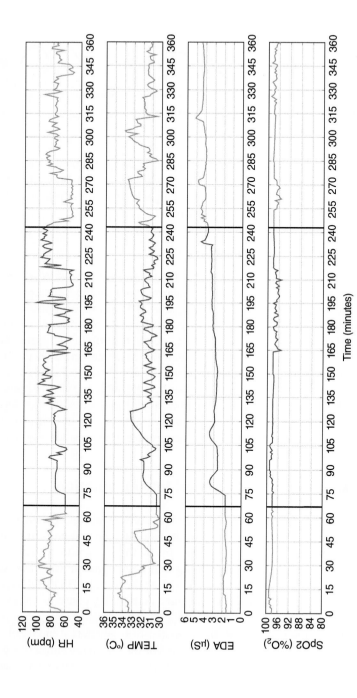

Figure 9.9 Hemodynamic variables after synchronization and preprocessing during a migraine episode (the curve between vertical bars).

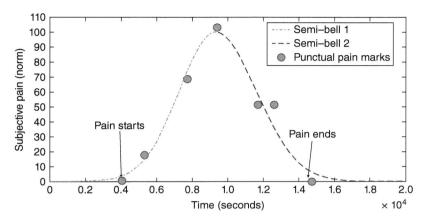

Figure 9.10 Modeling of subjective pain evolution curve using real data. The pain has been described as two semi-Gaussian curves.

than it recesses, so the symptomatic curve has been modeled as two semi-Gaussian curves, as they fit the patients subjective response. In addition to the discrete annotations of pain evolution, patients also indicate two time-points during the migraine attack. The first time-point indicates the onset of the pain when detected, and the second time-point indicates the end of pain. With all this information, two semi-Gaussian curves can be generated, as shown in Figure 9.10. $\{(\mu_1, \sigma_1), (\mu_2, \sigma_2)\}$ are the two semi-Gaussians parameters necessary to define a symptomatic curve. The symptomatic curve includes the pain period, as it reflects some changes in the migraine process. An example of the resulting function is shown in Figure 9.10, using actual data as well.

Now, we shall apply CERPS to this real environment. Following the best practices of the MBSE paradigm, the CERPS-Migraine system is simulated through a software model. Next, after validation, it is implemented in a physical device.

As a first step, we introduce a fine-grained migraine predictive model. This relation is expressed in Eq. (9.1).

$$\hat{y}\left[k + \Delta t\right] = f\left\{\text{TEMP}\left[k - p_1\right], \text{EDA}\left[k - p_2\right], \text{HR}\left[k - p_3\right], \text{SpO}_2\left[k - p_4\right]\right\}, \quad (9.1)$$

In Eq. (9.1), y is the signal that predicts the migraine pain in the future horizon $k + \Delta t$. This predictive model is a function f of the hemodynamic variables stated above, measured in a past windows defined by a parameter p_i. f is a function that can be implemented in different ways. In our research we have tested several predictive modeling alternatives. State-space models and Grammatical Evolution (GE) showed best results, and in the remaining of the chapter, we will present the implementation using both. State-space models relate changes in hemodynamic variables through an immeasurable state of the pain using matrices (Pagán et al., 2015), and GE predicts directly the pain value by means of mathematical functions applied over the hemodynamic variables (Pagán, Risco-Martín, Moya, & Ayala, 2016).

Figure 9.11 shows the block diagram of the actual implementation of the migraine prediction system. The following two sections provide details on how both the CERPS DEVS software model and the corresponding CERPS physical implementation are tackled.

9.3.1 Software Model

Figure 9.12 shows the details of the application of the CERPS architecture to a DEVS software model of this predictive system (Pagán et al., 2017). Next, after validating the system, it is adapted to a VHDL implementation (Henares et al., 2018). Although certain decisions must be made to adapt it, the process is remarkably simplified due to the modular nature shared by both DEVS and VHDL.

A general view of the DEVS simulation can be seen in Figure 9.12. This refers only to the coarse-grained migraine prediction using hemodynamic variables. As it can be seen, it is composed of four DASs, four DDs, four SSDs, a synchronizer, a predictor, and a decider. To analyze the operation of the system an additional module (EFgt) was added to the design. An extra module is used to provide the synchronization of the real pain values of each episode which will be compared with the generated predictions in the EFgt module.

One DAS were added by each monitored biometric variable. In the simulation, past data of migraine patients were used to replace the data entry flow that would correspond to the sensors. Also, four extra modules were added between the DASs and DDs to simulate the errors that can appear due to its use (ErrorInductors).

The four DDs read their inputs and introduce them into the system, associating a timestamp to each sample. The resulting pairs are directed to the SSD modules, to repair the possible errors in the signals. Saturation, fall, and noise errors are detected and are temporarily replaced by an estimation. In this case the repairing of the signal is done using a GPML procedure. The average fitting achieved using GPML ranges from 73.4% for SpO2, to 93.2% for EDA.

In a software simulation, it is suitable to run GPML for signal recovery. However, the implementation in real monitoring nodes is computationally expensive for signal repair in real time. In those cases, autoregressive (ARX) models are used. ARX models are time-series models with low computation requirements. GPML signal recovery needs only its own past data. On the contrary, ARX models use past data from the remaining variables, so if more than one signal is damaged, the system will not be able to recover any of them. Figure 9.13 shows how the signal recovery works using both alternatives applied over an HR signal with errors.

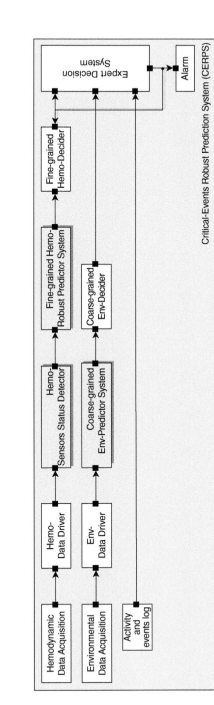

Figure 9.11 Overview of the implementation of the whole system architecture. There are three types of data sources: two of them perform fine and coarse predictions, while the third one serves as prediction support to the Expert Decision System.

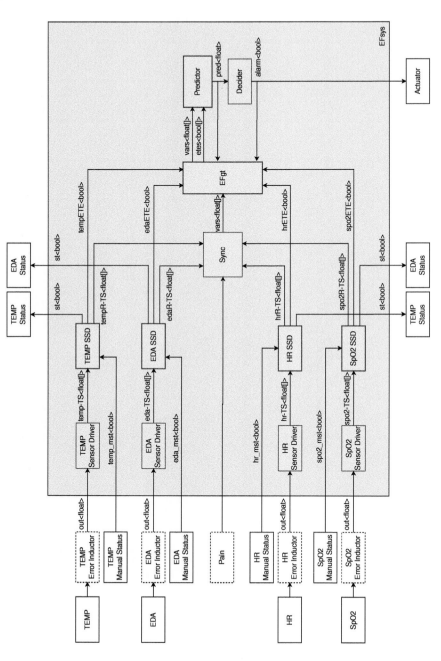

Figure 9.12 Root component of the migraine pruning system implemented in an FPGA with VHDL.

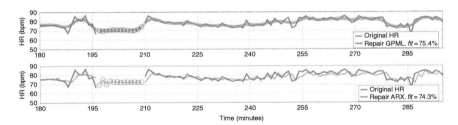

Figure 9.13 Example of signal repair using Gaussian process machine learning (GPML) and time series algorithms (ARX) for the HR signal.

The outputs of the four `SSD` are provided as input to the synchronizer. The `Sync` module synchronizes and buffers the data for simultaneously supplying the values for the four biometric variables to the coupled model `EFgt`. The pain value is also grouped with the other variables, if available. After this informative module, the synchronized information is processed in the `Predictor` module. In this case, it uses state-space models to generate predictions (solved using N4SID method) and it counts five different model sets (Pagán et al., 2015). One of them is intended to operate in the ideal situation, when all the variables are free of errors. The other four are trained to generate prediction using variable sets with a damaged signal (combinations of three operating variables). Hence, reliable predictions cannot be generated with two or fewer biometric variables. Each model set contains three N4SID models trained to generate predictions with three different time horizons. This was made to increase the effectiveness of the combined prediction, generated by a `Linear combiner` module. In this case, this module simply averages the three predictions (but depending on the case it can be suitable to weight them).

Finally, the output of the `Predictor` is sent to the `Decider`. This module is in charge of activating the alarm signal based on a binary function. The used threshold value is 32. This represents a 50% probability of the maximum pain level (see Figure 9.10). As it is the only decider of the system, there is no need of an `EDS` and this signal corresponds to the final alarm.

Once the software model is tested and validated (see Pagán et al. 2017), it is implemented in the hardware. As we will see, there are some modifications that must be done to the software model, because of some peculiarities of the physical implementation and the target device.

9.3.2 Physical Implementation

As seen before, simulation adds savings in terms of reduced implementation costs, time, or human and material resources. Simulation is the natural step in MBSE design, prior to a physical implementation because it accelerates error debugging phase and it allows the verification of the methodology. Before an actual device is implemented in hardware, a hardware/software (HW/SW) co-simulation that includes hardware-in-the-loop (HIL) is used. This will ensure that the system works in presence of actual hardware sensor failures and physical actuators, and triggers alarms accurately, as predicted by the simulation system. In this section, the changes and the considerations made in the process of translating the DEVS model into a VHDL implementation using an FPGA are discussed. The root component of this implementation is shown in Figure 9.14.

The first consideration to be made relates to the DASs. As this implementation works with real sensors, some interfaces are needed to link them to the real system. Both the TEMP and EDA sensors are analogical and need an analog-to-digital converter. Specifically, a Pmod AD2 has been selected. It counts with 4 channels, 12 bits of precision, and communicates through the Inter-Integrated Circuit (I2C) protocol. The remaining biometric variables, HR and SpO2, are read through a OEM III platform[2] (by NONIN®) using serial communication. It has several operating modes. The selected device streams one integer measure of each one of the variables each second.

As in this case only two DAS are there, we need Pmod AD2 converter and NONIN OEM III module for each of them. The system also counts two DD. Since in this case real sensor readings must be made, they have the additional task of interacting through suitable communication protocols. The first DD, that reads the TEMP and EDA values using the I2C protocol, has to send reading requests three times per second (sample rate of 3 Hz). The second DD, that reads HR and SpO2 values using serial communication, keeps waiting and read the data sent by the OEM III module (one data pack per second). Moreover, both modules must adapt the data format to match them to the data types used by the system.

As stated above, the Signal repair module operates in hardware based on an Auto Regressive model with eXogenous inputs (ARX). The evaluation of these models is more efficient and also offers acceptable results, but has the disadvantage that each module needs previous information of all the other variables of the system. For this reason, an additional module (BuffersHandler) was added to group that information and offer them as input to all the SSD modules of the system (see Figure 9.13).

Moreover, since the data obtained from the sensors have different sample rates, SSDs are placed after the synchronizer module. This causes the buffer to store the data by the minutes. In this way, the input of the ARX modules is already synchronized and consist of variables of the same data rate.

---

2 OEM-III: http://www.nonin.com/OEM-III-Module (accessed December 2018).

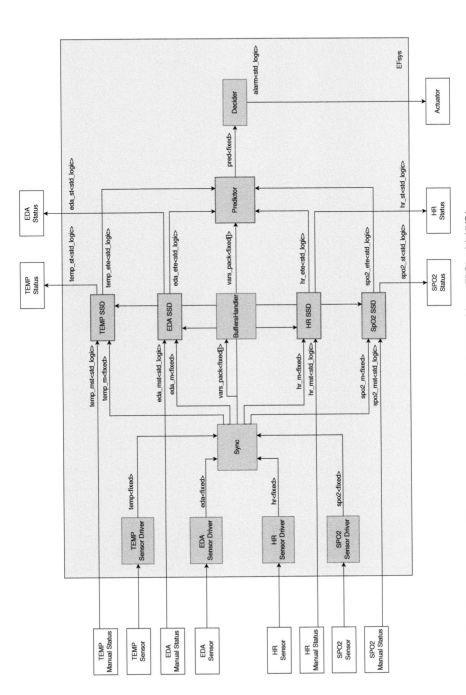

Figure 9.14 Root component of the migraine pruning system implemented in an FPGA with VHDL.

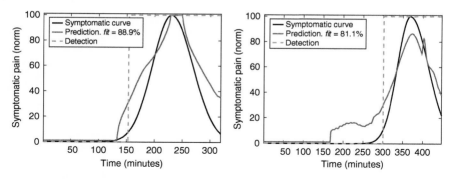

Figure 9.15 Test results: symptomatic periods for one of the trained patients.

As there are no large timing requirements in the generation of predictions (since a data set is processed per minute), the `Predictors` modules are joined into a single module. It loads the suitable models from a separately deployed memory module and generates three predictions in a sequential way. This process is controlled by the `SDSM2` module, with requests the predictions one by one, which are accumulated in the `Linear Combiner` as they are generated. This change allows the reuse of FPGA components and results in an optimized use of resources.

Models are trained to predict migraines 30 minutes ahead. A migraine event is detected when the *fit* metric reaches a 70% of accuracy and the average prediction rate (taking into account the availability of the sensors) is 76%, with a low rate of false positives.

The `Decider` module operates in the same way as in the SW DEVS simulation.

Figure 9.15 shows two prediction results and the corresponding alarms using N4SID models when all the sensors are available (Figure 9.15a), and when one sensor fails (Figure 9.15b) and how `SDMS2` actuates changing the predictive model used.

As floating-point numbers require higher processing in FPGAs (unless specialized modules are available), fixed-point numbers are used in this design. Fixed-point numbers assign a fixed number of bits for both the whole and the decimal part. Consequently, there is a precision loss that must be controlled to avoid introducing significant errors in the obtained results. In this way, two measures are taken: (i) values read from sensors are stored in data types with several decimals, enough to avoid precision loss, and (ii) the intermediate data types to perform the operations needed for signal repairing and prediction generation are analyzed to choose a suitable decimal length. For this, the RMSE is calculated with a range of decimal lengths and it was selected the minimum length with a RMSE lower than 2%. Results can be seen in Figure 9.16.

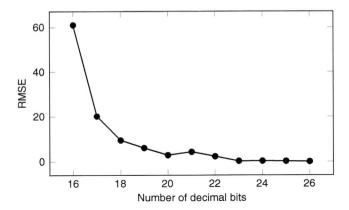

Figure 9.16 RMSE of the predictions depending on the number of decimal fixed bits used in the variables that hold the intermediate calculations in the `Predictor` module (compared with the results of the simulator, using floating-data types).

Once the feasibility of the procedure and methodology has been demonstrated, the next section shows how to make it energy efficient and how to tackle scalability issues.

9.4 Energy Consumption and Scalability Issues

This section deals with the energy consumption of the proposed system, as well as the scalability issues when the migraine detector is integrated into the health system.

9.4.1 Energy Consumption

The MCC scenario must be economically rewarding. In our implementation, we considered two interfaces where we can actuate: (i) from the WBSN to the gateway or smartphone, and (ii) from the gateway to the high computing facility or Data Center. Energy efficiency in the first interface means battery savings in the monitoring nodes and smartphones; on the other hand, energy efficiency in the second interface means saving money due to the reduction of the electricity bill of the power-hungry big computing facilities.

The use of wearable monitoring devices is becoming increasingly popular. For that reason, more and more research teams are focusing their efforts in reducing their power consumption to enlarge their battery life. Some of the common approaches are based on creating more efficient architectures (Braojos Lopez & Atienza, 2016), develop new on-data processing techniques (Ghasemzadeh, Amini, Saeedi, & Sarrafzadeh, 2015) or apply pre-processing techniques such as compressed sensing (Braojos et al., 2014). In this section, we present a methodology to optimize the energy consumption in the first of the interfaces (from the WBSN to the gateway), focused on the monitoring devices for prediction of migraine attacks in real time. This methodology is focused on the optimization of two main sources of energy consumption: (i) reducing the complexity of the processing in the physical device (our FPGA in this case), that leads to reduce the number of clock cycles required to execute the code and, (ii) reducing the consumption of peripherals using the minimum number of sensors. This is a multi-objective optimization problem, for which the multi-objective function can be formulated as follows:

$$\min\left(-fit, \#clk, E_{sensing}\right). \tag{9.2}$$

To minimize the multi-objective function, we use the developed migraine system with different GE predictive models m_i. Each m_i is defined as a mathematical expression. When the system is working with m_i, we extract three optimization objectives. These objectives, shown in Eq. (9.2), are:

1) the accuracy or *fit* of the predicted values for m_i (the *fit* is related to the normalized RMSE),
2) the number of clock cycles *#clk* that the predictive model m_i takes to be computed in the FPGA, and
3) the energy consumption $E_{sensing}$ of the sensors used.

The optimization process is based on the Non-dominated Sorting Genetic Algorithm II (NSGA-II) (Deb et al., 2002) that generates a set of non-dominated solutions (usually called Pareto front). The optimization is performed calculating the number of cycles consumed by the FPGA to compute the model m_i and the energy consumed by the sensors. The first parameter can be easily calculated taking into account the complexity of m_i, the second one is calculated taking into account the energy consumption of the sensors. In our tests, the skin surface temperature (TEMP) uses a thermistor (0.32 mJ/4.9 dBm), the electrodermal activity (EDA) uses two differential electrodes (0.32 mJ/−4.9 dBm), the electrocardiogram (ECG) to extract the Heart Rate (HR) uses two leads with one reference (396 mJ/26 dBm), and the blood oxygen saturation (SpO2) uses an 8000R SpO2 sensor.

With these parameters, several models of energy consumption of the MCC environment are obtained. As a result, we obtain two tri-dimensional pareto fronts. The solutions, since they are non-dominated solutions, have the same goodness. The ones with the maximum *fit* are selected. Figure 9.17 depicts 3D and 2D projections of the results obtained in the pareto fronts for two patients. They are globally convex, tending to minimize the error, the number of clock cycles and the energy consumed by the operation of the sensors. The chosen solutions are

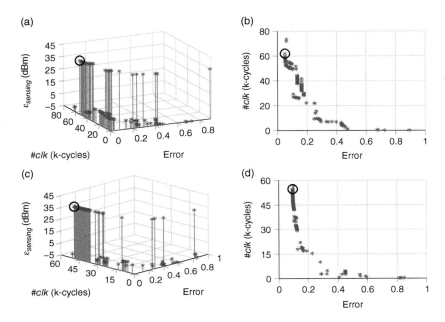

Figure 9.17 3D and 2D views of the Pareto Fronts, result of the optimization process. (a) Patient A. Idx = 2. 3D at 10′. (b) Patient A. Idx = 2. 2D at 10′. (c) Patient B. Idx = 5. 3D at 10′. (d) Patient B. Idx = 5. 2D at 10′.

painted with a circle. This methodology reflects savings up to 90% in the execution time. This directly reduces the energy consumption of the system, at the expense of barely degrading the accuracy of the models.

9.4.2 Scalability Issues

Large scale population monitoring systems in MCC scenarios are getting closer to reality. Smart cities combine ICT services and ICT resources to improve the urban environment and improve the citizen's life. In this context, there are examples of research areas that are applying monitoring and model generation to optimize processes and improve techniques. In the sports area, for instance, the monitoring is being used to generate predictions in team sports (Groh et al., 2014). In the medical area, multiple projects also follow this trend. Some applications are: remote diagnosis, disease alarms generation (Alemdar & Ersoy, 2010), prediction of atrial fibrillation (Milosevic et al., 2014), etc. Although these examples focus on specific cases, larger monitoring networks are a reality nowadays. Nevertheless, these networks must be supported by an adequate architecture that supports their progressive extension and improvement.

Table 9.1 Five scenarios for the workload balancing policies.

	Sensor device	Coordinator	Data center
SC1	Collect + transmit data	Receive + transmit data	Process data + perform predictions
SC2	Collect + transmit data	Receive + process + transmit data	Perform predictions
SC3	Collect + transmit data	Receive + process data + perform prediction + transmit data	—
SC4	Collect + process + transmit data	Receive + transmit data	Perform predictions
SC5	Collect + process + transmit data	Receive data + perform predictions + transmit data	—

This section explores the hypothetical deployment of a large-scale monitoring network centered in the migraine prediction, covering the 2% of European migraine sufferers. It is done extrapolating real data and simulation results of previous studies. As a result, it is calculated that the benefits of implant that network would report savings of €1272 million due to the benefits of the migraine prediction.

The proposed network consists of three main parts: the sensing nodes, the coordinator, and the Data Centers. Each one of the three network elements can operate in various modes. Sensing nodes can collect, transmit, and process data. The controller can receive, transmit and process data, and perform predictions, if necessary. The Data Center can process data and perform predictions. The energy efficiency policies take these possibilities into account to minimize the power consumption of the whole system. For the sake of clarity, among all combinations of scenarios, the five considered below are more significant from the energy perspective, shown in Table 9.1.

In these scenarios, the sensor nodes can act in two different modes:

- Streaming mode (SC1, SC2, SC3): they collect the raw information from sensors and relay it to the controller. S1 transmit ECG signal immediately, S2 transmit raw data from the OEM-III devices every second.
- Processing mode (SC4, SC5): HR is calculated from the ECG signal and it is transmitted every minute. SpO2 data are extracted from the OEM-III device and transmitted once a minute as well.

The Data Center that has been considered to develop the use case is composed of (i) a High-Performance Computing (HPC) cluster to train and validate the models, and (ii) a virtual Cloud computing cluster for online prediction.

All the online and offline tasks are characterized in terms of power and performance, taking into account the different network devices where they are executed. Offline tasks are data processing (in GPML), and model training and validation. Online tasks include data processing (in GPML), and online prediction. Model training and validation are CPU and memory intensive, whereas runtime prediction is a light-weight process (that can be performed by controllers).

In runtime processes, several additional policies are used to minimize the Data Center energy consumption. When the coordinators off-load computation from the Data Center, the numbers of VMs are reduced. This policy, when the processing charge decay, is combined with server turn-off policies and cooling optimization. Because of these techniques, the power usage in the Data Center is drastically reduced.

Further details about the Data Center features and the tasks to be performed are available in Pagán et al. (2018). With these features, HPC and Cloud clusters can handle up to 1 393 649 migraine patients (2% of the migraine sufferers in Europe). Its inclusion in the network is designed to be a gradual process. Four stages are planned, with rates of 50, 25, 15, and 10% of the total included patients. Total evaluation period is 10 weeks.

To analyze the differences derived from the use of controllers as intermediate elements, cases S4 and S5 have been studied. Specifically, the following variations are considered:

- SC4 (baseline): the coordinator simply forward computation to the Data Centers (there is no workload off-loading). The off-line phase is performed in the HPC cluster, and the online phase in the virtualized cluster.
- SC4 (optimized): energy minimization techniques are applied in the Data Center, but no workload off-loading policies are applied.
- SC5, 100% prediction: coordinators nodes perform data preprocessing (i.e., GPML), but all the predictions are generated in the virtualized Cloud. The off-line phase is performed in the HPC cluster.
- SC5, 30% prediction: coordinators execute both GPML and 70% of the predictions. The remaining 30% are computed in the virtualized cluster.

Figure 9.18a shows the utilization ratio of the proposed Data Center as new patients are introduced. Blue dots represent the gradual inclusion of patients (50, 25, 15, and 10%). The lower line corresponds to models that need to be retrained.

Figure 9.18b reflects the utilization of the virtualized clusters in the different scenarios. It can be seen how this charge decreases significantly as we off-load computation to the coordinators. This combined with the turn-off policies at the Data Center, results in a drastic reduction of power consumption.

As can be seen, MCC (and IoT in general) poses important challenges because of the large volumes of data that need to be gathered and analyzed. Among all the

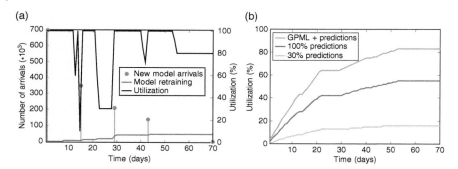

Figure 9.18 Utilization of the HPC Data Center and Cloud cluster. (a) HPC Data Center. (b) Cloud cluster.

MCC possible applications, population monitoring in eHealth define important constraints and demands energy minimization strategies to develop massive healthcare solutions in massive population scenarios. In this section we have shown a multilevel approach to reduce energy consumption of nodes, ensuring acceptable levels of scalability.

9.5 Conclusion

The recent advances in wireless sensor networks, medical sensors, and cloud computing are making CPS an excellent candidate to design predictive and proactive healthcare applications. In this chapter, we show a hardware-in-the-loop model-driven method, based on the DEVS formalism to apply system engineering on CPS oriented to prevent symptomatic events in chronic diseases or complex systems without an analytical description.

The architecture of the system has been presented. The derived methodology can be applied to chronic diseases cases where the prediction of symptomatic events is needed. Through its use, the resulting system significantly increases its robustness. On the one hand, temporary errors in the signal are detected and restored. On the other hand, the proposed system allows having several model sets and select the suitable one in real-time. In this way, different prediction algorithms and horizon times can be used, weighting them, if necessary, to improve the prediction quality.

As an example, the proposed architecture was applied to a migraine prediction system. The system is firstly developed in a DEVS M&S context that allows us to validate the design. Next, it was transformed into a VHDL implementation, taking advantage of the modularity of both the DEVS formalism and the VHDL language.

In addition, several techniques and policies for reducing energy have been presented. First, the impact of the algorithms and sensors used in the sensing nodes have been analyzed, reflecting savings up to 90% of energy consumption. Also, some techniques related to the management of both the Data Center and the virtualized servers have been discussed. The application of all these points in the developed workflow can result in a significant reduction in terms of energy consumption.

Finally, the deployment of a hypothetical large-scale network extending the migraine prediction case has been discussed. The different network elements and tasks are categorized and used to study the utilization ratio of the infrastructures, comparing approaches with and without intermediate coordinators.

We have shown how MBSE-based techniques and the use of appropriate M&S standard facilitates all the phases of the CPS design and the evaluation of objectives like reliability, performance, robustness, and energy consumption. This covers the full straightforward design method for CPSs, from the first conception, to the hardware design, taking care of the energy consumption and final massive deployment.

References

Alemdar, H., & Ersoy, C. (2010). Wireless sensor networks for healthcare: A survey. *Computer Networks*, *54*(15), 2688–2710.

Braojos Lopez, R., & Atienza, D. (2016). An ultra-low power NVM-based multi-core architecture for embedded bio signal processing. In *ICT-Energy Conference 2016*.

Braojos, R., Mamaghanian, H., Junior, A. D., Ansaloni, G., Atienza, D., Rincón, F. J., & Murali, S. (2014). Ultra-low power design of wearable cardiac monitoring systems. In *Proceedings of the 51st Annual Design Automation Conference* (pp. 1–6).

Deb, K., Pratap, A., Agarwal, S., & Meyarivan, T. (2002). A fast and elitist multiobjective genetic algorithm: NSGA-II. *IEEE Transactions on Evolutionary Computation*, *6*(2), 182–197.

Ghasemzadeh, H., Amini, N., Saeedi, R., & Sarrafzadeh, M. (2015). Power-aware computing in wearable sensor networks: An optimal feature selection. *IEEE Transactions on Mobile Computing*, *14*(4), 800–812.

Giffin, N., Ruggiero, L., Lipton, R., Silberstein, S., Tvedskov, J., Olesen, J., ... Macrae, A. (2003). Premonitory symptoms in migraine an electronic diary study. *Neurology*, *60*(6), 935–940.

Goadsby, P., Zanchin, G., Geraud, G. A., De Klippel, N., Diaz-Insa, S., Gobel, H., ... Fortea, J. (2008). Early vs. non-early intervention in acute migraine: 'Act when mild (awm)'. A double-blind, placebo-controlled trial of almotriptan. *Cephalalgia*, *28*(4), 383–391.

Groh, B. H., Reinfelder, S. J., Streicher, M. N., Taraben, A., & Eskofier, B. M. (2014). Movement prediction in rowing using a dynamic time warping based stroke detection. In *IEEE 9th International Conference on Intelligent Sensors, Sensor Networks and Information Processing (ISSNIP), 2014* (pp. 1–6).

Henares, K., Pagán, J., Ayala, J. L., & Risco-Martín, J. L. (2018). Advanced migraine prediction hardware system. In *Proceedings of the 50th Computer Simulation Conference* (pp. 7:1–7:12). San Diego, CA, USA: Society for Computer Simulation International. Retrieved from http://dl.acm.org/citation.cfm?id=3275382.3275389

Hu, X. H., Raskin, N. H., Cowan, R., Markson, L. E., & Berger, M. L. (2002). Treatment of migraine with rizatriptan: When to take the medication. *Headache: The Journal of Head and Face Pain, 42*(1), 16–20.

Linde, M., Gustavsson, A., Stovner, L., Steiner, T., Barré, J., Katsarava, Z., ... Andrée, C. (2012). The cost of headache disorders in Europe: The Eurolight project. *European Journal of Neurology, 19*(5), 703–711.

Milosevic, J., Dittrich, A., Ferrante, A., Malek, M., Quiros, C. R., Braojos, R., ... Atienza, D. (2014). Risk assessment of atrial fibrillation: A failure prediction approah. In *Computing in Cardiology Conference (CINC), 2014* (pp. 801–804).

OFSRC.UL. (2018). *Security Enabled Pervasive Biometric Sensors.* http://www.ofsrc.ul.ie/index.php/research/3-security-enabled-pervasive-biometric-sensors. (Online; accessed 28 December 2018).

Ordás, C. M., Cuadrado, M. L., Rodríguez-Cambrón, A. B., Casas-Limón, J., del Prado, N., & Porta-Etessam, J. (2013). Increase in body temperature during migraine attacks. *Pain Medicine, 14*(8), 1260–1264.

Pagán, J., Moya, J. M., Risco-Martín, J. L., & Ayala, J. L. (2017). Advanced migraine prediction simulation system. In *Proceedings of the 2017 Summer Simulation Multiconference (SummerSim 2017)*.

Pagán, J., Orbe, D., Irene, M., Gago, A., Sobrado, M., Risco-Martín, J. L., ... Ayala, J. L. (2015). Robust and accurate modeling approaches for migraine per-patient prediction from ambulatory data. *Sensors, 15*(7), 15419–15442.

Pagán, J., Risco-Martín, J. L., Moya, J. M., & Ayala, J. L. (2016). Grammatical evolutionary techniques for prompt migraine prediction. In *Proceedings of the Genetic and Evolutionary Computation Conference 2016* (pp. 973–980).

Pagán, J., Zapater, M., & Ayala, J. L. (2018). Power transmission and workload balancing policies in eHealth Mobile Cloud Computing scenarios. *Future Generation Computer Systems, 78*, 587–601.

Porta-Etessam, J., Cuadrado, M. L., Rodríguez-Gómez, O., Valencia, C., & García-Ptacek, S. (2010). Hypothermia during migraine attacks. *Cephalalgia, 30*(11), 1406–1407.

Research, H. T., & de Sevilla, U. (2018). *Impacto y situación de la Migraña en España: Atlas 2018* (pp. 1–183). Editorial Universidad de Sevilla.

Stovner, L. J., & Andree, C. (2010). Prevalence of headache in Europe: A review for the Eurolight project. *The Journal of Headache and Pain*, *11*(4), 289.

Vallejo, M., Recas, J., Del Valle, P. G., & Ayala, J. L. (2013). Accurate human tissue characterization for energy-efficient wireless on-body communications. *Sensors*, *13*(6), 7546–7569.

Zeigler, B. P., Praehofer, H., & Kim, T. G. (2000). *Theory of Modeling and Simulation. Integrating Discrete Event and Continuous Complex Dynamic Systems* (2nd ed.). Academic Press.

10

Model-Based Engineering with Application to Autonomy

Rahul Bhadani, Matt Bunting, and Jonathan Sprinkle

Department of Electrical and Computer Engineering, The University of Arizona, Tucson, AZ, USA

10.1 Introduction

Model-based design is a promising methodology in the design and deployment of autonomous cyber-physical systems. As the scale and complexity of such systems grow, designers and engineers have started employing a model-based approach towards model-based engineering (Al Faruque and Ahourai 2014). However, due to the inherent complexity of autonomous cyber-physical processes, it is imperative to account for dynamic behavior of the CPS under a wide variety of conditions. Complexity in CPS system arise due to its physical component being a function of continuous time while cyber component being a function of discrete-time. As a result, it is crucial to address the gap between modeling of autonomous CPS and validation and verification (V&V) cycle that takes into account the complex heterogeneous nature of such systems. Methods in model-based design enable CPS designers to implement tasks of V&V by abstracting large systems at a coarser level. These abstractions fall into different categories such as meta-modeling, interpreter design and language modeling, and structural and behavioral modeling (Nordstrom et al. 1999; Lédeczi et al. 2001c; Sprinkle and Karsai 2003; Emerson et al. 2004; Tolvanen and Kelly 2009; Jackson et al. 2011; Kelly et al. 2013).

While there are well-established methods of V&V in other domains of physical sciences, the consideration of heterogeneity and the external environment makes the V&V procedure in CPS a complicated task (Zheng et al. 2017). Often the possibility of

Complexity Challenges in Cyber Physical Systems: Using Modeling and Simulation (M&S) to Support Intelligence, Adaptation and Autonomy, First Edition. Edited by Saurabh Mittal and Andreas Tolk.

dependent component failure is ignored in cyber-physical systems when using traditional approaches, and in these cases models are easily invalidated when designers perform rudimentary integration testing with hardware in the loop or with real-world use cases. Therefore, merely relying on code-inspection, static analysis, module level testing and integration testing is not enough. In safety-critical systems such as autonomous vehicles, abstraction makes analysis of the design manageable. However, due to interaction with the physical world and deficiency in capturing all aspects of physical processes, design, or implementation inadequacies (i.e. bugs) are unveiled during run-time or deployment. The stochastic nature of physical processes and their interdependencies makes formal V&V indispensable, though not inexpensive, in the context of autonomous systems. This chapter explores model-based design techniques and architectures for their formal validation and verification in context of autonomous system, especially autonomous ground vehicle control.

10.2 Background

In this section, we present some background on traditional V&V approaches and progress in V&V with respect to Model-based design for CPS and hybrid systems.

10.2.1 Verification and Validation

The development of large-scale software-based systems has typically broken down the overall design into stages that eventually produce components. These components, then, are designed, tested, and integrated (with further testing) until the final product is approved by a customer. While such an approach can be brittle when it comes to building systems whose requirements are difficult to define, or whose stakeholders may not understand the cost consequences of the requirements they outline, the overall structure is introduced in order to see the component pieces – and in order to see the context in which this structure is modified when it comes to the approaches outlined in this chapter.

We treat verification as an answer to the question, "Is the hardware or software component under consideration behaving according to the specification?" In order to verify traditional software components, answers to this question are sometimes obtained through testing. These processes may include input/output testing, comparison to a functional equivalent or reference implementation, regression testing, etc. Testing is well-known to be a dynamic form of evaluation (which can be carried out only when the software under test is executed), in contrast to static evaluation techniques. These include design reviews whereby experts examine the code, and verification techniques by which the code is logically or symbolically executed in order to determine whether the requirements have been met (Tabuada 2009; Anta et al. 2010; Ammann and Offutt 2016; Clarke et al. 2018; Lonetti and Marchetti 2018).

We treat validation an answer to the question, "Is the component (hardware or software) specification likely to achieve the desired result?" Validation is considered both early and near the conclusion of the system design and implementation. In the early stages it is a means by which designers determine how high-level requirements should flow to low-level systems and (eventually) components. As the design concludes, validation is carried out on composed systems to determine whether individual component behaviors, when integrated, satisfy desired composed behaviors.

In summary, verification may be thought of as "are we building the product right?" and validation as "are we building the right product?" In the downward stroke of the "V" the design requirements flow down, and designs are finalized. In the upward stroke of the "V" the implementations are compared to design requirements (verification) and the integrated behaviors are tested to ensure that they satisfy higher-level requirements (Figure 10.1).

10.2.2 Model-Based System Approaches: The Tie to CPS

There are myriad texts that describe weaknesses in software processes for waterfall, agile, spiral, etc., when it comes to performing verification and validation. Rather than designing a new software design process to mitigate the

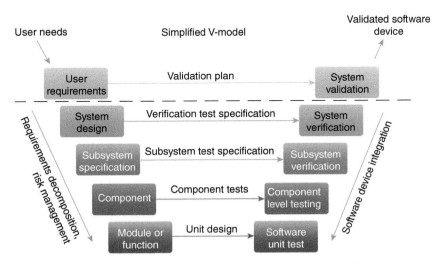

Figure 10.1 Simplified V-model Based on an Auto Tech Review Article "Verification and Validation Coverage for Autonomous Vehicle," Rath (2017).

risks associated with using conventional approach for CPS design, we apply model-based designs in which the reference specification is implemented as a model which can be executed in order to obtain the final implementation. In this modeling approach, often called Domain-Specific Modeling, the specification for how the system should be implemented is the same as the implementation (Sprinkle et al. 2009).

Let us explore how a model, as an abstract concept of a physical system, is created through a well-defined design process. These design processes are required to adhere to a set of rules in order to exploit them in a meaningful manner for a specific domain, e.g. in control engineering, manufacturing processing, and circuit design. The first technical step in the direction of creating a model that can be manipulated as part of a domain-specific language is metamodeling. Metamodels specify a set of rules that a model must adhere to (Sprinkle et al. 2010; Garitselov et al. 2012). This paves the way for creating models that can be manipulated by an end user, and also in creating a model that obeys certain syntactic rules – which restricts the kinds of models that can be created. By reducing the design space to the one in which models are more likely to be valid, it is likely that the models themselves can be verified and validated early in the design process. Once a metamodel is defined, it can be used to generate modeling artifacts that advance the design process. This may be software or code skeletons, or a configuration file generated via a transformation process. Rules of model transformation ensure that model analysis can be translated to model design and vice-versa.

After refining, models can be transformed into executable artifacts that can be deployed, a process referred to as Model Integrated Computing (MIC) (Sztipanovits and Karsai 1997; Karsai et al. 2003). Some commercially and open source tools are capable of performing the task of code-generation or configuration generated as a part of metamodeling such as IBM's Rational Rose (Quatrani 2002; Vidal et al. 2009), Eclipse Modeling Framework (Duddy et al. 2003), MathWork's Simulink (Dabney and Harman 2004), Generic Modeling Environment (Ledeczi et al. 2001b; Davis 2003; Molnár et al. 2007), MetaEdit+ (Tolvanen and Kelly 2009; Kelly et al. 2013), Modelio (Schranz et al. 2018). From this short discussion of metamodeling, we see that model-based engineering benefits heavily from metamodeling due to the structure that metamodeling provides to models, and the semantics that can be attached to metamodels. This has been proved to be an exciting development in Software Engineering for validation and verification of software systems specifically for physical systems.

Code generation in model-based design enables a model user to create models that could be implemented on a physical system. For systems like CPS, code generation is an indispensable tool where designers and engineers can build models of systems that are executable, harnessing the power of code-generators, also known as program synthesis environments (Ledeczi et al. 2001b; Tolvanen and

Kelly 2009). Such tools provide abstracted domain concepts and rules and a direct mapping from those concepts to code. A set of rules based on knowledge of domains may be developed to permit teams to specify relationships between heterogeneous blocks.

The key reflections of model-based design and its impact on V&V for CPS are closely related to the fact that for most CPSs, the final software component implementations must be tightly integrated with the dynamics of the physical system(s) being sensed and controlled. Thus designs will benefit greatly from domain-specific languages that are related to control and distributed systems, if those models can generate software. However, the integration testing required for CPSs frequently results in heterogeneous modeling languages that were not designed to work with one another: this ties back to the V&V approach of software-in-the-loop and hardware-in-the-loop testing, which provides a means by which pairwise component integration testing can be safely conducted before the final integrated system is tested.

10.2.3 Model-Based V&V

V&V of model-based designs is a well-developed area with a history of exploration that began with software systems. These encompass standards for abstraction of systems, formalizing state-machine models (Harel 1987), and creating specifications for systems and using them to verify properties of systems and relationships with interacting components of a system (Garland and Lynch 2001; Yilmaz 2017).

While cyber-physical systems need to incorporate both continuous and discrete time, conventionally progress has been made separately in the continuous and discrete domains. As a result, using a conventional V&V process for CPS may mean that the results do not deal with stochasticity, heterogeneity, and highly-coupled components of CPS systems that are safety-critical when their failure can lead to damage to lives and properties. When such results are available, it often comes at a high budget to implement, or a long time scale for development (Holt et al. 2017).

Verification of discrete-state systems is already well known in the area of hardware systems design and communication protocols (Clarke and Kurshan 1996). Extending these techniques to cyber-physical systems has been a challenge. A notable development in this area uses hybrid automaton and timed automaton are recent development in performing validation of a design of CPS systems (Sanfelice 2015).

One of the recent works in V&V include Statistical Model Checking (SMC) (David et al. 2012, 2015). SMC dictates that conduct simulation of the system, monitor them, and use statistical methods on gathered data to verify hypotheses with some degree of confidence on whether the system satisfies requirements or not. One such SMC tool called UPPAL-SMC uses networks of timed automata to conduct probabilistic analysis of system properties by monitoring the simulation of complex hybrid systems. In case of complex hybrid systems which are often multi-agent systems, formal verification methods are helpful in proving guaranteed safety and correctness of models. As an example, a platoon of connected vehicles can be considered a multi-agent system where a vehicle-to-vehicle (V2V) communication is needed for safe behavior of such systems. A hybrid automata approach is proposed for safety verification of such systems where continuous nature of systems is encapsulated in discrete states and discrete behaviors are represented by transitions between these states (Henzinger 2000). Improvement over hybrid automata model is Belief-Desire-Intention (BDI) model where agents act based on its beliefs, goals, and a set of event queue. As BDI models do not incorporate learnt behavior based on their past behavior (Kamali et al. 2017), proposes a verifiable BDI method in context of vehicle platooning. Another interesting work done in the area of formal verification for CPS is ModelPlex (Mitsch and Platzer 2016). Modelplex focuses on correct runtime validation that can be verified based on properties of models and ensures that those properties can provably be applied to CPS implementation. Modelplex utilizes correct-by-construction mechanism approach to monitor runtime behavior of a CPS system, which is also the core principle of Domain Specific Modeling Languages.

Another notable development towards V&V of cyber-physical systems is co-simulation. In co-simulation, subsystems from different domains that are known to be well-established are simulated together in a black box manner with suitable solvers (Neema et al. 2014; Zhang et al. 2014; Gomes et al. 2017). Co-simulation talks about theory and techniques to enable global simulation of coupled systems through the composition of the simulators. Each subsystem is developed and tested by respective domain experts and a composition is made through some algebraic constraints. Here, a master algorithm or an orchestrator is need to couple the system. This master algorithm specifies how data are moved between different subsystems. However, co-simulation denies the concepts of late coupling due to issues of large design gaps. In this case, co-simulations are required to contain model of all subsystems that can be integrated together at a later stage, models should be at appropriate level of abstractions so that simulation is not only accurate but efficient and it should allow model-based prototyping to improve usability (Zhang et al. 2014). This also enables verification of large scale system by sharing computational load among different co-simulators. The main challenge in co-simulation is for simulation to satisfy the property of correctness, i.e. if a co-simulation unit satisfies a property P, then how we should guarantee that overall simulation satisfies the property P. A full scale example of time-triggered automotive CPS using co-simulation has been discussed in Zhang et al. (2014) and interested reader are encouraged to refer to the article.

10.2.4 Application to Autonomy

In this chapter, we focus on aspects of model-based engineering with application to autonomous driving rather than dealing with theoretical analysis and formal methods. In Sections 10.3 and 10.4, we talk about the design cycle involving software-in-the-loop (SWIL) and hardware-in-the-loop (HWIL) testing. Section 10.5 presents use cases of autonomous driving from a perspective of SWIL and HWIL enabled model-based engineering. Section 10.6 presents Domain Specific Modeling Languages that can create a verifiable system, constraints to particular use cases with a pre-defined safety metrics to prevent failure of such systems under real-world operations. Section 10.7 summarizes the chapters with authors' findings and an insight on future development of model-based engineering.

10.3 Approaches to Model-Based Engineering

Model-Based Engineering (MBE) places models as the first class objects for system development. In MBE, the model is the core artifact of development practices, and encompasses everything from user requirement, system design, subsystem specification, component modeling, functional modeling with an aim to build right system and unit testing, component level testing, subsystem verification, and system validation to check whether whole process is realizing a right system. This property of MBE provides a well-defined relationship between different elements of a model through varying degree of abstraction, encapsulation, and functionalities. MBE provides an efficient methodology by integrating different steps of design cycle such as modeling of a system plant, synthesis and analysis of a controller for the plant, simulation of plant and controller, their implementation and deployment on the physical platform. With a model in the center of the design cycle, we formulate advance functional characteristics by using continuous-time and discrete-time computation building blocks. As an example, for developing a controller for a physical system, we can abstract the system as a black box with the system state defined by

$$x' = f\left(x, u\right) \tag{10.1}$$

$$y = g\left(x, u\right) \tag{10.2}$$

where x is the state of the system and u is the input to the system and y is the output, $f(\cdot)$ describes the system dynamics, and $g(\cdot)$ the output variables of the system.

In such a model, we are using abstraction to define the structure by which a system should be defined, but this kind of model provides the interfaces rather than the execution semantics. To make the model more executable, we define f and g, for brevity a model of $f(\cdot)$ for a car-like robot such as the one in Walsh et al. (1994) as

$$x' = f\left(x, u\right) = \begin{bmatrix} v\sin\theta \\ v\cos\theta \\ \dfrac{v\tan\delta}{L} \end{bmatrix} \tag{10.3}$$

where $x = (x_1, x_2, \theta)^T$ and $u = (v, \delta)^T$. θ represents the orientation of the vehicle, (x_1, x_2) is Cartesian position, input v is the vehicle's velocity, input δ is the steering angle, and the parameter L is the vehicle's wheelbase. When utilized with a simulation environment such as MATLAB/Simulink, this model is now executable, and can provide state update information for the location and orientation of the vehicle, with straightforward means by which the desired velocity and steering angle can be provided as an input.

This model, although executable and more specific than the one in Eq. (10.1), provides only the execution environment for design of a control algorithm that generates u: it does not provide means by which the environment can be sampled during testing, design of the actuators and their dynamics, etc. The application of model-based engineering aids in the complexity that comes with this overall process by enabling the composition of individual processes by which the system can be executed, tested, etc., through defined interfaces. The architecture also explicitly permits components within those interfaces to be replaced so that individual components can be replaced, redesigned, individually tested, etc., as part of the overall process.

There are various ways to measure the state of the system such as data-driven approach using system identification, and physics-based models; or we can use hardware-in-the-loop simulation and replace a computational model of the system by a physical platform/hardware. Once we abstract the system model of an autonomous vehicle, our task simplifies by one level with significant focus on requirements for the control input, output behavior, and controller development. This kind of model-based approach exploits system hierarchy and limits the complexity visible to the system designer at any time and reduces the search space during model interpretation (Abbott et al. 1993). A schematic of one such abstracted system is shown in Figure 10.2.

Each block developed as an abstracted element is called as a component. In MBE, component libraries are developed that can be assembled together to generate a system model at the platform level of abstraction. For example, MathWorks's modeling tool Simulink offers a rich set of component libraries with number of different interfaces for faster prototyping. These kinds of models with the support

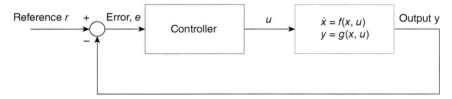

Figure 10.2 An abstracted system.

of simulation tools provide rapid prototyping, functional verification, software-testing, and software/hardware validation (Wilber 2006).

10.3.1 Workflows in Model-Based Engineering

Traditional workflows begin with requirement analysis as a documentation task either on a paper or in document files. Document based approaches experience limitations such as misinterpretation, ambiguity, and unstructured representation of information. It has been realized over the time that this leads to software and hardware teams finding themselves in a clear disconnect for requirements, implementations, and deployment (Hahn and Brunal 2017; Mall 2018; Ntanos et al. 2018). An example high-level workflow for such traditional approaches is illustrated in Figure 10.3.

During testing, gathered data are analyzed by engineers to figure out whether tests provided the expected result or something went wrong – was it a software problem or a fault in hardware? Were data out of the expected range, or is there a mismatch in the behavior when compared to the reference implementation? To resolve the issues, engineers update the model and re-run the test. In this whole process, it may be discovered very late that there are some flaws in the model, or perhaps even flaws in the requirements. It is evident that this is very time-consuming and expensive design technique.

The goal of MBE is to achieve a workflow that allows the system development to be expressed through models and translated to an executable system by means of software tools. The generated executable systems may consist of hardware or software elements, or both. They are often required to be computing-platform independent to target a wide variety of operating systems and architectures. The MBE approach is a plausible solution to manage the CPS development process involving inter-disciplinary topics by offering domain-specific tools and abstractions encompassing the system life-cycle. As CPS systems require knowledge of different disciplines, there is a growing desire to integrate these heterogeneous models. This kind of portability allows experts and designers to work with tools of their choices and preserve model compatibility with other domains. A standard MBE design process is illustrated in Figure 10.4. With this as one of the motivations for MBE, types of

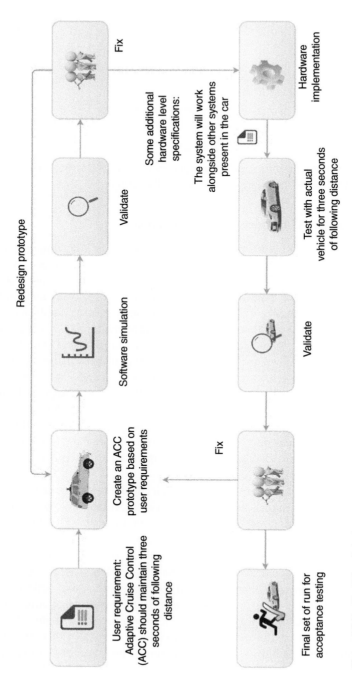

Figure 10.3 Traditional design process.

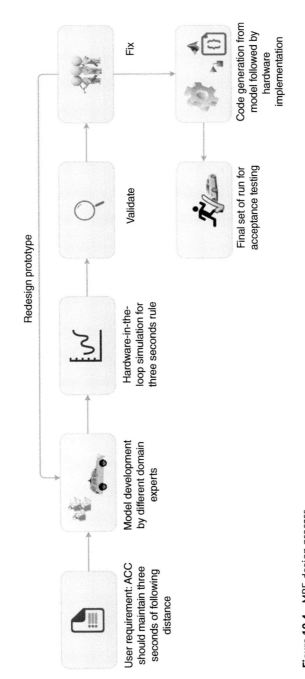

Figure 10.4 MBE design process.

models that may be employed in MBE process, ranging from highly abstract to domain-specific modeling languages (DSMLs), fall into two categories:

1) **The model of the system:** This represents an abstraction of the system of interest. This typically includes, but not limited to, definitions of behavior, performance, and structure of a specific system.
2) **A model about all systems:** This describes information or a set of rules that defines how models for a particular class of model may be created. This model is called a metamodel (Sztipanovits 2016).

10.3.2 Domain-Specific Modeling Environments

A Domain-Specific Modeling Environment (DSME) is a tool that can create custom DSMLs and allow for model creation (Nordstrom et al. 1999) (Ledeczi et al. 2001b). The use of such tools can greatly reduce the development time and use of a DSML through use of a model builder interface. Such tools also provide methods of model interpretation for code generation or analysis tools (Hemel et al. 2010). A language is defined by a syntax formal specification of component types and construction rules, and relationships between components. This syntax is the language formal specification and is the effective model of how domain models can be created, i.e. the metamodel. Figure 10.5 shows the relationship of various pieces involved in a DSME (Sprinkle and Karsai 2003). At the top left, the metamodel is defined by a domain expert that has extensive knowledge of how a DSML should be constructed. At this stage

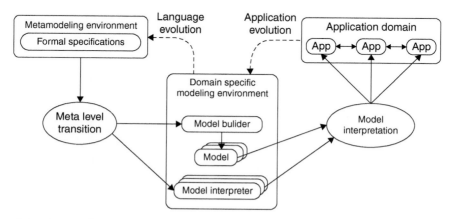

Figure 10.5 The process of DSML and model design using a DSME (Sprinkle and Karsai 2003).

they understand a model's possible set of high level components and how they are to be related. A metamodel is constructed using a DSME's own metamodel formal specification, known as a meta-metamodel (Álvarez et al. 2001). Different DSMEs may use a different schema for metamodel construction, sometimes in the form of XML schema or in the form of Unified Modeling Language (UML) class diagrams (Kobryn 1999; Routledge et al. 2002; Rumbaugh et al. 2004). Creation of a metamodel is then translated into components needed for model building and model interpretation. Often this stage is performed by a well-established DSME, where the metamodel is needed for model construction rules and constraints. The model builder provides the interface for a user to design models in the particular domain. This is similar to the usage of Simulink, where blocks are created and connected. During model building the model builder interface ensures that the rules of the metamodel are followed, thus enforcing models to be correct-by-construction. Important to note is that correct-by-construction in this context means that models will always fit into the metamodel and is therefore correct syntactically, however this does not mean that models are guaranteed to behave correctly. As an example, a DSML for a control system may have a metamodel that ensures that the edges are always connected between nodes, that there exists negative feedback, and that the dynamics are well-formed differential equations. The metamodel may however not ensure that the controller is stable, or that a particular constraint involving overshoot is not exceeded. Though the correct-by-construction ensures that a model follows a particular syntax, verification is an important step in meeting requirements. Part of the meta level transition can create a set of templates to aid in this process, but providing a set of model interpreters. These model interpreters are similar to compilers for code, where models are transformed into lower level artifacts at the top right of the figure. Such artifacts may include working output artifacts, or artifacts to be used for verification tools. Model interpretation is one of the key utility in a DSME due to producing artifacts from models, otherwise the models would exist purely as designs.

At the top of Figure 10.5, there are two dotted lines signifying Language Evolution and Application Evolution. Both of these are similar, and act as design feedback at different DSME stages. The application evolution is a model modification needed based on the feedback after generating output artifacts from a model. This may be the result of observing a system's behavior through a verification tool. Language evolution occurs when the DSML designer discovers that models in a particular domain are unable to be captured by the metamodel. The designer in this case must modify their metamodel to ensure a proper domain capture. Also, the interpreter may not be able to generate artifacts if they are unable to be modeled and thus the metamodel may need to be modified.

10.4   Modeling and Simulation in Model-Based Engineering

Modeling a dynamic behavior requires capturing a wide variety of use cases with number of different parameters and initial conditions. Simulation of model-based design ensures that those behaviors are captured in the model and tested for compliance with requirements and safety. As a result, a computational model is becoming increasingly popular approach that cut cost and reduces design time by facilitating numerical simulation as an alternative to building a hardware prototype. Once a computational model is finalized, a number of simulations are performed to test and refine the design before finalizing the implementation. This is a core principle of Software-in-the-loop (SWIL) simulation. SWIL allows multiple simulations with different parameters which is otherwise expensive and sometimes impossible with physical platforms. In safety-critical systems such as autonomous vehicles and medical CPS, simulation enables testing of unstable conditions without any risk to infrastructure or human lives. However, often times a part of the system is mathematically complicated to model or their dynamics are unknown. In such cases, we interface the sensor or hardware to be modeled with the computational model. This combined approach of modeling and simulation gives rise to the phenomenon of Hardware-in-the-loop (HWIL) simulation.

10.4.1   Computational Modeling in Model-Based Engineering

The use of computational modeling to simulate and study the behavior of complex CPS processes using mathematical equations and physics engine presents an ability to expedite system design and test new controllers and algorithms. Computational modeling contains numerous variables and simulations are performed by adjusting different variables often one or a combination of them and making inferences from observations. Simulation using computational modeling helps making predictions about system with a particular "what-if" scenario which is otherwise challenging and may pose financial, logistics, and life-threatening challenges.

This can be understood with an example of simulating an autonomous vehicle using ROS (Quigley et al. 2009) and the Gazebo simulator (Koenig and Howard 2004). Consider designing a vehicle-following algorithm for an autonomous vehicle (AV). To test a vehicle-following algorithm, we can conduct a real-life experiment consisting of an AV and a human-driven vehicle. The idea behind the vehicle-following algorithm is for the autonomous vehicle to follow a human-driven vehicle maintaining a safe-following distance. For the real-life experiment, we will also require a laser-based range-finder device such as a SICK LMS 291 or a radar mounted on the front bumper of the AV. The range-finder is required to estimate the distance between the AV and preceding human-driven vehicle. This distance information can be fed into a vehicle-following algorithm to adjust the speed to maintain a safe-following distance.

Conducting a real-life experiment requires a confidence about vehicle-following algorithm to ensure that the AV doesn't collide with the preceding vehicle at any given point. The guaranteed-safety of the algorithm demands that we find an alternative approach to develop confidence well before testing. One such alternative is employing computational modeling in the Model-based Engineering to design and test control algorithms for individual component processes. Now the question is, what are the things we require to model to design and implement vehicle-following algorithm? The following discussion presents an overview of environment modeling, component modeling, and sensor modeling necessary for model-based engineering of autonomous vehicle applications.

10.4.1.1 Environmental Modeling

We begin with a broad understand of environment where agents involved in the experiment are going to interact with each other and external surroundings. There are number of commercial and open-source tools available for environment modeling such as Gazebo (Koenig and Howard 2004), Carsim (Benekohal and Treiterer 1988), Webots (Michel 2004), V-rep (Rohmer et al. 2013), to name a few. In the example presented in this chapter, we use the Gazebo simulator. A snapshot of environmental modeling simulator from Gazebo is shown in Figure 10.6. In most cases, the environmental setup is created in order to exercise the test cases that drive the functional simulations, rather than a realistic simulator for any use case.

With the environmental model in place, the reader can see that portions of the earlier components for vehicle state (e.g. (x, y, θ), v, as described in Eq. (10.3)) can be gathered from the simulator's execution of vehicle dynamics. The environmental model is additionally required to enable realistic data gathering from simulated sensors, for distances between the vehicle and other objects, perhaps even simulated motion of those objects.

10.4.1.2 Component Modeling

A computational model is required for modeling and simulation of an autonomous vehicle. An oversimplified model of a vehicle uses a 3D box shaped body and a kinematic or dynamical equation such as that described in Eq. (10.3). Alternatively, an actual 3D model of vehicle chassis developed using 3D modeling software such as Solidworks (2019) or Blender can be used. Steering mechanism is based on revolute joints and follows Ackermann principle of differential steering on its inner and outer tires. For a realistic appearance, the visual component of main body is provided with a 3D chassis model of Ford Escape vehicle developed in Solidworks. Using Gazebo simulator to create 3D model of the autonomous

Figure 10.6 An example of environment modeling in Gazebo.

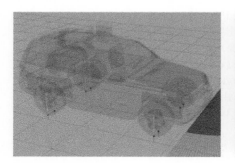

Figure 10.7 Computational model of an autonomous vehicles; left side shows skeleton of the model and right shows solid view.

enables a designer to specify physical properties of a model such as inertia, coefficient of friction, damping coefficient etc. Figure 10.7 presents an example of such model used for designing vehicle-following algorithm in a realistic setting. In this model, tire has been modeled as a cylinder to reduce the computational cost that otherwise comes from calculating contact forces with an actual tire models which can be thousands of mesh triangles to construct tread of tires.

Additionally developed components will include controllers designed to issue control commands to the vehicle, and to interpret data from sensors. When using domain-specific approaches, these can be designed using control software such as Simulink to prototype the design, and then code generation to provide the executable behavior.

10.4.1.3   Sensor Modeling

For autonomous driving applications, sensor modeling is as significant as the modeling of a vehicle. For a RangeFinder sensor, a point-cloud model based on ray-tracing and ray-casting is used in Gazebo that has the capability to include the stochastic nature of a sensor for realistic data-collection. Sensor modeling for the application of computer vision is a wider topic of synthetic vision systems which is out of the scope of this chapter. Interested readers may refer to (Lohani and Mishra 2007; Peinecke et al. 2008; Davar et al. 2017). Here, we will briefly discuss modeling of LiDAR sensor that is one of the widely used component in autonomous and semi-autonomous vehicle. Testing an autonomous vehicle application requires realistic LiDAR data but due to logistical issues, having a computational model of LiDAR increases the use-case coverage.

Computational modeling of LiDAR uses ray-casting technique. In ray-casting, a 3-D vector is used to check intersections with other geometries in the virtual environment. A coordinate of the intersected point is recorded. Using multiple 3-D vectors, a three-dimensional image of intersecting rays from the LiDAR response is constructed. A collection of coordinate points generated from ray-casting is called a point cloud. As ray-casting is not done in parallel in all physics engines, the engine must wait to produce the visualization until a 360° ray-casting is complete. As a result, modeling performance is tied to the available computing power. Some physics engines take the approach of updating the visualization of point-cloud as often as coordinate points from ray-casting are available. Some open-source libraries are available for simulating the computational model of sensors such as Velodyne 3D LiDAR and SICK LMS 291 2D LiDAR. One such open source project is offered under the ROS platform. Figure 10.8a and 10.8b display simulation of raytracing and points clouds in visualization.

10.4.2   Software-in-the-Loop Simulation

Cyber-Physical Systems are heterogeneous by their nature and interact with physical world via myriad sensing and actuation devices. Even though we manage to accomplish interfacing physical devices with computer system during the development process, critical and risky use cases may not be handled.

Early in the design stage and/or to capture use-cases with varying conditions more difficult to create in reality, simulations are used. These simulations must be

(a)   (b)

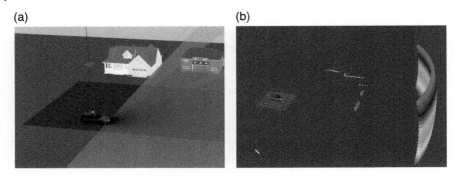

Figure 10.8 (a) Raytracing of simulated LiDAR sensor in Gazebo World (b) Point cloud visualization obtained from simulated LiDAR sensors in rviz.

tightly integrated with the software infrastructure to accelerate productivity and reduce ambiguity, in order to ensure that the functional interactions work through the interfaces that will be used in deployment.

Within the example of developing controllers for an autonomous vehicle, the development of software components is substantially complicated when each component needs to interact with a physical device or a sensor. Early in the development cycle interaction may take place through purely software testing. Once the functional behavior is tested using traditional software engineering approaches, it is now necessary to transition to interfacing with external emulators and simulators.

Due to these limitations, virtual prototypes (developed in software) provide an alternative for developing and testing full-scale applications through a process called software-in-the-loop simulation or SWIL. These virtual prototypes can be readily modified and tested under wide variety of inputs and conditions without any risk. SWIL allows testers to create a number of test scenarios such as different traffic conditions, road conditions, presence or absence of external factors, etc.

10.4.3 Hardware-in-the-Loop Simulation

Software-in-the-loop simulation offers great capability in reducing system-level bugs and reduces the cost that comes from troubleshooting in later stages otherwise. Once SWIL testing has completed, it is necessary to transition some of those SWIL prototypes to physical hardware. This process of replacing one or more software components by a representative physical device is known as hardware-in-the-loop simulation or HWIL.

We can explore this concept with consideration of components that require sensor data. Having simulated a sensor in early stages of development is good, but sensors such as LiDAR, radar, and cameras suffer from stochastic and shot noises which are difficult to model computationally with precision without sacrificing performance. In that case, we may want to replace a simulated sensor with an actual hardware device, keeping the rest of software component (for example, the vehicle model), connected in place.

In some cases, a newer sensor model may not have a simulated version available, since developing a computational model of a sensor is a daunting task and requires thorough understanding of physics and electronics. In HWIL, a virtual prototype can be replaced by its equivalent physical device, and real data obtained from physical devices can be injected to reduce dependence on a synthetic dataset. In our vehicle-follower algorithm example, the velocity controller requires data from LiDAR mounted on the front bumper of the AV to estimate the distance and relative velocity of the target.

In the SWIL approach, a vehicle model in simulation receives distance information from a simulated LiDAR which doesn't represent stochastic nature of the data and doesn't exhibit perturbation in dataset due to absence of random non-flat road conditions. By injecting data gathered from a real LiDAR mounted on a real vehicle, we can fill this gap and present new scenario to the controller under development.

10.5 Use Case: Velocity Control of Automotive CPS

Vehicles have gone beyond mechanical devices to become automotive cyber-physical systems, equipped with electronic components and control systems. There are a number of limitations and challenges in automotive CPS in the current state-of-art, for example the ability to collect global information about the environment in which they operate. On-board sensors are typically designed for in-vehicle information gathering, and the data they gather are valid for that particular execution only, since closing the loop with a different control would likely alter the data in the next sample period.

The goal to provide high confidence automotive CPS that are self-driving has potential advantages for mobility as well as safety and reliability (Work et al. 2008). Some of the problems discussed here have been an active area of research that incorporates existing and new technologies such as Road-Side Units (Lochert et al. 2008), DSRC (Shin et al. 2019), use of LTE (Mazzola et al. 2016) and 5G network (Mavromatis et al. 2018) for dissemination of information, formation and control of vehicular platooning, etc. In this section, we present how MBE is proving indispensable tool in modeling, verifying and validation of components used in autonomy for self-driving cars, which will be required to bring the next generation of automotive CPS to people.

We next discuss a very specific use case of controlling velocity of a car-like vehicle, in order to provide a well-defined system that can be used to validate the reaction of a following vehicle with adaptive cruise control (ACC). We are interested in determining the behavior of an ACC vehicle whose feedback control law is unknown. To achieve the objective, we set up several tests with the help of two ACC vehicles, a leader and a follower. The lead vehicle can execute velocities consistently, and its controller is designed to exercise the follower vehicle in order to determine its (thus far unknown) control law. We start with specifying a set of inputs, and how inputs change with respect to time.

This requirement is translated into a state model (implemented in Simulink Stateflow) as shown in Figure 10.9. The outputs of this model are the inputs to the

```
1) Start from rest.
2) Accelerate to 20 mph at 0.5 m/s².
3) Maintain speed at 20 m/s for 30 seconds.
4) Increment speed by 10 mph.
5) Maintain speed for 30 seconds.
6) Repeat steps 4-7 until speed is 60 mph.
7) Decelerate to 0 mph at 1.0 m/s².
8) If speed is ε≈0, terminate.
```

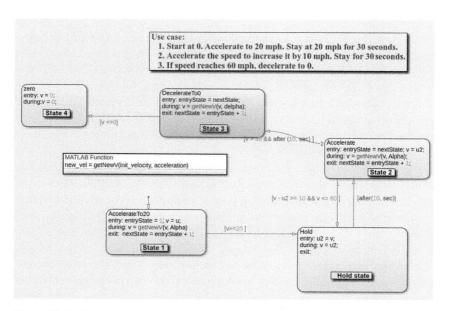

Figure 10.9 A state machine with set specified inputs for ACC example, implemented in Stateflow.

vehicle's velocity controller. First, these are tested on a vehicle model within Simulink using the model developed in Eq. (10.3) to check the traces of the output. These software tests are integrated within the Simulink execution environment, and do not require explicit code generation to take place in order to test: the execution semantics of Simulink are sufficient.

Once this is verified through input/output testing, the system test progresses to SWIL through a simulated environment as illustrated in Figure 10.10. In order to execute the SWIL tests, the Stateflow model is interpreted into a C++ based ROS component, using the Robotic System Toolkit and its interface to our Gazebo environment (Bhadani et al. 2018). With SWIL, the vehicle dynamics and sensor data are generated in Gazebo using its physics model, the sensor data (elapsed time, as well as position and velocity) come through ROS and are generated in Gazebo, and the controller's output comes from the generated C++ code as a commanded velocity, which is shared to ROS which sets rotational velocities for each of the

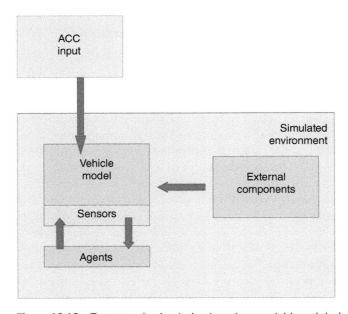

Figure 10.10   Test setup in simulation based on model based design. ACC input modeled state machine is provided to vehicle model residing within simulated environment. Vehicle's sensors interact with agents to provide information to vehicle. External components shown above are like agents but there is one-way interaction only as contrary to two-way interaction with agents. Further exact physical phenomenon of interaction with external components is modeled as unknown to represent behaviors such as road-grade, snow, etc.

wheels to achieve the desired vehicle velocity. With SWIL testing, the code is not verified per se, but integration errors can be discovered.

Once we build enough confidence about the inputs generated from our controller, we perform tests with the leader vehicle without the follower. This HWIL testing begins with driving the vehicle under fully human control, but checking the output from the controller to compare those outputs to the ones produced by the human driver. This kind of HWIL testing determines whether the data coming from the physical hardware are of sufficient fidelity and frequency, and may also uncover minor errors in accumulated noise or in disturbances that are not covered by the SWIL system. Following successful HWIL integration tests, we process to a HWIL test with the entire system, with the output of our controller hooked up to the physical system. As we run our customized velocity controller, we record the outputs of the controller in order to apply those HWIL outputs as HWIL inputs to our later designs of the inferred ACC controller of the follower vehicle, to determine whether our estimates of the follower ACC model are accurate.

10.6 Use Case: Domain Specific Modeling Language for CPS Design

Often in the case of a CPS, requirements may span over multiple design domains. For example, a traditional controller may be designed to operate a vehicle's steering for lane following based on camera images, but delay needs to be considered due to the delay between networked computers or algorithm computation time. Should an issue exist in this system where the initial design behaves incorrectly, a designer may attempt to either redesign the controller to be more robust, redesign the network system to reduce latency, or optimize computational algorithms. Design can be done independently across such multiple domains; however, this may require a domain expert in each particular field. Furthermore, individual experts in each domain may be required to translate the design from the high level design into the low level working artifacts. Such a process is prone to human error and can be very time consuming.

Domain experts can instead codify their knowledge for a specific platform by designing a Domain-Specific Modeling Language (DSML) (Kelly et al. 2013). DSML creation involves the design of a syntax for a high level models (Chen et al. 2005). Along with a formal syntax a semantic mapping provides the necessary representation of models needed for producing low level artifacts. A model constructed using a DSML may be represented visually, and closely match generic modeling methodologies (Lédeczi et al. 2001a). As an example, Figure 10.2

demonstrates a domain of control systems engineering. This simple model of a control system is commonly used model of a feedback control system and abides by a particular syntax and semantic mapping. Each graph node represents a math formula, each edge has a directional arrow, and some edges have a label. Semantically, a controls engineer understands how there exists a system input and output, the difference between the plant and controller, and the existence of a feedback loop. Simulink is an example use case of a DSML where control systems may be designed. Expert knowledge has also been implemented to both simulate Simulink models and generate low level artifacts in the form of C code.

Code generation and verification are two key utilities that a well-constructed DSML can provide for the development cycle of a complex, safety-critical system. An interpreter designed for model verification can transform the model into artifacts necessary to ensure that requirements are met. If a model is verified to have a safe behavior through valid verification tools, then proper code generators will only produce safe code. Some implementations of a DSML with verification interpreters can even specifically prevent the execution of code generation until verification is passed (Bunting et al. 2016). This manner of preventing unsafe code from being generated ensures that the iterative design process loop in Figure 10.4 is complete before feeding the models forward for deployment.

10.6.1 An Example DSML in WebGME

The Web-Based Generic Modeling Environment (WebGME) is an example tool that allows for the metamodel to be defined visually using the UML class-like diagrams (Maróti et al. 2014; Gray and Rumpe 2016). WebGME uses a web browser and a web-server with a database backend to store DSMLs, models, and interpreters. Being web-based, the interpreters and other plugins are written using Javascript. Plugins other than model interpretation may be used to decorate model objects to represent components in a domain in visually appealing manner.

Figure 10.11 is an example DSML of a simple hybrid controller designer using WebGME. This is a very simple hybrid controller representation and is specifically intended for creating paths for the autonomous vehicle shown in Figure 10.7. The metamodel of the hybrid controller is on the left. This is the appearance of metamodel construction using the class-like diagrams in WebGME. Software engineers familiar with UML design will quickly understand how to construct metamodels with little instruction, though there are differences between WebGME metamodels and class diagrams. Since WebGME is built specifically for model building, some of the relationships between classes appear differently than in UML. In this design, a diagram is the top-most class that contains modes and transitions. Transitions, following from state diagram design, represent transitions from one control mode to another, and thus has relationships designating a source mode src and destination mode dst. There are two special modes, start and end that inherit from a standard mode. Lastly, each mode contains a controller with settable parameters. Using this metamodel, WebGME's model builder can let users design a simple hybrid controller model shown in the middle pane of Figure 10.11. The model is a simple sequence of modes, and one mode's parameters are shown. Due to the model builder and metamodel, the modes and transitions are guaranteed to be constructed correctly, but the tuning of controller may exhibit unsafe behavior.

The right pane in Figure 10.11 shows both generated MATLAB and verification output artifacts from the model. For each controller mode, the interpreter generates a verification code for ensuring certain safety constraints are not violated. This language has two parts to be verified, the first is to check that a particular goal is reached and the second is to ensure that controller tunings are not overly aggressive. Upon a successful verification, code generators can produce artifacts for a vehicle to be controlled based on the model of the hybrid controller.

This language was constructed to be simple and not include all generic parts of hybrid control design since the end users are non-experts in control theory. The intention of these models is for a known autonomous vehicle with only times transition and a specific controller type, thus the language is highly domain-specific. By reducing the design complexity through simplifying the design space, non-experts may still design behaviors for complex CPSs, including students as young as in grade 4 (Bunting et al. 2016). In this case, a non-expert will only be able to specify modes with properly connected transitions, and will only be able to set particular parameters.

10.7 Conclusion and Insight

In this chapter, we have explored the roles of model-based simulation and design in autonomous cyber-physical systems, and how they differ from traditional design of software systems. The heterogeneity of CPSs motivates the use of software-in-the-loop and hardware-in-the-loop testing to reduce errors often uncovered through integration testing, and to have realistic error and noise conditions inserted into software components early in the design process. Importantly, the application of domain-specific languages – either well-known examples such as Simulink/Stateflow, or customized languages such as those developed through WebGME – permits end-users to develop references implementations in a language much closer to the design specification, and code generation reduces the time needed to transform those references implementations into the final software components.

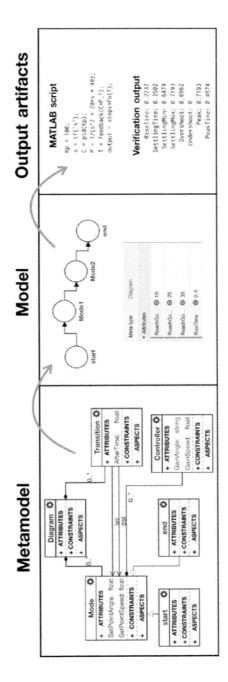

Figure 10.11 An example DSML of a simple hybrid control system with verification script generation.

Designers of autonomous systems are faced with important questions when it comes to addressing the scope and scale of societal-scale systems, in terms of the safety of those systems and the need to have rigorous examination of the software that composes their behavior. The continued use of models to build those behaviors enables verification tools to have defined behaviors that can be examined, rather than depend upon black-box implementations where the corner cases are less likely to be discovered until catastrophic failure.

Acknowledgments

This work was supported by the National Science Foundation under awards 1446435, 1253334, and by a gift from The MathWorks. Additional thanks are given to Nancy Emptage for her support of hardware-in-the-loop experiments that permitted the many models of the system to be tested.

References

Abbott, B., Bapty, T., Biegl, C., Karsai, G., & Sztipanovits, J. (1993). Model-based software synthesis. *IEEE Software, 10*(3), 42–52.

Al Faruque, M. A., & Ahourai, F. (2014). A model-based design of cyber-physical energy systems. In *Proceedings of the 19th Asia and South Pacific Design Automation Conference (ASP-DAC)* (pp. 97–104). IEEE.

Álvarez, J. M., Evans, A., & Sammut, P. (2001). Mapping between levels in the metamodel architecture. In *Proceedings of the 4th International Conference on the Unified Modeling Language* (pp. 34–46). Springer.

Ammann, P., & Offutt, J. (2016). *Introduction to software testing.* Cambridge University Press.

Anta, A., Majumdar, R., Saha, I., & Tabuada, P. (2010). Automatic verification of control system implementations. In *Proceedings of the 10th ACM International Conference on Embedded Software* (pp. 9–18). ACM.

Benekohal, R., & Treiterer, J. (1988). Carsim: Car-following model for simulation of traffic in normal and stop-and-go conditions. Transportation Research Record, no. 1194.

Bhadani, R., Sprinkle, J., & Bunting, M. (2018, April). The CAT Vehicle Testbed: A Simulator with Hardware in the Loop for Autonomous Vehicle Applications. In *Proceedings 2nd International Workshop on Safe Control of Autonomous Vehicles (SCAV 2018)*, Electronic Proceedings in Theoretical Computer Science (vol. 269). Porto, Portugal.

Blender Online Community (n.d.). Blender: A 3D modelling and rendering package. Blender Foundation, Blender Institute, Amsterdam [Online]. Available: http://www.blender.org

Bunting, M., Zeleke, Y., McKeever, K., & Sprinkle, J. (2016). A safe autonomous vehicle trajectory domain specific modeling language for non-expert development. In *Proceedings of the International Workshop on Domain-Specific Modeling* (pp. 42–48). ACM.

Chen, K., Sztipanovits, J., & Neema, S. (2005). Toward a semantic anchoring infrastructure for domain-specific modeling languages. In *Proceedings of the 5th ACM International Conference on Embedded Software*. ACM, pp. 35–43.

Clarke, E. M., & Kurshan, R. P. (1996). Computer-aided verification. *IEEE Spectrum*, *33*(6), 61–67.

Clarke, E. M., Jr., Grumberg, O., Kroening, D., Peled, D., & Veith, H. (2018). *Model checking*. MIT Press.

Dabney, J. B., & Harman, T. L. (2004). *Mastering Simulink*. Pearson.

Dassault Systemes SolidWorks Corporation (2019). SolidWorks 2019 Proven Design to Manufacture Solution [Online]. Available: https://www.solidworks.com/

Davar, S. A., Chemander, T., Jansson, M., Jonathan Laurenius, R., & Tibom, P. (2017). Virtual generation of LiDAR data for autonomous vehicles.

David, A., Du, D., Larsen, K. G., Legay, A., Mikučionis, M., Poulsen, D. B., & Sedwards, S. (2012). Statistical model checking for stochastic hybrid systems, arXiv preprint arXiv:1208.3856.

David, A., Larsen, K. G., Legay, A., Mikučionis, M., & Poulsen, D. B. (2015). Uppaal SMC tutorial. *International Journal on Software Tools for Technology Transfer*, *17*(4), 397–415.

Davis, J. (2003). GME: The generic modeling environment. In *Companion of the 18th annual ACM SIGPLAN Conference on Object-Oriented Programming, Systems, Languages, and Applications* (pp. 82–83). ACM.

Duddy, K., Gerber, A., & Raymond, K. (2003). Eclipse Modelling Framework (EMF) import/export from MOF, JMI. Technical report, CRC for Enterprise Distributed Systems Technology (DSTC).

Emerson, M. J., Sztipanovits, J., & Bapty, T. (2004). A MOF-Based Metamodeling Environment. *Journal of Universal Computer Science*, *10*(10), 1357–1382.

Garitselov, O., Mohanty, S. P., & Kougianos, E. (2012). A comparative study of metamodels for fast and accurate simulation of nano-CMOS circuits. *IEEE Transactions on Semiconductor Manufacturing*, *25*(1), 26–36.

Garland, S. J., & Lynch, N. A. (2001, September 11). Model-based software design and validation.*US Patent 6,289,502*.

Gomes, C., Thule, C., Broman, D., Larsen, P. G., & Vangheluwe, H. (2017). Co-simulation: State of the art, arXiv preprint arXiv:1702.00686.

Gray, J., & Rumpe, B. (2016). The evolution of model editors: Browser-and cloud-based solutions.

Hahn, C., & Brunal, E. M. (2017). Matlab and simulink racing lounge: Vehicle modeling. MATLAB and Simulink Racing Lounge [Online]. Available: https://web.archive.org/web/20190109221627/https://www.mathworks.com/videos/matlab-and-simulink-racing-lounge-vehicle-modeling-part-1-simulink-1502466996305.html

Harel, D. (1987). Statecharts: A visual formalism for complex systems. *Science of Computer Programming, 8*(3), 231–274.

Hemel, Z., Kats, L. C., Groenewegen, D. M., & Visser, E. (2010). Code generation by model transformation: A case study in transformation modularity. *Software & Systems Modeling, 9*(3), 375–402.

Henzinger, T. A. (2000). The theory of hybrid automata. In *Verification of digital and hybrid systems* (pp. 265–292). Springer.

Holt, S., Collopy, P., & DeTurris, D. (2017). So it's complex, why do I care? In *Transdisciplinary Perspectives on Complex Systems* (pp. 25–48). Springer. [Online]. Available: https://doi.org/10.1007/978-3-319-38756-7_2

Jackson, E. K., Levendovszky, T., & Balasubramanian, D. (2011). Reasoning about metamodeling with formal specifications and automatic proofs. In *Proceedings of the International Conference on Model Driven Engineering Languages and Systems* (pp. 653–667). Springer.

Kamali, M., Dennis, L. A., McAree, O., Fisher, M., & Veres, S. M. (2017). Formal verification of autonomous vehicle platooning. *Science of Computer Programming, 148*, 88–106. special issue on Automated Verification of Critical Systems (AVoCS 2015). [Online]. Available: http://www.sciencedirect.com/science/article/pii/S0167642317301168

Karsai, G., Agrawal, A., & Ledeczi, A. (2003). A metamodel-driven MDA process and its tools. In *Workshop in Software Model Engineering*.

Kelly, S., Lyytinen, K., & Rossi, M. (2013). Metaedit+ a fully configurable multi-user and multi-tool case and came environment. In *Seminal contributions to information systems engineering* (pp. 109–129). Springer.

Kobryn, C. (1999). UML 2001: A standardization odyssey. *Communications of the ACM, 42*(10), 29–37.

Koenig, N. P., & Howard, A. (2004). Design and use paradigms for gazebo, an open-source multi-robot simulator. In *IROS* (vol. 4, pp. 2149–2154). CiteSeerX.

Lédeczi, Á., Bakay, A., Maroti, M., Volgyesi, P., Nordstrom, G., Sprinkle, J., & Karsai, G. (2001a). Composing domain specific design environments. *Computer, 34*(11), 44–51.

Ledeczi, A., Maroti, M., Bakay, A., Karsai, G., Garrett, J., Thomason, C., … Volgyesi, P. (2001b). The generic modeling environment. In *Workshop on Intelligent Signal Processing* (vol. 17, p. 1). Budapest, Hungary.

Lédeczi, Á., Nordstrom, G., Karsai, G., Volgyesi, P., & Maroti, M. (2001c). On metamodel composition. In *Proceedings of the 2001 IEEE International Conference on Control Applications (CCA)* (pp. 756–760).

Lochert, C., Scheuermann, B., Wewetzer, C., Luebke, A., & Mauve, M. (2008). Data aggregation and roadside unit placement for a VANET traffic information system. In *Proceedings of the 5th ACM International Workshop on VehiculAr Inter-NETworking* (pp. 58–65). ACM.

Lohani, B., & Mishra, R. (2007). Generating LiDAR data in laboratory: LiDAR simulator. *International Archives of Photogrammetry an Remote Sensing and Spatial Information Sciences, 52*, 12–14.

Lonetti, F., & Marchetti, E. (2018). Emerging software testing technologies. In *Advances in computers* (Vol. *108*, pp. 91–143). Elsevier.

Mall, R. (2018). *Fundamentals of software engineering*. PHI Learning.

Maróti, M., Kereskényi, R., Kecskés, T., Völgyesi, P., & Lédeczi, A. (2014). Online collaborative environment for designing complex computational systems. *Procedia Computer Science, 29*, 2432–2441.

Mavromatis, I., Tassi, A., Rigazzi, G., Piechocki, R. J., & Nix, A. (2018). Multi-radio 5g architecture for connected and autonomous vehicles: Application and design insights, arXiv preprint arXiv:1801.09510.

Mazzola, M., Schaaf, G., Stamm, A., & K¨urner, T. (2016). Safety-critical driver assistance over LTE: Toward centralized ACC. *IEEE Transactions on Vehicular Technology, 65*(12), 9471–9478.

Michel, O. (2004). Cyberbotics Ltd. Webots: Professional mobile robot simulation. *International Journal of Advanced Robotic Systems, 1*(1), 5.

Mitsch, S., & Platzer, A. (2016). Modelplex: Verified runtime validation of verified cyber-physical system models. *Formal Methods in System Design, 49*(1–2), 33–74.

Molnár, Z., Balasubramanian, D., & Lédeczi, Á. (2007). An introduction to the generic modeling environment. In *Proceedings of the TOOLS Europe 2007 Workshop on Model-Driven Development Tool Implementers Forum*.

Neema, H., Gohl, J., Lattmann, J., Sztipanovits, J., Karsai, G., Neema, S., ... Sureshkumar, C. (2014). Model-based integration platform for FMI co-simulation and heterogeneous simulations of cyber-physical systems. In *Proceedings of the 10th International Modelica Conference* (pp. 235–245); March 10–12; 2014; Lund; Sweden, no. 096. Linköping University Electronic Press.

Nordstrom, G., Sztipanovits, J., Karsai, G., & Ledeczi, A. (1999). Metamodeling-rapid design and evolution of domain specific modeling environments. In *Proceedings of the IEEE Conference and Workshop on Engineering of Computer-Based Systems, 1999 (ECBS'99)* (pp. 68–74). IEEE.

Ntanos, E., Dimitriou, G., Bekiaris, V., Vassiliou, C., Kalaboukas, K., & Askounis, D. (2018). A model-driven software engineering workflow and tool architecture for servitised manufacturing. *Information Systems and e-Business Management, 16*, 1–38.

Peinecke, N., Lueken, T., & Korn, B. R. (2008). LiDAR simulation using graphics hardware acceleration. In *Digital Avionics Systems Conference* (pp. 4–D). DASC 2008. IEEE/AIAA 27th. IEEE.

Quatrani, T. (2002). *Visual modeling with rational ROSE 2002 and UML*. Addison-Wesley Longman Publishing.

Quigley, M., Conley, K., Gerkey, B., Faust, J., Foote, T., Leibs, J., ... Ng, A. Y. (2009). ROS: An open-source Robot Operating System. In *ICRA Workshop on Open Source Software* (vol. 3, no. 3.2, p. 5). Kobe, Japan.

Rath, M. (2017). Verification and validation coverage for autonomous vehicle. Auto Tech Review [Online]. Available: https://web.archive.org/web/20190109204743/ https://autotechreview.com/opinion/guest-commentary/verification-and-validation-coverage-for-autonomous-vehicle

Rohmer, E., Singh, S. P., & Freese, M. (2013). V-rep: A versatile and scalable robot simulation framework. In *Proceedings of the IEEE/RSJ International Conference on Intelligent Robots and Systems (IROS)* (pp. 1321–1326). IEEE.

Routledge, N., Bird, L., & Goodchild, A. (2002). UML and XML schema. *Australian Computer Science Communications*, *24*(2), 157–166. Australian Computer Society.

Rumbaugh, J., Jacobson, I., & Booch, G. (2004). *The Unified Modeling Language reference manual*. Pearson Higher Education.

Sanfelice, R. G. (2015). Analysis and design of cyber-physical systems: A hybrid control systems approach. In *Cyber-Physical Systems* (pp. 3–31). CRC Press.

Schranz, M., Bagnato, A., Brosse, E., & Elmenreich, W. (2018). Modelling a CPS Swarm System: A Simple Case Study. In *Proceedings of the 6th International Conference on Model-Driven Engineering and Software Development (MODELSWARD 2018)*.

Shin, D., Kim, B., Yi, K., Carvalho, A., & Borrelli, F. (2019). Human-centered risk assessment of an automated vehicle using vehicular wireless communication. *IEEE Transactions on Intelligent Transportation Systems*, *20*(2), 667–681.

Sprinkle, J., Mernik, M., Tolvanen, J.-P., & Spinellis, D. (2009). Guest editors' introduction: What kinds of nails need a domain-specific hammer? *IEEE Software*, *26*(4), 15–18. Available: http://dx.doi.org/10.1109/MS.2009.92

Sprinkle, J., Rumpe, B., Vangheluwe, H., & Karsai, G. (2010). *Metamodelling*. Lecture Notes in Computer Science. (Vol. *6100*, ch. 4, pp. 59–78). Springer. [Online]. Available: http://www.springer.com/computer/swe/book/978-3-642-16276-3

Sprinkle, J. M., & Karsai, G. (2003). Metamodel driven model migration. (PhD dissertation). Citeseer.

Sztipanovits, J. (2016). Metamodeling [Online]. Available: https://web.archive.org/web/20190109231001/http://w3.isis.vanderbilt.edu/Janos/CS388/Presentations/Metamodeling.pdf

Sztipanovits, J., & Karsai, G. (1997). Model-integrated computing. *Computer*, *30*(4), 110–111.

Tabuada, P. (2009). *Verification and control of hybrid systems: A symbolic approach.* Springer Science & Business Media.

Tolvanen, J.-P., & Kelly, S. (2009). Metaedit+: Defining and using integrated domain-specific modeling languages. In *Proceedings of the 24th ACM SIGPLAN Conference Companion on Object Oriented Programming Systems Languages and Applications* (pp. 819–820). ACM.

Vidal, J., De Lamotte, F., Gogniat, G., Soulard, P., & Diguet, J.-P. (2009). A co-design approach for embedded system modeling and code generation with UML and MARTE. In *Proceedings of the Conference on Design, Automation and Test in Europe. European Design and Automation Association* (pp. 226–231).

Walsh, G., Tilbury, D., Sastry, S., Murray, R., & Laumond, J.-P. (1994). Stabilization of trajectories for systems with nonholonomic constraints. *IEEE Transactions on Automatic Control, 39*(1), 216–222.

Wilber, J. (2006). Bae systems proves the advantages of model-based design [Online]. Available: https://web.archive.org/web/20190115071038/https://www.mathworks.com/company/newsletters/articles/bae-systems-proves-the-advantages-of-model-based-design.html

Work, D., Bayen, A., & Jacobson, Q. (2008). Automotive cyber physical systems in the context of human mobility. In *Proceedings of the National Workshop on High-Confidence Automotive Cyber-Physical Systems* (pp. 3–4).

Yilmaz, L. (2017). Verification and validation of ethical decision-making in autonomous systems. In *Proceedings of the Symposium on Modeling and Simulation of Complexity in Intelligent, Adaptive and Autonomous Systems* (p. 1). Society for Computer Simulation International.

Zhang, Z., Eyisi, E., Koutsoukos, X., Porter, J., Karsai, G., & Sztipanovits, J. (2014). A co-simulation framework for design of time-triggered automotive cyber physical systems. *Simulation Modelling Practice and Theory, 43*, 16–33.

Zheng, X., Julien, C., Kim, M., & Khurshid, S. (2017). Perceptions on the state of the art in verification and validation in cyber-physical systems. *IEEE Systems Journal, 11*(4), 2614–2627.

Part IV

The Cyber Element

11

Perspectives on Securing Cyber Physical Systems

Zach Furness

INOVA Health System, Sterling, VA, USA

11.1   Cyber Physical Systems

This section will provide a brief overview of Cyber Physical Systems and discuss their relationship to related systems such as embedded systems and Internet of Things (IoT).

11.1.1   CPS Definition

Cyber-physical systems (CPS) are "smart systems that include engineered interacting networks of physical and computational components" (NIST 2017). The term "smart" refers to the fact that CPS generally involves "sensing, computation, and actuation." CPS typically bridge connections between operational technologies (those that perform control and actuation functions) and traditional information technologies (such as those that perform sensing and computation). The ability to automate transactions between those domains and create faster sense/response times is one of the primary advantages of CPS. CPS are not limited to single systems that bridge Information Technology (IT) and Operational Technology (OT) domains. A CPS will often be part of a "system-of-systems" that has multiple sensor, computation, and actuator connections that share data directly, or act on the results of aggregated data across multiple sensors (NIST 2017).

The enhanced functionality provided by automated response in these systems also comes with a tradeoff in ensuring the trustworthiness of the underlying algorithms and data in these systems. Imagine an autonomous vehicle (and example of a CPS) that has been trained to detect as stop sign and reacts by

Complexity Challenges in Cyber Physical Systems: Using Modeling and Simulation (M&S) to Support Intelligence, Adaptation and Autonomy, First Edition. Edited by Saurabh Mittal and Andreas Tolk.
© 2020 John Wiley & Sons, Inc. Published 2020 by John Wiley & Sons, Inc.

applying brakes to the vehicle and bringing the car to rest. The ability of the CPS to correctly sense the input (stop sign) and react accordingly (stop) without any human interaction across a variety of possible environmental conditions (weather, obscured view, etc.) is fundamental to the operation of the CPS. Any manipulation of the input data or modifications to the algorithms could compromise this system and the safety of its passengers. There needs to be emphasis placed on the security, privacy, safety, reliability, and resilience of such systems, given the degree of connectedness and autonomy being introduced (NIST 2017).

11.1.2 Related Systems

In this section, we'll cover some of the similarities and differences among systems related to CPS including embedded systems and Internet of Things (IoT).

Embedded systems are "collections of computers and networks that monitor and control physical processes" (Lee and Seshia 2017). Embedded systems are often part of a CPS, providing the functionality required to sense and react directly with the physical environment. They also tend to be applied in cases requiring real-time response in systems. Examples are numerous, ranging from avionics systems in aircraft to control systems in automobiles that provide anti-lock braking or fuel efficiency. Embedded systems tend to be characterized by their direct coupling to physical systems (via microcontrollers and other devices) as opposed to CPS which may be made up of larger numbers of connections that perform automated sensing and response among multiple components.

Internet of Things (IoT) refers to collections of computing devices used to sense, compute, and respond to physical systems connected via larger networks such as the internet. IoT devices share many similarities with CPS systems and are often used interchangeably with CPS. IoT devices are characterized by four primary capabilities (Boeckl et al. 2018):

1) **Transducer capabilities.** Transducer capabilities act as the means for computing devices to interact directly with the physical systems they are attached to. Two primary types of transducer capabilities are those related to sensing and actuating.
2) **Data capabilities.** These include data storage and processing capabilities.
3) **Interface capabilities.** Interface capabilities may exist at multiple levels such as application-level interfaces, human user interfaces, and network interfaces.
4) **Supporting capabilities.** These may include device management, cybersecurity, and privacy functions.

11.2   CPS Security Challenges

Because of the unique nature of Cyber-Physical Systems, they introduce several challenges for security. Their coupling to physical systems creates vulnerabilities that are not typically present in traditional IT systems. Such challenges are not limited to a specific industry or domain – numerous examples will be discussed in the following section.

11.2.1   Transportation: Automobile Security

As mentioned earlier, today's automobiles are examples of Cyber-Physical Systems. They are outfitted with a variety of automated functions including driver assist technology, automatic braking, and sense and avoid technology. Each of these functions senses conditions around the automobile and responds automatically with minimal human intervention. And as a result, these functions introduce vulnerabilities that were not originally intended in the design.

In July of 2015, Charlie Miller and Chris Valasek demonstrated the ability to hack into a car remotely and control key functions of the car through the vehicle's internet enabled entertainment system (Greenberg 2015). The demonstration and subsequent publication of the exploit has raised awareness of the problem across the automotive industry. The compromise of a network-enabled automobile is an example of how a CPS can be compromised. It also highlights a major concern of CPS – that such manipulation of those systems could lead to human injury or even loss of life.

Miller and Valasek were ultimately able to gain access to the Jeep Cherokee's controller access network (CAN) bus where the major functions of the car (steering, braking, acceleration, etc.) are controlled. While there is no direct connection between the CANBUS and the internet, the Jeep's entertainment system serves as a pivot point. A service known as "Uconnect," which provides internet connectivity for thousands of Fiat Chrysler automobiles, provides connectivity between the car's entertainment system and the internet. Miller and Valasek were able to access the entertainment system through that connection and rewrite the firmware that controlled the entertainment system to allow it to send commands over the CANBUS network. Once they had gained access to the CANBUS, they were able to disable or engage components such as steering, braking, or the engine itself.

The vulnerabilities documented in the Jeep car hack highlight one of the primary challenges in securing CPS. While creating enhanced functionality by connecting the physical systems of the car to the internet, unforeseen vulnerabilities are introduced that can disable the automobile and impact driver safety. The inability to fully predict and understand the unintended consequences of design decisions can have catastrophic impacts once fielded. And as the complexity of

CPS grows, understanding the full range of possible vulnerabilities will continue to be a challenge.

Vulnerabilities in connected automobiles are not limited to the Jeep car hack. In 2018, researchers demonstrated how they could gain access to a Tesla Model S through cloning of the Tesla key fob (Greenberg 2018). By precomputing and cataloging all potential cryptographic keys for the fob (a database of over 6-TB) they were able to correctly respond to the Tesla's cryptographic challenge – thereby unlocking the car and turning on the engine. While such an attack does not directly gain access to critical functions of the car, it highlights how the physical access control of the automobile is vulnerable based on network-enabled functionality (the wireless key fob). By exploiting such physical access to CPS, attackers can gain access to such systems.

The future of transportation is already moving beyond single network-enabled automobiles. Future systems will be composed of networks of connected vehicles that will interact with other cars as well as with sensors in the road. The implications for CPS are significant – creating an entire network of multiple, connected CPS entities. Information on individual automobiles such as position, speed, and heading, will be transmitted in messages using Vehicle-to-Vehicle (V2V) communications – requiring that the integrity and authenticity of those messages is maintained (McGurrin and Gay 2016). As in other transportation CPS examples, the inability to properly engineer security controls during the design of V2V and Vehicle-to-Infrastructure (V2I) communications could introduce vulnerabilities that have significant safety impacts.

11.2.2   Health IT: Medical Device Security

A network-enabled medical device is a Cyber-Physical System. These devices have direct physical interaction with a patient as either diagnostic tools (monitoring devices), or drug delivery systems (wireless infusion pumps) while also being network-enabled. And as with other CPS, their network enabled nature provides a potential for attackers to access such system and inflict harm to patients.

In Rajkumar et al. (2016), the authors highlight how the "combination of the embedded software controlling the devices, the new networking capabilities, and the complicated physical dynamics of the human body" warrant characterizing these devices as a distinct class of CPS – Medical CPS (MCPS). They also go on to cite the increasing use of intelligence in these devices as another aspect that makes them distinct. For example, the devices may automatically change their behavior (dosage levels, etc.) based on monitoring changes in the physiology of the patient. This characteristic of CPS creates another potential security concern as the "AI" can be influenced by corrupted input data to take action thereby inducing harmful effects in the patient.

In September 2017, the Industrial Control Systems (ICS) Cyber Emergency Response Team (CERT) issued an advisory citing vulnerabilities discovered by researcher Scott Gayou in Smiths Medical 4000 Wireless Syringe Infusion Pumps (ICS-CERT 2017). These pumps provide intravenous drugs to patients and use software on the device to manage delivery of such drugs. Many infusion pumps can be accessed and controlled remotely via a wireless network, creating a means for attackers to access the pump and inflict harm on an unsuspecting patient (Donovan 2018). This vulnerability is another example of how functionality added to a CPS through network connectivity created an unforeseen vulnerability that could jeopardize safety. The National Cybersecurity Center of Excellence (NCCoE) recently published guidance on ways to protect such infusion pumps through their Special Publication 1800-8, "Securing Wireless Infusion Pumps" (O'Brien et al. 2018).

Infusion Pumps are not the only class of MCPS that is vulnerable to attacks. Heart devices such as pacemakers have been shown to be vulnerable to cyber-attacks, as demonstrated by Kevin Fu in 2008 (Halperin et al. 2008). In 2011, Jerome Radcliffe demonstrated how vulnerabilities in an Insulin Pump could be exploited in a presentation at the 2011 Black Hat Conference (Radcliffe 2011). Similar vulnerabilities have been documented across a wide range of medical devices including anesthesia devices, ventilators, defibrillators, and lab equipment (Larson 2017). Beyond the concern over directly tampering with these devices, there is an additional concern over how data from these devices could be intercepted and modified in ways that could falsify patient medical records and lead to "unnecessary procedures or medical prescriptions" (Rapaport 2018). In Almohri et al. (2017), the authors discuss security concerns across a broad range of MCPS.

These reports also caution that while the vulnerabilities in these systems exist, there has yet to be a documented case in which a patient has been harmed as a result of an MCPS cyber-attack. Nonetheless, the increasing complexity of these systems as well as the rapid increase of cyber-attacks against medical devices, suggests that security must be engineered into these devices from the beginning and not addressed once the device has been brought to market. Over-engineering security into these devices can also be a concern. The introduction of security controls – such as complex passwords, multi-factor authentication, and other controls – could potential hamper patient safety by limiting access to devices involved in life-saving procedures.

11.2.3 Energy Systems: Smart Grid

Energy systems are rapidly evolving as complex systems that rely on vast networking and connectivity, sophisticated software that monitors operations and adapts to conditions to effectively deliver energy to end users. The scale at which energy CPS operate is quite large – in fact, energy systems are really just collections of hundreds or thousands of individual CPS. Understanding and securing such large CPS networks is a significant challenge.

The term "Smart Grid" is used to describe the evolution of the current energy grid to a system of even greater complexity, with advanced monitoring, control, communications, and delivery systems enabled by advances in computer and networking technologies (Shabanzadeh and Moghaddam 2013). The employment of such technology will ideally lead to advances in the generation and delivery of energy to end users. However, the application of such advances may also create vulnerabilities in these systems that could allow attackers to shut down power to users.

One of the most notable cyberattacks against an electric utility involved the use of the malware "CRASHOVERIDE" in an attack on the Ukrainian electric grid operations in December of 2016. This December 2016 attack was the second attack in as many years but the main difference in this attack was that "the 2016 attack was fully automated. It was programmed to include the ability to 'speak' directly to grid equipment, sending commands in the obscure protocols those controls use to switch the flow of power on and off" (Greenberg 2017). This type of attack could foreshadow more sophisticated attacks against energy utilities in the near future. And as elements of the Smart Grid emerge, the opportunity to gain access through greater connectivity and automation is a concern.

The National Institute of Standards and Technology (NIST) has issued guidance on securing the Smart Grid (NIST 2014; Stouffer et al. 2015). In Stouffer et al. (2015), differences between securing Industrial Control Systems (ICS) and traditional IT security systems are discussed. Among those differences are the following:

- ICS systems tend to be more time-critical, and many are built on real-time operating systems.
- Availability of ICS systems is usually much greater than that of traditional IT systems. Outages that impact service are not acceptable. Traditional measures of patching or rolling in software fixes by taking down and rebooting a component are typically not acceptable with ICS systems.
- Unlike IT systems where data confidentiality and integrity are paramount, the focus of ICS is on human safety and public health. Any security control that impacts public safety is unacceptable.
- ICS components – such as Programmable Logic Controllers (PLCs) – are directly tied to physical operations.
- Unlike IT systems which typically have a lifespan of 3–5 years, an ICS may be in use for 10–15 years. In many cases, it may be infeasible to employ security controls due to the legacy nature of the equipment or inability to convince the manufacture to make changes.

Stouffer et al. (2015) goes further in identifying potential ICS security architecture elements that can be employed to help secure ICS systems.

11.3 Challenges and Opportunities for M&S in CPS Security

Modeling and Simulation (M&S) technologies have shown to be an effective tool in improving the overall design, development, integration, and testing of systems. Examples of using M&S to improve design, and create efficiencies in the systems engineering process are prevalent. The design and construction of the Boeing 777 made use of significant M&S to fully evaluate performance of design tradeoffs prior to "bending metal" (Kossiakoff et al. 2011). M&S has also been used extensively in the engineering of complex systems (Tolk and Rainey 2015).

The idea of using M&S as a means of identifying potential security vulnerabilities, *during the design phase of the system*, has tremendous appeal to security engineers. As was seen in the examples mentioned earlier, one of the underlying challenges in securing CPS systems is the inability to identify the unintended consequences of design decisions. M&S of CPS is still relatively immature and requires representation of the physical dynamics of the target system as well as the network connectivity of that system and the influence of such connections on that target system. As such capability matures, we could foresee a number of instances in which M&S could be used in effective ways to secure CPS in the future.

11.3.1 Incorporating M&S into Systems Security Engineering and Resiliency

Too often, little thought is put into securing systems while they are being developed (as has been noted previously). More recently, the idea of incorporating security into the formal systems engineering process has begun to gain prominence. The NIST publication 800-160 is evidence of that increased focus (Ross et al. 2016). The insights of that document apply to CPS as well as traditional IT systems. Using that guide as a reference, there are numerous opportunities to apply M&S to the Systems Security Engineering process. NIST 800-160 specifically cites the use of M&S in the cyber resiliency verification process to "evaluate correctness before a system element is implemented, based on design artifacts." NIST 800-160 also mentions the potential for M&S application during the design phase to "support analysis of potential systemic consequences stemming from the compromise of a given resource."

A related topic to system security engineering is Cyber resilience – defined as the ability to "anticipate, withstand, recover from, and adapt to adverse conditions, stresses, attacks, or compromises on cyber resources" (Bodeau and Graubert 2017). Unlike traditional methods which look to close vulnerabilities within a system, cyber resilience focuses more on building in redundancy into systems which will allow it to function in the face of a security compromise. M&S methods could help to advance approaches in understanding CPS resiliency, through modeling of system behavior to a variety of threats and helping to develop resilient approaches. In addition to the resiliency applications for M&S mentioned in 800-160 above, the more recent volume, NIST Special Publication 800-160, Volume 2, titled "Systems Security Engineering: Cyber Resiliency Considerations for the Engineering of Trustworthy Secure Systems" offers additional opportunities on potential applications of M&S (Ross et al. 2018).

11.3.2 Digital Twin Concept for CPS Security

In many CPS examples, gaining access to the actual physical system is not an option either due to the ongoing use of the system in a critical application (energy, health) or concerns on behalf of the manufacturer. Having access to a digital "twin" of a CPS would provide the ability to investigate responses of the system to threats while not taking these systems offline.

A digital twin is a "virtual representation of a physical object or system across its life-cycle. It uses real-time data and other sources to enable learning, reasoning, and dynamically recalibrating for improved decision making" (Mikell and Clark 2018). Because of the connected nature of CPS systems, they are always evolving and changing as new software features are incorporated into the system and new connections created. The digital twin concept ensures that as the actual CPS changes, sensors on that system communicate those changes to the digital representation so it too, can maintain consistency. Digital twins are already gaining prominence in applications such as manufacturing by allowing analysis of production processes and machinery (Castellanos 2018). In 2017, Gartner named digital twin technology as one of their top strategic technology impacts (Pettey 2017).

The digital twin concept would allow a security engineer to monitor a digital replicate of a CPS system and create digital approaches to security without impacting the actual device. The application of security controls, evaluation of vulnerabilities, and understanding of response of a CPS to threats could all be done in the digital domain and validated prior to applying to the actual CPS system. Simulations of device behavior could be running continually with recommended actions provided to the security engineer.

11.3.3 M&S for CPS Risk Assessment

At its heart, cybersecurity is all about managing risk. Blindly applying security controls without understanding what risks exist within an enterprise will not lead to improved security. Instead, it will potentially lead to a misalignment of security resources against the highest priority vulnerabilities. The Risk Management Framework, developed by NIST, is the most widely accepted approach for organizations to document and manage risk. It outlines a seven step process for assessing and managing risk through the system development life cycle (NIST Joint Task Force 2018).

One of the tasks outlined within the RMF is the requirement to perform a system-level risk assessment. Such as assessment typically relies upon "identification of threat sources and threat events affecting assets, whether and how the assets are vulnerable to threats, the likelihood that an asset vulnerability will be exploited by a threat, and the impact (or consequence) of the loss of the asset" (NIST Joint Task Force 2018). M&S applications that can not only represent the detailed complexity of a CPS but also how a CPS is coupled to the operational environment, and affected by external threats, could be incredibly useful as tools for performing risk assessments. Such tools could be updated continuously with inputs from CPS and other systems and continuously run to generate outputs with unexpected outcomes – linking the concept of the digital twin to performing risk assessment. The likelihood of those outcomes based on the current security posture of the organization and expected threats could help to automate the risk assessment process.

11.3.4 CPS Cyber Ranges

A major concern in securing CPS is understanding how vulnerabilities within systems can lead to cascading effects among large numbers of systems at scale. Because a CPS may rely on many networked connections across multiple systems, the potential for unanticipated behavior at scale is a real possibility. Cyber ranges – that consist of large numbers of emulated endpoints deployed through virtualization – offer an opportunity to investigate such behavior in a controlled environment. The National Cyber Range, run OSD's Test Resources Management Center (TRMC), is one of the best known cyber ranges with thousands of nodes able to assess complex interactions of systems as large scales (Ferguson et al. 2014). However, one of the limitations of many of these ranges is their inability to represent CPS. As noted earlier, most CPS differ from traditional IT systems and the OS that are used so simply virtualizing thousands of OS instances is not sufficient to address CPS analysis at scale. As M&S becomes better at representing such CPS, deploying instances of those simulations on a large range would allow for more meaningful analysis of CPS interactions.

References

Almohri, H., Yao, D., Cheng, L., & Alemzadea, H. (2017). On Threat Modeling and Mitigation of Medical Cyber-Physical Systems. In *2017 IEEE/ACM International Conference on Connected Health: Applications, Systems, and Engineering Technologies (CHASE)*.

Bodeau, D., & Graubert, R. (2017). *Cyber Resiliency Design Principles. Selective Use Throughout the Lifecycle and in Conjunction with Related Disciplines*. The MITRE Corporation.

Boeckl, K., Fagan, M., Fisher, W., Lefkovitz, N., Megas, K., Nadeau, E., ... Scarfone, K. (2018). *Draft NIST 8228, Considerations for Managing Internet of Things (IoT) Cybersecurity and Privacy Risks*. National Institute of Standards and Technology.

Castellanos, S. (2018). Digital Twins Concept Gains Traction Among Enterprises. *Wall Street Journal*, 12 September 2018.

Donovan, F. (2018). Wireless Infusion Pumps Could Increase Cybersecurity Vulnerability. http://HealthITSecurity.com, 28 August 2018.

Ferguson, B., Tall, A., & Olsen, D. (2014). National Cyber Range Overview. In *2014 IEEE Military Communications Conference (MILCOM)*.

Greenberg, A. (2015). Hackers Remotely Kill a Jeep on the Highway – With Me In It. *Wired Magazine*, 21 July 2015.

Greenberg, A. (2017). 'Crash Override' the Malware That Took Down a Power Grid. *Wired Magazine*, 12 June 2017.

Greenberg, A. (2018). Hackers Can Steal a Tesla Model S in Seconds by Cloning it's Key Fob. *Wired Magazine*, 10 September 2018.

Halperin, D., Heydt-Benjamin, T., Ransford, B., Clark, S., Defend, B., Morgan, W., ... Maisel, W. (2008). Pacemakers and Implantable Cardiac Defibrillators: Software Radio Attacks and Zero-Power Defenses. In *2008 IEEE Symposium on Security and Privacy*.

ICS-CERT Advisory ICSMA-17-250-02A (2017). Smiths Medical Medfusion 4000 Wireless Syringe Infusion Pump Vulnerabilities, 7 September 2017.

Kossiakoff, A., Sweet, W., Seymour, S., & Biemer, S. (2011). *Systems Engineering Principles and Practice*. Development of the Boeing. (Vol. 777). Wiley, page 278

Larson, J. (2017). Medical Device Security Considerations – Case Study. In *RSA Conference 2017*. www.rsaconference.com.

Lee, E., & Seshia, S. (2017). *Introduction to Embedded Systems – A Cyber-Physical Systems Approach*. MIT Press.

McGurrin, M., & Gay, K. (2016). *USDOT Guidance Summary for Connected Vehicle Pilot Site Deployments, Security Operational Concept*. US Department of Transportation.

Mikell, M., & Clark, J. (2018). Cheat Sheet: What Is a Digital Twin. IBM Internet of Things Blog.

NIST, Cyber-Physical Systems Public Working Group (2017). Framework for Cyber-Physical Systems: Volume 1, Overview. June 2017.

NIST Joint Task Force (2018). *Risk Management Framework for Information Systems and Organizations*, October 2018. Draft NIST Special Publication. (Vol. 800-37). National Institute of Standards and Technology.

NIST Smart Grid Interoperability Panel Smart Grid Cybersecurity Committee (2014). *Guidelines for Smart Grid Cybersecurity, Vol 1–3*, September 2014. NISTIR 7628 Revision. (Vol. 1). National Institute of Standards and Technology.

O'Brien, G., Edwards, S., Littlefield, K., McNab, N., Wang, S., & Zheng, K. (2018). *Securing Wireless Infusion Pumps in Healthcare Delivery Organizations*. NIST Special Publication. (Vol. 1800-8). National Cybersecurity Center of Excellence (NCCoE).

Pettey, C. (2017). Prepare for the Impact of Digital Twins. Gartner.com, 18 September 2017.

Radcliffe, J. (2011). Hacking Medical Devices for Fun and Insulin: Breaking the Human SCADA System. In *2011 Black Hat Conference*.

Rajkumar, R., de Niz, D., & Klein, M. (2016). *Cyber Physical Systems*. Addison Wesley.

Rapaport, L. (2018). *Pacemakers, Defibrillators, are Potentially Hackable. Reuters*, 20 February 2018.

Ross, R., Graubart, R., Bodeau, D., & McQuaid, R. (2018). *Systems Security Engineering: Cyber Resiliency Considerations for the Engineering of Trustworthy Secure Systems*. NIST Special Publication 800-160, Volume 2. NIST.

Ross, R., McEvilley, M., & Oren, J. C. (2016). *Systems Security Engineering: Considerations for a Multidisciplinary Approach in the Engineering of Trustworthy Secure Systems*. NIST Publication 800-160 Volume 1. NIST.

Shabanzadeh, M., & Moghaddam, M. P. (2013). What is the Smart Grid? Definitions, Perspectives, and Ultimate Goals. In *International Power System Conference*, November 2013.

Stouffer, K., Pillitteri, V., Lightman, S., Abrams, M., & Hahn, A. (2015). *Guide to Industrial Control Systems (ICS) Security*. NIST Special Publication. (Vol. 800-82). NIST.

Tolk, A., & Rainey, L. (2015). *Modeling and Simulation Support for Systems of Systems Engineering Applications*. Wiley.

12

Cyber-Physical System Resilience

Frameworks, Metrics, Complexities, Challenges, and Future Directions

Md Ariful Haque[1], Sachin Shetty[1], and Bheshaj Krishnappa[2]

[1] *Computational Modeling and Simulation Engineering, Old Dominion University, Norfolk, VA, USA*
[2] *Risk Analysis and Mitigation, ReliabilityFirst Corporation, Cleveland, OH, USA*

12.1 Introduction

Cyber-physical systems (CPSs) are engineered systems built from the integration of computation, networking, and physical processes. Researchers often generalize "CPS" as an integrated system of cyber and physical systems, where embedded computers and networks are used to compute, communicate, and control the physical processes (Baheti and Gill, 2011; Wang, 2010). Advances in CPS make them crucial in most of the industries, e.g. energy delivery systems (EDS) (McMillin et al., 2007), healthcare systems (Cheng, 2008), transportation systems (Xiong et al., 2015), or smart systems (smart grid, smart homes, smart cities, etc.) (Amin, 2015; Yu and Xue, 2016). The advancements in engineering processes also bring the risk of cyberattacks because of the integration of the cyber and physical domains – in other terms information technology (IT) and operational technology (OT) domain. Therefore, the cyber resiliency of such systems considering the vulnerabilities of different components within the networked system is an integral part of CPS security analysis.

12.2 Cyber Resilience: A Glimpse on Related Works

The National Academy of Sciences (NAS) (Cutter et al., 2013) defined resilience as *the ability to prepare and plan for, absorb, recover from, or more successfully adapt to actual or potential adverse events*. The authors (Linkov, Eisenberg,

Complexity Challenges in Cyber Physical Systems: Using Modeling and Simulation (M&S) to Support Intelligence, Adaptation and Autonomy, First Edition. Edited by Saurabh Mittal and Andreas Tolk.
© 2020 John Wiley & Sons, Inc. Published 2020 by John Wiley & Sons, Inc.

Plourde, et al., 2013) used the resilience definition provided by NAS to define a set of resilience metrics spread over four operational domains: physical, information, cognitive, and social. In another work (Linkov, Eisenberg, Bates, et al., 2013), the authors applied the previous resilience framework (Linkov, Eisenberg, Plourde, et al., 2013) to develop and organize useful resilience metrics for cyber systems. Bruneau et al. (2003) proposed a conceptual framework initially to define seismic resilience, and later the authors (Tierney and Bruneau, 2007) introduced the R4 framework for disaster resilience. The R4 framework comprises *robustness* (ability of systems to function under degraded performance), *redundancy* (identification of substitute elements that satisfy functional requirements in event of significant performance degradation), *resourcefulness* (initiate solutions by identifying resources based on prioritization of problems), and *rapidity* (ability to restore functionality in timely fashion).

MITRE presents a framework for cyber resiliency engineering (Bodeau and Graubart, 2011). The framework identifies the cyber resiliency goals, the threat model for cyber resiliency, and structural layers to which cyber resiliency could be applied. Most of these frameworks discuss standard practices and provide guidance from different angles of resilience study, but lack of clear explanation on the quantitative resilience metrics formulation. Another issue with these frameworks is that they are most suitable for information technology system (ITS) rather than the CPS. The CPSs have unique requirements that make them different from typical ITS: First, for CPS, real time, safety, and continuity of service are essential, while for ITS the confidentiality and integrity of data are important, and momentary downtime can be tolerated (Macaulay and Singer, 2016). Second, it is common to apply anti-malware software in ITS, and they often automatically download and apply security patches. The CPS uses control systems that are designed for functionality rather than security and with limited memory and processing capacity (Macaulay and Singer, 2016). The installations of anti-malware solutions that consume a lot of memory and processor capacity for the automatic updates are not applicable to the CPS.

Lots of research works are done on resilience study of industrial control system (ICS), which is mainly a type of CPS. The National Institute of Standards and Technology (NIST) provides a framework (Sedgewick, 2014) for improving the cybersecurity and resilience of critical infrastructures that are supported by both ITS and ICS. The NIST framework identifies five functions that organize cybersecurity at the highest levels: *identify* (develop understanding of and manage risk to systems, assets, data, and capabilities), *protect* (develop and implement appropriate safeguards to ensure delivery of critical infrastructure services), *detect* (identify the occurrence of a cybersecurity event), *respond* (take action regarding a detected cybersecurity event), and *recover* (maintain plans for resilience and to restore any capabilities or services that were impaired due to a cybersecurity event).

In another work (Stouffer, Falco, and Scarfone, 2011), NIST provides detailed guidelines for ICS security. Collier et al. (2016) outline the general theory of performance metrics and highlight examples from the cybersecurity domain and ICS. Bologna, Fasani, and Martellini (2013) define the necessary measures to be taken to make ICS and critical infrastructures resilient. The above works present different useful insights from the cybersecurity standpoint but analyze resilience considering a specific section of the CPS network rather than incorporating the complete CPS cyber threat scenario. We are concerned about the estimation of cyber resilience for the CPS considering the whole system domains of security concerns. Thus, a comprehensive resilience metrics formulation is necessary for the CPS.

Yong, Foo, and Frazzoli (2016) present a state estimation algorithm that is resilient to sparse data injection attacks on the CPS. On the survey side, Humayed, Lin, Li, and Luo (2017) present an excellent overall security survey on the CPS by considering security, cyber-physical components, and CPS-level perspectives. Koutsoukos et al. (2017) present a modeling and simulation integrated platform for the evaluation of CPS resilience with an application to the transportation systems. Although the resilience assessment irrespective of the types of the CPS is the goal of the chapter, we limit our focus on the CPSs involving ICSs, EDS, and oil and gas systems. The following subsections are going to discuss the CPS architecture in brief and relate the complexities to handle to make the system cyber-resilient.

12.3   Cyber-Physical System Resilience

There are different applications of CPS, and the network system architecture is varied based on the functional area. Here we present a generic CPS architecture by considering the applications related to the ICS to explain the cyber resilience concepts as illustrated by Haque, De Teyou, Shetty, and Krishnappa (2018). In Figure 12.1, we present an ICS architecture to discuss CPS in general. An ICS is a set of electronic devices to monitor, control, and operate the behavior of interconnected systems. ICSs receive data from remote sensors measuring process variables, compare those values with desired values, and take necessary actions to drive (through actuators) or control the system to function at the required level of services (Galloway and Hancke, 2013; Macaulay and Singer, 2016). Industrial networks are composed of specialized components and applications, such as programmable logic controllers (PLCs), Supervisory Control and Data Acquisition (SCADA) systems, and distributed control systems (DCS) (Cardenas et al., 2009). There are other components of ICS such as remote terminal unit (RTU), intelligent electronic devices (IED), and phasor

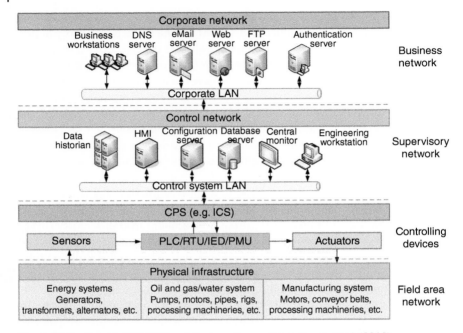

Figure 12.1 Generic CPS (ICS) architecture. *Source:* From Haque et al. (2018).

measurement units (PMU). Those devices communicate with the human–machine interface (HMI) located in the control network.

The risk of cyberattack comes into play when the corporate and control network have communications for regular business operations because part of the corporate network system is open to the Internet to communicate with stakeholders and business entities outside ICS network. Threat vectors also come from the heterogeneity (Humayed et al., 2017) of the different building blocks of the CPS and their control and monitoring hardware and software systems. Because of the complex interconnections and interactions among various components within the CPS, it is difficult to identify or trace any attack vector that may involve exploitation of multiple CPS components in sequence. Thus, to make the CPS cyber-resilient, it needs a wide range of efforts including the study of the system-level intrusion detection and prevention mechanisms as well as the system capabilities. Such capabilities can be the ability to divert the attack event, use redundant resources, respond and recover within the defined timeframe with minimal impact, keep learning the vulnerabilities and attack vectors, and evaluate and update security and privacy policies.

12.3.1 Resilient CPS Characteristics

Wei and Ji (2010) present the resilient industrial control system (RICS) model where the authors have identified the following three characteristics of the ICS to be resilient:

- Ability to minimize the undesirable consequence of an incidence.
- Ability to mitigate most of the undesirable incidents.
- Ability to restore to normal operation within a short time.

All the above characteristics are a repetition of robustness, resourcefulness, redundancy, and rapidity as illustrated in the R4 resilience framework (Tierney and Bruneau, 2007). As CPS or ICS works closely with the field devices with interfaces to the control centers and there need to be connections between control and the corporate network, a wide range of efforts spanning the system level and organizational level are necessary to make the CPS cyber-resilient. A detail resilience graph-based analysis is presented in Section 12.6 to illustrate the challenges in modeling and simulation of CPS resilience.

12.3.2 Need for Resilience Metrics

There are diverse applications of CPS, e.g. ICS, smart grid systems, medical devices, autonomous automobiles (intelligent cars), etc. Based on the field of application, there are variations in the cyber and physical components as well as in the cyber-physical interconnections and protocol communications within the cyber-physical devices. Within the scope of this chapter, we limit our focus on the ICS to explain the required aspects of resilience. As illustrated in the related works section, there are a lot of research works going on with the development of standard practices and guidelines to make the CPS cyber-resilient, which have lack of specific quantitative cyber resilience metrics. Therefore, we feel the need for the development of the quantitative cyber resilience metrics for the CPS.

The availability of quantitative cyber resilience metrics would assist the concerned industry operators to assess and evaluate the CPS and focus on the weak points to improve. Thus, one of the objectives of this work is to derive quantitative resilience metrics and development of a simulation platform that can handle the network architecture, scan the vulnerabilities, generate useful quantitative resilience metrics, and provide recommendations to improve the overall network resilience posture. There is no doubt that there is a high need for the resilience metrics automation across various industries. Thus, an approach to quantify the resilience metrics and development of a simulation platform to automate the metrics generation process is the need of the time. Also, the inclusion of the modeling and simulation paradigm in the CPS resilience study is a crucial research aspect to consider.

12.4 Resilience Metrics and Framework

Resilience metrics derivation is one of the goals of this chapter. In this section, we cover the CPS cyber threat landscape, CPS cyber resilience metrics and sub-metrics, and cyber resilience framework for the CPS. To have a smooth transition, we provide qualitative resilience metrics computation methodology in Section 12.5 and a detailed discussion on the quantitative modeling of resilience metrics in Section 12.6.

12.4.1 CPS Cyber Threat Landscape

The threat landscapes that CPSs are facing today are coming from different threat vectors. We use ICS to describe the threat landscape to CPS in general; in other terms we use ICS in some cases to represent the CPS. As the ICS industry grows larger and complex, the types and severity of targeted threats increased. Some of the threats that are identified to be a part of the CPS or ICS threat landscape are put together below. A mapping of the resilience metrics analysis domains with the threats is presented in Figure 12.2 where some of the attacks are collected from the discussion provided by Andrew Ginter (2017) and Cardenas et al. (2009):

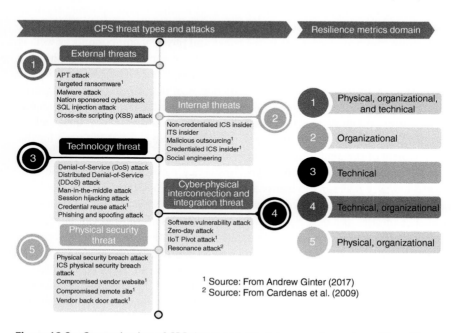

Figure 12.2 Categorization of CPS threat and attack types and mapping with resilience metrics domain.

- *External threats:* These threats arise from adversaries such as nation-sponsored hackers, terrorist groups, and industrial competitors through espionage activities. Cyber intruders may launch an advanced persistent attack (APT) attack, where the goal is to steal some valuable information on the network's assets without getting detected. One example of such an attack in recent times is the Stuxnet attack on the Iranian nuclear centrifuges. Some other threats such as targeted ransomware are discussed by Andrew Ginter (2017).

- *Internal threats:* Today, as the work processes in any industry are segmented and done by contractors or third-party vendors, ICS companies need to share system access information to outside business partners. That makes the ICS vulnerable to potential cyber threats. There also exists direct insider threat from the employees of the ICS company itself who are provided with legitimate access to the ICS network for regular operation and maintenance related tasks, which falls in the category of credentialed ICS insider attack.

- *Technology threats:* Many of the ICS networks run on legacy technology where the most concern part is the protocol-level communication among different ICS products within the network. Thus, many of them are lack of strong authentication or encryption mechanism (Laing, 2012). Even if they use authentication procedure, the weak security mechanisms (e.g. weak password, default user accounts) are not enough to protect the system from smart adversaries. Some of the attacks that can take place due to the CPS technology itself are presented in Figure 12.2.

- *ICS and ITS integration threats:* Due to the integration and interconnection of the ICS network with the control system network and corporate network for business operations, ICS devices become vulnerable to cyberattacks as part of the corporate network is open for communication over the Internet. Only putting the ICS devices behind the firewalls does not necessarily safeguard them, because today smart intruders are expert enough to launch a multi-host multi-stage cyberattack by exploiting several stepping stones on the way to the valued ICS assets.

- *Physical infrastructure security threats:* Sometimes lack of proper infrastructure security to the ICS devices poses severe threats to the ICS network. One such example of poor physical security is sharing of the floor space of ICS devices with the ITS devices (e.g. employees' computers, routers, switches, etc.) and thus providing easy access to the ICS devices (e.g. PLC, IED, RTU, PMU, etc.). Other sorts of threats may arise from compromised vendor website or compromised remote sites.

Each of the above categories of threats falls under any or a combination of the three major domains of resilience assessment: physical, organizational, or technical. To handle the external threats, it needs a comprehensive effort from all three areas: physical, organizational, and technical. To control the cyber-physical inter-connectivity threats although more emphasis should be given on the technical side, it also needs to consider the policies that mostly depend on the management decisions. Thus, it falls under technical and organizational domains. Similarly, physical security threats are part of physical security and organizational policy, and therefore, resilience metrics dealing with the physical security threats need to consider the physical and organizational security posture. We incorporate the categorization while defining the sub-metrics for the cyber resilience assessment of the CPS, which is discussed in the following subsections.

12.4.2 CPS Resilience Metrics

We decompose the four R4 metrics (*robustness*, *redundancy*, *resourcefulness*, and *rapidity*) proposed by Bruneau et al. (2003) into a hierarchy of several domains and sub-metrics, each of which can be analyzed independently. Each of the broad R4 metrics is subdivided into three domains – *physical*, *organizational*, and *technical* – to cover most of the threat vectors of the CPS. There are sub-metrics under each of the areas that are organized in a tree structure as given in Figure 12.3 and can effectively contribute to the cyber resilience assessment for ICS.

The metrics illustrated in Figure 12.3 are self-explanatory and easy to under-stand. Within the scope of this chapter, we analyze two different approaches to estimate cyber resilience metrics for CPS: qualitative approach and quantitative approach. We aim to assess the broad R4 metrics using both methods. The quali-tative approach discussion is presented in Section 12.5, while the quantitative modeling and simulation approaches are discussed in Section 12.6. Within the scope of the chapter, the detail definition of each of the sub-metrics seems unnecessary. Here we explain some of the sub-metrics of the resourcefulness metrics.

Physical resourcefulness is having two sub-metrics: physical monitoring and protective technology. Physical monitoring includes all sorts of devices and systems that allow monitoring the physical ICS, which includes but not limited to the surveillance systems, alarm monitoring systems, etc. Protective technology refers to the efforts to protect information and asset security, phys-ical solutions or policies to safeguard CPS assets, documentation, etc. This may also involve automatic action initiation based on the alarm or security breach, e.g. automatic door lock because of consecutive unauthorized access attempts to control centers.

Organizational monitoring and detection refer to monitoring of the employee actions and audit of the system command logs to identify the pres-ence of potential insider threat. Organizational response and recovery refer to

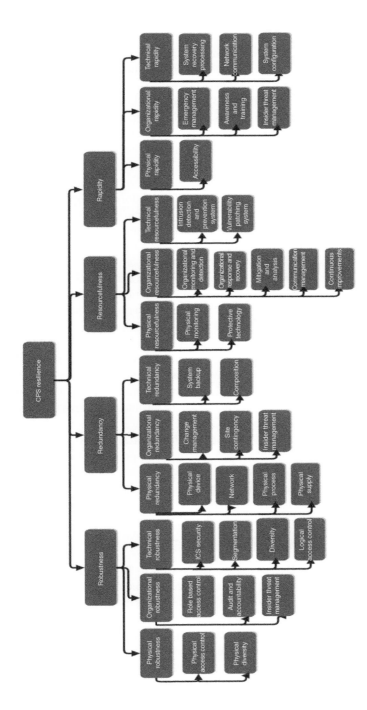

Figure 12.3 CPS cyber resilience metrics structural hierarchy.

the organizational policies to handle any cyberattack scenario, the performance of documented actionable policies, and evaluation of the actions performed. Mitigation and analysis refer to the organizational capabilities and efforts in mitigating any potential cyberattack, analyzing the attack types and originating points, and learning to prevent the similar attack in the future. This may also include training programs with a simulated attack in a controlled environment to train the operators better. Communication management refers to the organization hierarchy-wise communication that is necessary for detecting, protecting, and preventing any cyberattack. It may include the organizational communication policies in case of a cyberattack event. Continuous improvement is the resourcefulness area where CPS vendors and operators need to evaluate what measures should be in the cybersecurity detection and protection mechanism to cope up with the changing technological environment.

In the technological resourcefulness, the sub-metrics are intrusion detection and prevention system and vulnerability patching system. Intrusion detection and prevention system deals with the intrusion detection systems (IDS) and intrusion prevention systems (IPS). IDS detect and IPS prevent any potential information security breach in the ICS and ITS areas. The organization may have strict policies and regulations regarding the use and application of the IDS and IPS. The policy or the rule set in IDS and IPS need to keep updated constantly. Some vulnerabilities may not be detected by the IDS because there are zero-day vulnerabilities. In those cases, the system should have some sorts of abnormal behavior alarm generation in place. Vulnerability patching system makes sure if the computers and servers in the ITS domain are updated with the latest antivirus patches, automatic update of the system is enabled.

Unlike the ITS, ICS or CPS do not have traditional antivirus software. It is necessary to keep these systems updated by the vendor recommended system updates or patches. The resilience metrics and sub-metrics illustrated in Figure 12.3 are used to derive a cyber resilience framework for the CPS as presented in Section 12.4.3.

12.4.3 Cyber-resilient Framework for CPS

We provide a detailed CPS resilience framework in Table 12.1, as presented by Haque, De Teyou, Shetty, and Krishnappa (2018). The framework is designed to assess the cyber resilience of the CPS (i.e. ICS in this discussion) and is based on the CPS resilience metrics hierarchy presented in Figure 12.3. The framework can serve as a platform to create secure and resilient CPS/ICS across different industries (energy, oil and gas, manufacturing, etc.). Again, the resilience is assessed in terms of robustness, redundancy, resourcefulness, and rapidity in three domains: physical, organizational, and technical. The details provided in the framework make it self-explanatory.

Table 12.1 Cyber resilience framework for CPS.

	Robustness (R_1)	Redundancy (R_2)	Resourcefulness (R_3)	Rapidity (R_4)
Physical	*Access control* • Physical barrier policy (guards, walls, rooms, gates) • Identification and authentication (biometric, smart card, PIN code) • Physical ports protection and electronic device policy *Segmentation* • Physical isolation of ICS sites from corporate sites • Physical isolation of storage site from the processing site *Diversity* • Product and vendor diversity *Risk mitigation* • Threat identification, characterization, and mitigation	*Contingency* • Alternate storage/processing site, power supply, and communication network • Protection of alternate sites and power supply • PLC and RTU redundancy (shadow or separate mode) *Composition* • Capabilities to deploy new PLC/RTU interoperable with the ICS to ensure continuity of the process	*Monitoring and detection* • Capabilities to monitor the physical environment to detect cybersecurity events (video cameras, motion detectors, sensors, and various identification systems) *Response and recovery* • Capabilities to investigate and repair physical devices	*Communication latency* • The delay between adverse event and detection *Restoration delay* • The delay to access damaged devices (debugging ports, remote access) • The delay between detection and restoration (mean time to repair) • Switching delay for backup operations (hot, cold, warm) *Learning* • Update device configuration in response to recent events and performs better in the future

(Continued)

Table 12.1 (Continued)

	Robustness (R_1)	Redundancy (R_2)	Resourcefulness (R_3)	Rapidity (R_4)
Organizational	*Access control* • Visitor escort and access agreements policy • Restriction of physical access to ICS to authorized employee only • Personnel designation, screening, termination and transfer policy • Terms of employment policy *Segmentation* • Employee-specific roles and responsibility *Diversity* • Diverse group of employees to mitigate insider attacks *Risk mitigation* • Planning, implementation, and progress monitoring	*Contingency* • Business continuity planning and coordination *Composition* • Capabilities to deploy new manufacturing processes	*Monitoring and detection* • Capabilities to monitor personnel activity to detect cybersecurity events (account management, configuration change control) • Record and classification of incident *Response and recovery* • Collection of evidences for civil and criminal actions • Audit policy and change management policy	*Communication latency* • Cyber events are reported timely to appropriate personnel • Responsibilities and procedures are clearly defined for adequate personnel to ensure quick response *Restoration delay* • Timely availability of dedicated resources for service restoration (budget, manufacturer support, and tools) *Learning* • Registration to association and security conferences • Review security policy during and after adverse events • User awareness and training (penetration tests, role-based training)

Technical

Access control
- Port, protocol, and traffic filtering in CS
- Wireless and remote access policy (authentication and encryption policy)
- Email and browser policy (URL filtering, attachment, supported email clients and browser, plugins)

Segmentation
- Firewalls/gateways between corporate and CS network
- Firewalls/gateways between ICS and third-party network
- DMZ for control systems (CS) and corporate network

Diversity
- Disjoint technologies between CS and corporate network
- Software, firmware, and hardware diversity

Risk mitigation
- Continuous vulnerability scanning and patching
- Identification and mitigation of ICS weaknesses due to old technology design (clear communications, poor coding practices, low CPU/memory)
- Identification and mitigation of ICS weaknesses due to implementation (weak logging/ authentication, weak scripting interface, malfunction devices)
- Default configuration avoidance in CS and corporate network (default account/password, unused services/components)

Contingency
- Backup secured on protected servers
- Backup copies of information system and software
- Adequate resource to ensure availability of data

Composition
- Capabilities to deploy new protocols and technologies interoperable with the ICS to ensure continuity of the process

Monitoring and detection
- Capabilities to monitor network logs to detect cybersecurity events
- ICS protocol attacks detection (Modbus, DNP3, ICCP, etc.)
- Network-based IDS between CS and corporate
- Host-based IDS in both CS and corporate

Response and recovery
- Anti-malware tools, IRS
- Reduction of malware spread from corporate to control system network
- Dynamic reconfiguration

Communication latency
- Fault isolation latency
- Measurement reading latency
- Routine automation latency

Restoration delay
- Intrusion response frequency

Learning
- Logs correlation with vulnerability databases
- Dynamic reconfiguration after cyberattacks
- Online learning of attacker strategies

12.5 Qualitative CPS Resilience Metrics

One of the objectives in this work is to make the resilience metrics operational, i.e. the metrics should be analytically measurable and quantifiable with the available data and resources. One challenge of developing useful, generalizable resilience metrics for CPS/ICS is that the data regarding the cybersecurity of CPS/ICS is not available mostly due to regulation that requires reporting only a subset of cyberattacks as stated in McIntyre, Becker, and Halbgewachs (2007). Consequently, ICS companies with established cybersecurity policies prefer to keep their data confidential, mainly due to privacy and the proprietary nature of information. By taking into consideration the lack of system data availability, resilience metrics can be evaluated with a qualitative approach using cautiously chosen sets of questionnaires. The questionnaires should be designed in a way so that it can address each sub-metric and therefore capture qualitative information about the system resilience posture.

We derive the resilience metrics by aggregating the individual sub-metrics using a multi-criteria decision-making approach such as the analytical hierarchy process (AHP) (Saaty, 2008). N data sets are collected from N cybersecurity experts to compute the resilience metrics. Here data sets represent the response of questionnaires related to the ICS. For our example case as in Figure 12.4, we have used $N = 10$. This type of data collection from security experts is used to provide a scoring formula for cybersecurity metrics (Tran, Campos-Nanez, Fomin, and Wasek, 2016; Wilamowski, Dever, and Stuban, 2017), and we adopt the same methodology to collect the data.

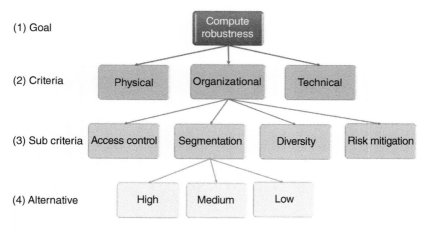

Figure 12.4 Decomposition of robustness using analytical hierarchy process.

The steps involved in implementing the AHP methodology are presented below:

Step 1: Each resilience metric is decomposed into a hierarchy of four levels: *goal* (maximize the corresponding resilience metric), *criteria*, *sub-criteria*, and *alternatives* (possible values that the sub-criteria can take). Figure 12.4 illustrates the hierarchy that we build for the robustness metric as an example. Each criterion at lower-level criteria affects the overall effort of maximizing the robustness. In the same way, each sub-criterion (respective to alternative options) affects its corresponding criterion.

Step 2: Data were collected from N cybersecurity experts (subject matter experts [SMEs]) corresponding to the hierarchy of Figure 12.4, $N = 10$ for our sample illustration. The data collection process was based on a pairwise comparison implementing the qualitative scale explained in Saaty (2008).

Step 3: The pairwise comparisons of all criteria and alternatives constructed in step 2 are organized into a square matrix. Mathematically, the pairwise comparison matrix A for m factors requires $m \times m$ elements. Each entry in A denoted by a_{ij} represents the comparison between factor i and factor j. The pairwise comparison matrix can be determined by using Eq. (12.1):

$$A = \begin{bmatrix} 1 & a_{12} & \cdots & a_{1m} \\ \dfrac{1}{a_{12}} & 1 & \cdots & a_{2m} \\ \cdots & \cdots & \cdots & \cdots \\ \dfrac{1}{a_{1m}} & \dfrac{1}{a_{2m}} & \cdots & 1 \end{bmatrix} \tag{12.1}$$

In addition, the individual responses of the N SMEs were aggregated by using the geometric mean, which yielded to one unique comparison matrix:

$$a_{ij} = \left[\prod_{k=1}^{N} (a_{ij})_k \right]^{1/N} \tag{12.2}$$

where $(a_{ij})_k$ is the response obtained by the kth cybersecurity expert.

Step 4: When all judgments are made, the relative weight of each criterion with respect to the goal is calculated. These weights are obtained with the normalized right eigenvector of the pairwise comparison matrix A. The relative weights of all the sub-criteria and alternative options are generated using the same process.

Step 5: A consistency ratio (CR) is calculated to measure how accurate or consistent are the judgments of the N expert responses. For the kth cybersecurity

Table 12.2 Values of the random index for small problems ($m<10$).

m factors	2	3	4	5	6	7	8
RI	0.00	0.58	0.9	1.12	1.24	1.32	1.41

expert response, if $CR_k<0.1$, then the judgment of this expert can be accepted; otherwise it should be excluded from the analysis for inconsistency. The consistency ratio can be evaluated by comparing the consistency index (CI) with the random index (RI) (Saaty, 2008):

$$CR = \frac{CI}{RI} \qquad (12.3)$$

The values of the random index are shown in Table 12.2 for small problems ($m<10$), and the consistency index CI is given by

$$CI = \frac{\lambda_{max} - m}{m - 1} \qquad (12.4)$$

In Eq. (12.4), λ_{max} is the maximum eigenvalue of the expert judgment.

Step 6: The resilience metric is calculated by combining the total of the weights of the elements of each level multiplied by the weights of the corresponding lower-level elements.

12.6 Quantitative Modeling of CPS Resilience

CPSs are highly interconnected systems with heterogeneity in different system levels. Thus, any attempt to quantitative resilience modeling and estimation needs to consider the network topology, system vulnerabilities, asset criticality, interconnectivity, and underlying physical processes. It is hard to model and estimate the cyber resilience considering all the areas of security concerns in CPS without impacting system performance and stability. In this section, we discuss in high level the step-by-step modeling and simulation approaches that need to be carried out for the quantitative modeling and estimation of CPS cyber resilience.

12.6.1 Critical Cyber Asset Modeling

Determining asset criticality is one of the fundamental security analysis techniques used by researchers (Haque, Shetty, and Kamdem, 2018; Kellett, 2016).

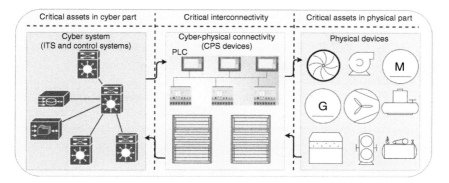

Figure 12.5 Critical assets in different system levels.

Identifying the critical assets provides network administrator a direction to focus on the most critical network components and thus facilitates the best use of resources concerning cost and time. In the CPS, asset criticality comes into play in three different layers: critical assets in the cyber and control part, essential inter-connectivity between the cyber and the physical part, and the critical assets in the physical part as shown in Figure 12.5. Critical assets in the cyber domain are those assets that upon exploitation can lead to the exploitation of control devices or cause significant damage to the physical processes. Critical interconnections are those cyber-physical interconnections whose disturbances can result in the una-vailability of the services and may cause physical processes to shut down.

While determining the critical asset in the cyber part is not difficult, considera-tion of all the three parts of criticality and combining them within a single plat-form undoubtedly need a lot of efforts. As we are concerned about the CPS resilience, we focus mostly on the first two parts, which include the cyber part and the interconnectivity part. We plan to use the graph-theoretical approach consid-ering the network topology and vulnerabilities of the network nodes to assess the asset criticality in a CPS environment. Graph-theoretical approach facilitates to analyze the shortest paths, in other terms most probable attack paths by modeling the cost of the attacker along the paths by using established shortest path algo-rithms. Also, different multi-criteria decision analysis (MCDA) methods assist in taking the optimum decision in identifying the asset criticality in a complex CPS.

12.6.2 Stepping Stone Attack Modeling

Different CPS components within the IT and OT systems have specific function-alities. Thus, because of the distinct nature and behavior of the systems at differ-ent parts such as in the cyber part, interconnectivity part, and physical part as in Figure 12.5, it makes hard for an intruder to exploit a vital asset in the CPS in one single step. However, today smart intruders are expert enough to launch a multi-host multi-stage cyberattack by utilizing several intermediate stepping stones on the way to the valued ICS or CPS assets. NIST has provided several guidelines (Stouffer et al., 2011) considering the vulnerabilities that arise because of ITS and ICS integration. Modeling the stepping stone attacks is a difficult problem to address because on the one hand the intruder gets more knowledge and expertise by exploiting different hosts in sequence and on the other hand the defender may apply dynamic network configuration based on the detected exploitation scenario. Thus, modeling the stepping stone attack paths is an important aspect in CPS resilience assessment to consider the dynamic nature of the attacker and defender actions. There are several attempts to detect and model the stepping stone attacks in ICS or CPS (Gamarra et al., 2018; Nicol and Mallapura, 2014; Zhang and Paxson, 2000).

Figure 12.6 illustrates an approach to model the stepping stone attack paths using the vulnerability graph where the different network layers are based on the NIST recommended defense-in-depth architecture (Stouffer et al., 2011). The graph model (Haque, 2018) uses different layers to represent different IT and OT layers, where it considers the physical devices such as the power generation systems and the field devices as the potential targets for cyberattacks. The layers corporate demilitarized zone (DMZ) and corporate local area network (LAN) belong to the pure cyber domain or the IT domain. Control system DMZ and control system LAN belong to the interconnectivity layer between the cyber (IT) and physical device layer (ICS/CPS). The network is drawn for application to the EDS where the physical device layer consists of power station networks and other field communication devices. The edge weights in the vulnerability graph model are coming from the exploitability and impact metrics of the Common Vulnerability Scoring System (CVSS) by Mell, Scarfone, and Romanosky (2007). In the probabilistic modeling of such stepping stone attacks in the CPS, game theory and Markov decision process (MDP) are useful to model the attacker and defender actions and to determine the most suitable actions for the defender in different states of the network.

12.6.3   Risk and Resilience Modeling and Estimation

One of the main objectives of CPS resilience assessment is to estimate the risk and resilience of the CPS. In the resilience estimation of CPS and other systems, the resilience graph (Bruneau et al., 2003; Wei and Ji, 2010) is utilized. Figure 12.7 presents a generic resilience graph for the CPS resilience modeling and estimation in the event of a cyberattack incident. The parameter notations are given in Table 12.3. We aim to model the CPS resilience using a combination of the vulnerability graph analysis and the resilience curve estimations and derive the broad R4 resilience

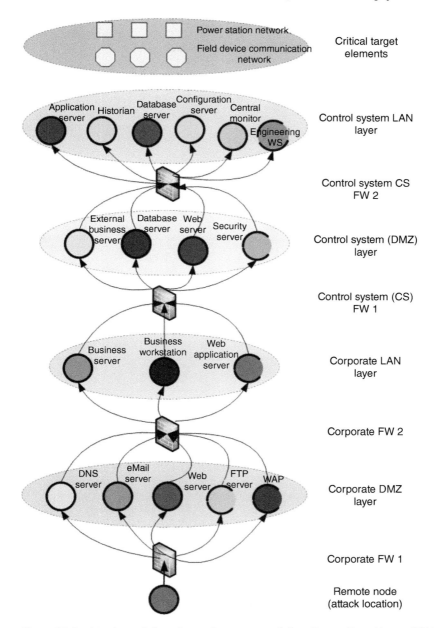

Figure 12.6 Attack graph-based stepping stone modeling. *Source:* From Haque (2018).

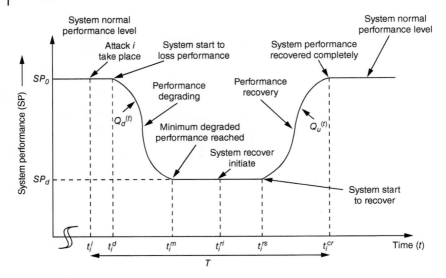

Figure 12.7 System performance graph in the event of cyber incident.

Table 12.3 Notations and parameter description for resilience graph.

Notation	Parameter description
SP_0	System normal performance level before any cyberattack incidence
SP_d	System minimum degraded performance level after a cyberattack incidence
t_i^i	Time when attack incidence i starts taking place
t_i^d	Time when system overall performance starts degrading due to attack incidence i
t_i^m	Time when system reaches to minimum performance level due to attack incidence i
t_i^{ri}	Time when operators identify the attack event and start initiating system recovery
t_i^{rs}	Time when system starts to recover performance after recovery attempt initiation
t_i^{cr}	Time when system completely recover performance and reach to pre-incident performance level
T	Time period of attack initiation to system complete recovery
$Q_d(t)$	Time-dependent system performance degradation behavior
$Q_u(t)$	Time-dependent system performance recovery behavior

metrics as illustrated in Figure 12.3. Here we present a brief discussion on the modeling of resilience and risk using the system performance graph of Figure 12.7.

The system performs at the performance level SP_0 before a cyber incident takes place. This performance is the overall system performance considering the normal stable operation and required services availability. At time t_i^i, a cyberattack event i takes place on the CPS, and the system starts showing degraded performance at time t_i^d. We assume the attack incidence causes damage to part of the systems and the complete system is not taken out of service. Thus, the system continues to perform in reduced performance and reach to the minimum degraded performance SP_d at time t_i^m. Because there are IDS and intrusion prevention mechanisms present in the CPS network, the system administrators and operators can identify the attack incident, analyze it, and initiate recovery at time t_i^{ri}. Because of the recovery initiation, the system performance starts recovering at time t_i^{ts} and performance reach to normal operation level as like pre-attack incidence at time t_i^{cr}. The time between t_i^i and t_i^{cr} is the total period T of resilience due to an attack event i as given in Eq. (12.5):

$$T = t_i^{cr} - t_i^i \tag{12.5}$$

To estimate the resilience, the only challenge is to estimate the nature of time-dependent performance degradation $Q_d(t)$ and time-dependent performance recovery $Q_u(t)$, which are functions of attack severity, the fraction of the system under adverse event, nature and severity of service losses, attack recover capability, etc. In general, the two time-dependent performance parameters are expressed in Eq. (12.6) as follows:

$$\left.\begin{aligned} Q_d(t) &= f\left(attack\ severity, system\ fraction\ under\ attack, service\ loss\right) \\ Q_u(t) &= f\left(attack\ diagnostic, system\ fraction\ out\ of\ service, recover\ ability\right) \end{aligned}\right\} \tag{12.6}$$

Using Figure 12.7, the resilience $\mathfrak{R}$ for the CPS is estimated as follows:

$$\begin{aligned} \mathfrak{R} = \frac{1}{T * SP_0}\Bigg[& SP_0\left(t_i^d - t_i^i\right) + \int_{t_i^d}^{t_i^m}\left\{SP_0 - SP_d - Q_d(t)\right\}dt \\ & + \int_{t_i^{ts}}^{t_i^{cr}}\left\{SP_0 - SP_d - Q_u(t)\right\}dt + \int_{t_i^d}^{t_i^{cr}}\left(SP_0 - SP_d\right)dt \Bigg] \end{aligned} \tag{12.7}$$

Resilience $\mathfrak{R}$ is normalized by the system normal performance level SP_0. As risk and resilience are treated as having opposite nature in evaluating a system performance during a cyber incident, risk $\mathcal{R}$ is defined as follows:

$$Risk, \mathcal{R} = 1 - \mathfrak{R} \tag{12.8}$$

We consider risk and resilience analysis as complementary to each other. While the computation of resilience is the goal, determining risk cannot be ignored. Both risk and resilience would give a complete posture of the security analysis for the CPS.

12.6.4 Modeling and Design of Attack Scenarios

Most of the cyberattacks that take place in the ITS domain are also applicable in the CPS domain. Figure 12.2 illustrates some of the attacks and threats that are directly applicable to the CPS domain. Often the cyberattack in the CPS is composed of multiple combinations of attack types. One of the challenges in CPS resilience modeling and estimation is to model different attack scenarios and their possible impacts on the CPS to get an estimate of the CPS security and resilience posture.

We present an illustration of the CPS attack scenario in Figure 12.8 using the NIST recommended defense-in-depth architecture by Stouffer et al. (2011) where

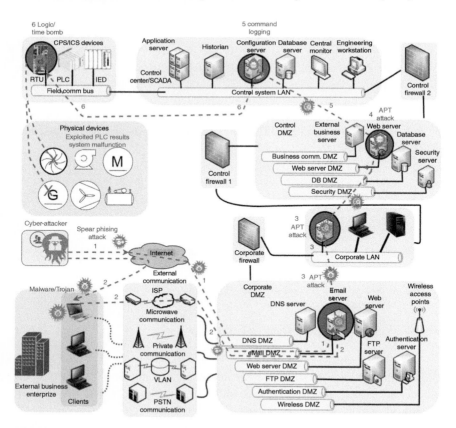

Figure 12.8 CPS attack scenario illustration using NIST defense-in-depth architecture.

we consider the ICS or SCADA systems as the target CPS. In this example, an attacker, having no information whatsoever about the network architecture and intellectual intelligence, uses two different attack mechanisms to penetrate the corporate DMZ network. On the one hand, the attacker installs malware or Trojan in one of the client computers through Internet malicious website browsing (marked as 2 in Figure 12.8), and on the other hand, he tries continuously spear phishing attacks through the Internet communication by sending malicious email attachments to the DMZ users' email accounts (marked as 1).

Using the above mechanisms, the attacker is successful in getting access to one of the email servers in the corporate DMZ, launches APT (marked as 3), and steals system access information and gets important idea about the systems communications. From there he gets successful in penetrating one of the desktop computers attached to the corporate LAN and installs backdoors to bypass the firewalls. Following a similar way, the attacker gets successful in accessing one of the web servers in the control DMZ and start logging the commands and communications (marked as 4).

Using the knowledge gathered from control DMZ, he becomes successful in getting access to the file systems in the configuration servers of the control LAN bypassing the firewalls (marked as 5). Then it is all about a suitable time to set and launch the logic time bomb attack (marked as 6), which may initiate at low traffic hours (generally at late nights when limited operators are available in the control centers to monitor the abnormality). Depending on the nature of the attack, the logic bomb may initiate destructive commands, delete important configuration file, delete backup files so that restoration of the system becomes difficult, and execute other actions depending on the information the attacker is able to gather about the CPS network. The regular lines on the diagram show the network internal connectivity, and the dashed lines on the diagram show the attack propagation.

The above example points to the necessity of modeling and design of the attack scenarios in a simulated environment to be able to make the system resilient from different types of cyberattacks. Again, to model the attacks, the network topology and the vulnerability information are necessary. We plan to use the directed multigraph as illustrated in Figure 12.6 to model and design the attack scenarios. One of the essential requirements to address the designing of attack scenarios is to map the vulnerabilities with the attack types. It is one of the areas we are currently working on.

12.6.5 Modeling Underlying Physical Processes and Design Constraints

CPS is the crucial part of the ICS, SCADA, smart grids, medical devices, and intelligent autonomous vehicles. Because of the diverse applications of the CPS, it is difficult to create a single cyber-resilient model considering the underlying physical processes and design constraints. The goal of this work is not to model the physical processes or bring any change to the current established physical systems, but to consider up to certain levels the physical processes in the cyber-resilient architecture modeling. The objective is not to contradict the modeling processes with each other, but to make them complement each other. Thus, there is a need for research works in the cyber-resilient modeling of the physical processes. Examples of such can be to incorporate the PLC ladder logic of the ICS or IEEE bus system analysis of the EDS into the cyber-resilient modeling architecture development. Often machine learning (ML) techniques are useful in designing the intrusion detection and prevention systems. ML can also be used in the complex decision-making processes where there does not exist enough data to support the underlying processes due to system confidentiality and regulatory requirements. Qualitative data with decision tree analysis generate convincing results in such cases.

12.7 Simulation Platform for CPS Resilience Metrics

The CPSs are complex. As there is lack of visibility of the underlying interconnections and message communications due to heterogeneous nature of the system, a simulation tool can assist to assess the CPS security and resilience posture to some extent without making any change to the existing system and physical processes. A simulation platform is proposed in this chapter to handle the complex CPS resilience assessment. The simulation tool has two broad approaches: qualitative approach and quantitative approach. Both approaches generate from the resilience framework discussed in Section 12.4, follow some form of mathematical analysis, and produce resilience metrics and analysis from two different angles as presented in Figure 12.9. The details of both simulation tools and their integration process are discussed in the following subsections.

12.7.1 Qualitative Simulation Platform

Figure 12.10 presents a high-level architecture of the simulation engine for the CPS qualitative simulation platform as illustrated by Haque, Shetty, and

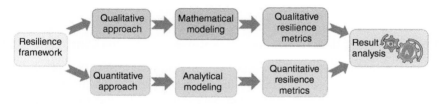

Figure 12.9   Process flowchart of cyber resilience tool for CPS.

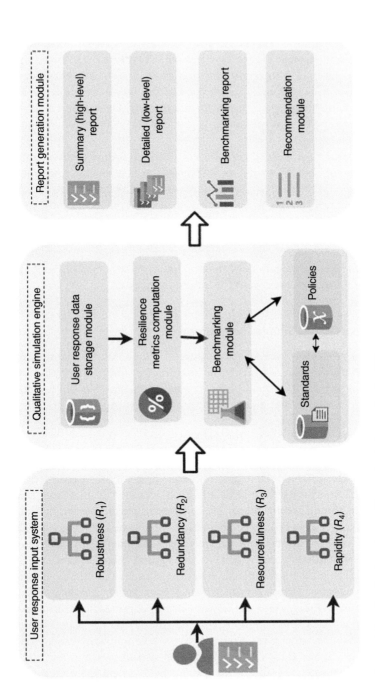

Figure 12.10 Qualitative cyber resilience simulation engine for CPS.

Krishnappa (2019). The user response system is the input systems where the users (CPS operators, engineers, cybersecurity experts, SMEs, etc.) are asked to respond to the questionnaires regarding the systems security posture. As illustrated in the framework in Table 12.1, the questionnaires are spanned across physical, organizational, and technical domain under each of the broad R4 resilience metrics: robustness, redundancy, resourcefulness, and rapidity. There are significant challenges in designing the questionnaires as some of the technical survey questions are application specific.

The responses provided by the users are sent to the qualitative simulation engine. The simulation engine is responsible for analyzing the qualitative responses, generating qualitative resilience metrics, and performing benchmarking with the industry standards and recommended best practices by different regulatory bodies (NIST, ICS-CERT, NERC-CIP, ISA, etc.). Lastly, the report generation module presents the analysis results in the form of a high-level summary report, a detailed low-level report, benchmarking report, etc. The report generation module also provides necessary recommendations to improve the overall resilience posture by comparing the assessment with the industry standards.

12.7.2 Quantitative Simulation Platform

Figure 12.11 presents a generic architecture of the quantitative simulation platform for CPS resilience assessment. The quantitative simulation platform is analogous to the testbed. The quantitative simulation platform is entirely different from the quantitative simulation platform and works independently. There are several management systems and modules to capture the resilience assessment perfectly. A brief description is provided here:

- *Interface management system (IMS):* IMS of the quantitative simulation platform is divided into admin and user interfaces. The admin interface can make changes in the simulation platform (e.g. the creation of new user accounts, modification in the input systems, etc.). The user interface is not allowed to create new users, but it can make changes in the network topology as required. Network topology input is the first step in simulation of the CPS resilience because the simulation engine will produce a graph-theoretic network based on the user inputs of the network components, software, and other applications.
- *Database management system (DMS):* DMS is the local repository of the commonly known vulnerabilities. It downloads the vulnerability information provided by the National Vulnerability Database (NVD) (NIST). The data are stored in a local database designed for the tool where it contains necessary vulnerability information and their CVSS exploitability and impact metrics (Mell et al., 2007). These metrics are used as the weights in the graph-theoretical models in the vulnerability graph module of the quantitative simulation engine.

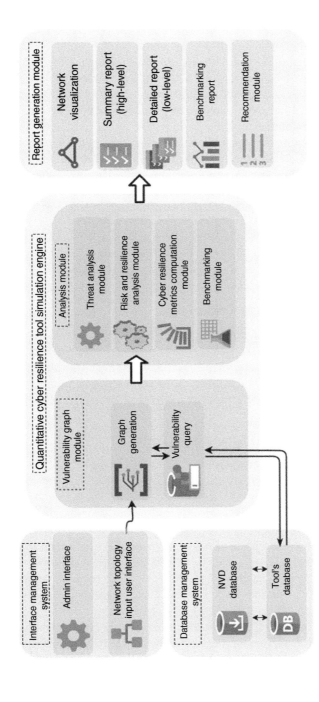

Figure 12.11 Quantitative cyber resilience simulation engine for CPS.

- *Quantitative simulation engine:* There are two major parts in the quantitative simulation engine. One is the vulnerability graph generation module, and the other is the analytical module. The vulnerability graph module takes the network topology input from the user, extracts the corresponding vulnerabilities for the network nodes by communicating with the vulnerability repository in DMS, and generates the graph model for the simulated network like what we present in Figure 12.6. The graph consists of edges where the edges represent the detected vulnerabilities, and the weights are quantitative exploitability and impact scores found from the exploitability and impact scores of confidentiality, integrity, and availability provided by the CVSS. As such, the simulation engine can only handle the known vulnerabilities, and it cannot handle the zero-day vulnerabilities. The analysis module analyzes the threats, computes the risk and resilience metrics, and performs the benchmarking for the resilience assessment of the CPS.
- *Report generation module:* The report generation module provides the network visualization, high-level summary report, a detailed low-level report, benchmarking report, and important recommendations for resilience improvement.

12.7.3 Verification and Validation Plan

The analytical models that are under development will be verified by the simulation platforms. Because of the lack of CPS security and resilience data, it is hard to validate the analytical results. We plan to use different statistical tools to validate our models. One such method is the multivariate analysis technique, which can be used to illustrate the relationships among the broad R4 metrics of resilience.

12.7.4 A Use Case of the Simulation Platform

A use case of the proposed simulation platform for the CPS of the oil and gas sector is presented using Figure 12.12. In this illustration, the physical layer is the bottommost layer where the physical processes occur. There are different sensors to monitor the system and the environment, such as temperature sensor, smoke detectors, fire alarm systems, security surveillance camera, ventilation system, lighting system, etc. The physical layer is transferring sensor data to the control center through the PLC, IED, or RTU in the cyber-physical layer. We consider the I/O processing devices (such as PLC, RTU, IED, etc.) and actuators to be the cyber-physical layer devices because these devices receive physical sensor data from field devices and controls the physical systems through the actuators using the commands initiated from the HMI in the control station. The control layer contains the control servers, master terminal units (MTU), HMI, application servers, data historian, etc. Control layer is responsible for monitoring and controlling

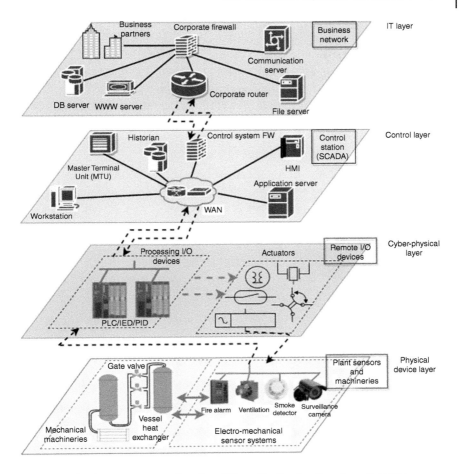

Figure 12.12 A use case example using CPS application in oil and gas industry.

the system performances. The IT layer communicates with the control layer to receive system data to evaluate the performance of the physical devices. We illustrate here the use case using both qualitative and quantitative approaches of the simulation platform.

Use Case of Qualitative Simulation Platform: The qualitative simulation platform is designed with a set of system-specific questionnaires to serve the security and resilience assessment for the use case. In the qualitative tool, the users are expected to provide the answers to the systems security-related questions.

The tool then aggregates the results and provides quantitative metrics that would not only help the executives but also help the system operators to analyze the security and resilience of the system using their assessment. Here we provide some of the usages of the proposed qualitative tool:

- *In-depth insights from resilience sub-metrics:* Figure 12.13 shows a sample simulation output generated from the proposed qualitative tool for the sub-metrics of robustness and resourcefulness. The tool produces the metrics directly from

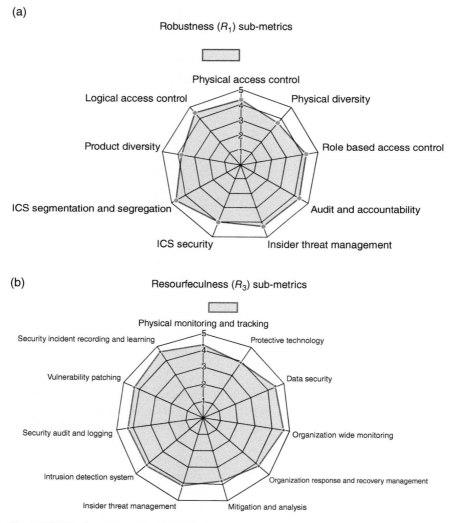

Figure 12.13 Sample qualitative metrics of (a) robustness and (b) resourcefulness generated from the qualitative tool.

the users' qualitative responses. In both the robustness and resourcefulness metrics, we see that most of the resilience areas are working fine with values ≥ 4.0 on a scale of 1.0–5.0, which means $\geq 75\%$ of resilient performances in those areas.

- *Overall insights on cyber resilience:* The qualitative tool computes the cyber resilience metrics R and its major components R_1, R_2, R_3, R_4. It also provides physical, technical, and organizational metrics under each of the R4 resilience metrics. Thus, it could give a satisfactory overall insight on the cyber resilience of the CPS based on the users' assessment.

Use Case of Quantitative Simulation Platform: In the quantitative simulation platform, the users need to set up the network topology up to the cyber-physical layer starting with the IT layer. In each of the layers, the user needs to provide the host, software, or applications running on the host, patch information, and the physical or logical connectivity. The tool then generates the vulnerability multigraph using the information provided by the user where the vulnerability information is extracted using the DMS of the tool as illustrated in Figure 12.11. The tool performs the analysis in the simulation engine and is expected to provide (but not limited to) the following information:

- *Network vulnerability visualization:* The tool provides a complete visualization of the system with the connectivity and vulnerability information, which would assist the operators to analyze the security of the system.
- *Cyber critical paths:* The tool would identify the critical assets in the network and provide the cyber critical paths with scores or probability of the attack using the underlying modeling and simulation techniques.
- *Risk and resilience metrics:* The tool would provide the cyber risk and resilience metrics in both high level and low level to facilitate further analysis by the network operators and security experts.
- *Benchmarking and recommendations:* The quantitative tool is designed to perform benchmarking with the industry and regulatory standards provided by the authorized bodies and provide meaningful recommendations for improving overall CPS network security and resilience posture.

12.8 Complexities, Challenges, and Future Directions

CPS is an active research field with numerous complexities and challenges from the cybersecurity perspective. The complications come from the complex nature of the system design, components heterogeneity, complex interconnections, lack of overall visibility, the trade-off between security and reliability of physical processes, etc. This section discusses some of the selected complexities and challenges of CPS security and future directions:

- *Security by design:* CPSs (e.g. ICS) are legacy systems where security is not considered as a design consideration because of the isolated command and control mechanisms and isolation from direct Internet communication. Physical security is thought to be enough for safeguarding the physical devices and underlying connections. Considering the widespread use of CPS and the growing need for different layers of interconnectivity, it is now the high time to include the security of the devices as an essential concern in the design processes.
- *Real-timeliness nature:* CPSs are real-time systems. The operational requirement of real-time availability makes CPS hard to assess a security threat and implement a preventive mechanism in real time within the tolerable delay limit. As pointed out by Humayed et al. (2017), cryptographic mechanisms could cause delays in the operations of real-time devices. Thus, the design of the CPS should consider inclusion of security mechanisms that recognize the real-timeliness nature and security processes without significant delay.
- *Heterogeneity and interconnectivity:* CPSs are heterogeneous in nature having complex interconnections of the cyber and physical layer devices. The proprietary protocols (e.g. Modbus and DNP3 in ICS or smart grids) are not free of vulnerabilities due to the isolation consideration during the design of the protocols. Thus, the communication due to the interconnections of the devices and protocol-level security need to be taken into account in the design process. The authors (Fovino, Carcano, Masera, and Trombetta, 2009; Majdalawieh, Parisi-Presicce, and Wijesekera, 2007) focus on this in details.
- *Underlying physical processes:* CPSs are designed to operate closely with the embedded physical processes. The complex underlying processes reduce the visibility of the overall security of the CPS, and thus, the underlying processes and dependencies should be considered in the resilient network topology design process.
- *IDS and IPS:* There are some fundamental differences of the IDS and IPS design between the ITS and the CPS. In ITS, the security is ensured by software patching and frequent updating, which is not suitable for the CPS because of limited memory resources. Another concern in the IDS and IPS design for the CPS is that CPSs are real time, and thus, patching of those devices that needs to take them offline by suspending operations is difficult to justify economically and operationally. Therefore, design of IDS and IPS solutions specific to the CPS and design of novel attack-detection algorithms are the need of the time. The article by Mitchell and Chen (2014) focuses on the development of the IDS solutions applicable to the CPS.

- *Secure integration:* The integration of new components with the existing CPS should perform security testing before putting online. There is a need for vulnerability assessment of the components to be added to the existing systems.

- *Understanding the consequences of an attack:* Often it is difficult to visualize the consequences of an attack in the ICS or CPS. It is essential to perform penetration tests and assess the impacts of a cyberattack on the CPS. Also, the prediction of the attacker strategy is crucial in defending the CPS network. The organizational and technical security policy should consider performing penetration tests up to certain security levels.

- *Malicious insider:* It is one of the most difficult challenges in securing the CPS against cyberattacks. Often disgruntled employees can be an attack vector for the targeted attacks on the CPS, which are hard to identify. There is also social engineering, which may have temptation effect on an employee with ICS credentialed access. Often unintentional attacks are also possible such as the use of USB sticks. Thus, organizations should have clearly defined policies and conduct security training for the employees to be aware of the situation and responsibilities.

- *Resilience, robustness, security, and reliability:* Resilience, robustness, and security are critical challenges from the cybersecurity perspective for the CPS devices. Often CPSs are designed to be reliable as the stability of operation is considered most important during the design process. Although the reliability is utmost necessary for the physical devices, security and resilience cannot be ignored. The uncertainty in the environment, cyberattacks, and errors in physical devices threaten the overall CPS security and reliability. Therefore, the resilience and security of the devices need to be considered equally important as reliability during the design phase.

The big question that comes naturally is how the proposed simulation platform to assess the resilience metrics for the CPS would handle the above challenges. As we have explained in Section 12.3, the availability of the resilience metrics would provide guidelines and directions in the different stages of CPS operations (e.g. design process, monitoring, recovery, etc.) and ensures the overall security by pointing to the improvement areas with essential recommendations. The qualitative and quantitative methods of resilience assessment are complementary to each other, and they are designed to handle the security challenges in the CPS by providing quantitative operational metrics of cyber resilience for the CPS. Including the physical processes and cyber-physical interconnectivity in the simulation platform would be a big challenge because of system connectivity changes based on the application area. Thus, the authors would like to include partial if not complete physical and cyber-physical components into the simulation platform. It is also possible to add multiple simulation modules where each module would serve specific CPS application area (such as one module for the smart grid, another module for oil and gas, etc.). Modeling and simulation of the complete CPS regardless of the application area would be a mammoth task and could be saved for future works.

12.9 Conclusion

CPSs play a vital role in critical infrastructures, ICSs, and many other applications. The growing interdependencies and interconnections among different parts of CPS make it vulnerable to cyber threats than any time before. The complex nature of CPS in conjunction with the lack of clear visibility due to the integration of different cyber, cyber-physical, and physical components make it difficult to handle the security concerns and resilience challenges. The chapter aims to present essential cyber resilience metrics for the CPS and proposes a simulation platform to help in the automation of the resilience computation process. Among the other discussions, the authors try to explain the need for the resilience metrics in the CPS and the way the proposed resilience metrics can help in identifying technical areas for improvement and generate recommendations. The authors discuss the complexities and challenges to secure the CPS from cyberattacks and answer the big question of how the resilience metrics can assist in handling those challenges. Overall, the authors believe that the chapter provides the necessary guidance in the ongoing CPS resilience and security analysis and would also provide directions for future research.

Acknowledgment

This material is based upon work supported by the Department of Energy under Award Number DE-OE0000780.

Disclaimer

This report was prepared as an account of work sponsored by an agency of the U.S. government. Neither the U.S. government nor any agency thereof nor any of their employees makes any warranty, express or implied, or assumes any legal liability or responsibility for the accuracy, completeness, or usefulness of any information, apparatus, product, or process disclosed, or represents that its use would not infringe privately owned rights. Reference herein to any specific commercial product, process, or service by trade name, trademark, manufacturer, or otherwise does not necessarily constitute or imply its endorsement, recommendation, or favoring by the U.S. government or any agency thereof. The views and opinions of authors expressed herein do not necessarily state or reflect those of the U.S. government or any agency thereof.

References

Amin, M. (2015). Smart grid. *Public Utilities Fortnightly*, March 2015.

Baheti, R., & Gill, H. (2011). Cyber-physical systems. *The Impact of Control Technology*, *12*(1), 161–166.

Bodeau, D., & Graubart, R. (2011). Cyber resiliency engineering framework. *MTR110237, MITRE Corporation*.

Bologna, S., Fasani, A., & Martellini, M. (2013). Cyber security and resilience of industrial control systems and critical infrastructures. In *Cyber Security* (pp. 57–72). Springer.

Bruneau, M., Chang, S. E., Eguchi, R. T., Lee, G. C., O'Rourke, T. D., Reinhorn, A. M., ... Von Winterfeldt, D. (2003). A framework to quantitatively assess and enhance the seismic resilience of communities. *Earthquake Spectra*, *19*(4), 733–752.

Cardenas, A., Amin, S., Sinopoli, B., Giani, A., Perrig, A., & Sastry, S. (2009). *Challenges for securing cyber physical systems.* Paper presented at the Workshop on Future Directions in Cyber-Physical Systems Security.

Cheng, A. M. (2008). *Cyber-physical medical and medication systems.* Paper presented at the 28th International Conference on Distributed Computing Systems Workshops (ICDCS).

Collier, Z. A., Panwar, M., Ganin, A. A., Kott, A., & Linkov, I. (2016). Security metrics in industrial control systems. In *Cyber-Security of SCADA and Other Industrial Control Systems* (pp. 167–185). Cham: Springer.

Cutter, S. L., Ahearn, J. A., Amadei, B., Crawford, P., Eide, E. A., Galloway, G. E., et al. (2013). Disaster resilience: A national imperative. *Environment: Science and Policy for Sustainable Development*, *55*(2), 25–29.

Fovino, I. N., Carcano, A., Masera, M., & Trombetta, A. (2009). Design and implementation of a secure modbus protocol. In *International Conference on Critical Infrastructure Protection* (pp. 83–96). Berlin, Heidelberg: Springer.

Galloway, B., & Hancke, G. P. (2013). Introduction to industrial control networks. *IEEE Communications Surveys and Tutorials*, *15*(2), 860–880.

Gamarra, M., Shetty, S., Nicol, D. M., Gonazlez, O., Kamhoua, C. A., & Njilla, L. (2018). *Analysis of stepping stone attacks in dynamic vulnerability graphs.* Paper presented at the 2018 IEEE International Conference on Communications (ICC).

Ginter, A. (2017). The top 20 cyber attacks on industrial control systems. *Waterfall Security Solutions*.

Haque, M. A. (2018). *Analysis of bulk power system resilience using vulnerability graph* (Master of Science (MS) thesis). Modeling Simulation and Visualization Engineering, Old Dominion University. doi:https://doi.org/10.25777/fqw2-xv37.

Haque, M. A., De Teyou, G. K., Shetty, S., & Krishnappa, B. (2018). *Cyber resilience framework for industrial control systems: Concepts, metrics, and insights.* Paper presented at the 2018 IEEE International Conference on Intelligence and Security Informatics (ISI).

Haque, M. A., Shetty, S., & Kamdem, G. (2018). Improving bulk power system resilience by ranking critical nodes in the vulnerability graph. In *Proceedings of the Annual Simulation Symposium* (pp. 8). Society for Computer Simulation International.

Haque, M. A., Shetty, S., & Krishnappa, B. (2019). *ICS-CRAT: A cyber resilience assessment tool for industrial control systems.* Paper presented at the 4th IEEE International Conference on Intelligent Data and Security (IDS).

Humayed, A., Lin, J., Li, F., & Luo, B. (2017). Cyber-physical systems security: A survey. *IEEE Internet of Things Journal, 4*(6), 1802–1831.

Kellett, M. (2016). Ranking assets based on criticality and adversarial interest. *Defence Research and Development Canada.*

Koutsoukos, X., Karsai, G., Laszka, A., Neema, H., Potteiger, B., Volgyesi, P., ... Sztipanovits, J. (2017). SURE: A modeling and simulation integration platform for evaluation of secure and resilient cyber-physical systems. *Proceedings of the IEEE, 106*(1), 93–112.

Laing, C. (Ed.) (2012). *Securing Critical Infrastructures and Critical Control Systems: Approaches for Threat Protection.* USA: IGI Global.

Linkov, I., Eisenberg, D. A., Bates, M. E., Chang, D., Convertino, M., Allen, J. H., ... Seager, T. P. (2013). *Measurable Resilience for Actionable Policy* (pp. 10108–10110). USA: ACS Publications.

Linkov, I., Eisenberg, D. A., Plourde, K., Seager, T. P., Allen, J., & Kott, A. (2013). Resilience metrics for cyber systems. *Environment Systems and Decisions, 33*(4), 471–476.

Macaulay, T., & Singer, B. L. (2016). *Cybersecurity for Industrial Control Systems: SCADA, DCS, PLC, HMI, and SIS.* UK: Auerbach Publications.

Majdalawieh, M., Parisi-Presicce, F., & Wijesekera, D. (2007). DNPSec: Distributed network protocol version 3 (DNP3) security framework. In *Advances in Computer, Information, and Systems Sciences, and Engineering* (pp. 227–234). Springer.

McIntyre, A., Becker, B., & Halbgewachs, R. (2007). Security metrics for process control systems. *Sandia National Laboratories* (Sandia Report SAND2007-2070P).

McMillin, B., Gill, C., Crow, M., Liu, F., Niehaus, D., Potthast, A., & Tauritz, D. (2007). Cyber-physical systems distributed control: The advanced electric power grid. *Proceedings of Electrical Energy Storage Applications and Technologies.*

Mell, P., Scarfone, K., & Romanosky, S. (2007). *A complete guide to the common vulnerability scoring system version 2.0*. Paper presented at the Published by FIRST-Forum of Incident Response and Security Teams.

Mitchell, R., & Chen, I. R. (2014). A survey of intrusion detection techniques for cyber-physical systems. *ACM Computing Surveys (CSUR)*, *46*(4), 55.

Nicol, D. M., & Mallapura, V. (2014). *Modeling and analysis of stepping stone attacks*. Paper presented at the Proceedings of the 2014 Winter Simulation Conference.

Saaty, T. L. (2008). Relative measurement and its generalization in decision making why pairwise comparisons are central in mathematics for the measurement of intangible factors the analytic hierarchy/network process. *RACSAM-Revista de la Real Academia de Ciencias Exactas. Fisicas y Naturales. Serie A. Matematicas*, *102*(2), 251–318.

Sedgewick, A. (2014). *Framework for improving critical infrastructure cybersecurity, version 1.0*. NIST-Cybersecurity Framework.

Stouffer, K., Falco, J., & Scarfone, K. (2011). Guide to industrial control systems (ICS) security. *NIST Special Publication*, *800*(82), 16–16.

Tierney, K., & Bruneau, M. (2007). Conceptualizing and measuring resilience: A key to disaster loss reduction. *TR News* (250).

Tran, H., Campos-Nanez, E., Fomin, P., & Wasek, J. (2016). Cyber resilience recovery model to combat zero-day malware attacks. *Computers and Security*, *61*, 19–31.

Wang, F. Y. (2010). The emergence of intelligent enterprises: From CPS to CPSS. *IEEE Intelligent Systems*, *25*(4), 85–88.

Wei, D., & Ji, K. (2010). *Resilient industrial control system (RICS): Concepts, formulation, metrics, and insights*. Paper presented at the 3rd International Symposium on Resilient Control Systems (ISRCS), 2010.

Wilamowski, G. C., Dever, J. R., & Stuban, S. M. (2017). Using analytical hierarchy and analytical network processes to create Cyber Security Metrics. *Defense Acquisition Research Journal: A Publication of the Defense Acquisition University*, *24*(2), 186–221.

Xiong, G., Zhu, F., Liu, X., Dong, X., Huang, W., Chen, S., & Zhao, K. (2015). Cyber-physical-social system in intelligent transportation. *IEEE/CAA Journal of Automatica Sinica*, *2*(3), 320–333.

Yong, S. Z., Foo, M. Q., & Frazzoli, E. (2016). *Robust and resilient estimation for cyber-physical systems under adversarial attacks*. Paper presented at the American Control Conference (ACC), 2016.

Yu, X., & Xue, Y. (2016). Smart grids: A cyber-physical systems perspective. *Proceedings of the IEEE*, *104*(5), 1058–1070.

Zhang, Y., & Paxson, V. (2000). *Detecting stepping stones*. Paper presented at the USENIX Security Symposium.

13

The Cyber Creation of Social Structures

E. Dante Suarez[1] and Loren Demerath[2]

[1] *School of Business, Department of Finance and Decision Sciences, Trinity University, San Antonio, TX, USA*
[2] *Department of Sociology, Centenary College of Louisiana, Shreveport, LA, USA*

13.1   Introduction

The use of computer simulation has changed the way that science is done in the twenty-first century. Computational models are currently being used to evaluate the effects of climate change, the possibility of a financial meltdown, or the probability of you buying your holiday gifts online. We argue in this chapter that the trend towards increasing reliance on these methodologies will continue apace, not only in areas where such methodologies are already commonplace, but in all aspects of research, work, and our daily lives. In this work we attempt to marry some of the insights of traditional social science with those of complexity studies and systems thinking to provide the basis for a holistic paradigm that can inform the incoming wave of Cyber Physical Systems (CPS) that will surely inundate our lives in years to come.

The popular conception of this incoming wave of computer-oriented solutions to model our world is often encapsulated in the general and open term of Big Data (see Akoka et al. (2017) for a comprehensive view on current literature on the topic). Big Data is a concept that is usually applied loosely to reference approaches that involve vast amounts of data, complex numerical techniques, and complicated and descriptive models of human behavior. This growing body of literature comes well equipped with hordes of scholars, practitioners, and college graduates ready to implement these novel approaches to any phenomenon they encounter, particularly to situations that are not well served by more mundane, linear methodologies. A common factor in some of these applications is a lack of sufficient attention to the underlying theory relating the model to the observed behavior, or to the versatility of the model

Complexity Challenges in Cyber Physical Systems: Using Modeling and Simulation (M&S) to Support Intelligence, Adaptation and Autonomy, First Edition. Edited by Saurabh Mittal and Andreas Tolk.
© 2020 John Wiley & Sons, Inc. Published 2020 by John Wiley & Sons, Inc.

and its range of possible applications. A simulation that is produced by a data-mining process may contain multiple topological aspects that are particular to the way that the model is built, and therefore not a true reflection of the real-world phenomenon at hand. However, it is important to stress that the novelty of Big Data does not come only from the large number of observations available. After all, most statistical methodologies are asymptotic, and are at least in theory built on the assumption of an infinite number of data points. The novelty of Big Data mostly stems from the fact that it is able to marry different types of data, from qualitative to quantitative, with different formats and granularities, even incorporating variables that are hard to describe and measure in normal circumstances – the so called "fuzzy variables." We argue that this matches well with the multiple levels of reality that come about in a world full of emergent phenomena, and that this is an essential feature of the cyber/human systems that this book explores.

The next step in the evolution of computer-aided descriptions of our reality involves a direct interaction between the machine and our world. Enter the new world of Cyber Physical Systems (CPS), generally defined as systems with integrated physical and computational capabilities that can interact with humans through a variety of modalities (Baheti and Gill 2011). This broad definition seems to fit most aspects of the modern world we live in, where computers have become an integral part of our daily life. The types of CPS that this article is making reference to, however, often involve a large combination of heterogeneous and distributed elements, such as sensors and robots, as well as computers and humans. CPS provide a new arena where computers can directly interact with humans and other actors to create fully integrated systems where the boundaries between human interaction, social dynamics, and artificial systems are blurred. In an upcoming paper proposing human simulation as a lingua franca for the computational social sciences and humanities, Tolk et al. (2018c) describe these types of interactions in the context of modeling and simulation. In the abstract of this paper, they put forth the following strong manifesto:

> Models are well established in the scientific community as mediators, contributors and enablers of scientific knowledge. We propose a potentially revolutionary linkage between the social sciences, humanities and computer simulation, forging what we call "human simulation." We explore three facets of human simulation, namely: (1) the simulation of humans, (2) the design of simulations for human use, and (3) simulations that include humans as well as simulated among the actors.

If a simulation includes human actors integrated into the processes of the system, then it represents one of these CPS. Such simulation has the potential to become an integral part of our world, providing us an opportunity to evaluate different possible outcomes for even a minor change in the variables, configuration, or components in the model. The CPS as a whole can subdivide the tasks and then explore aspects of reality where computational capabilities can be of most use. In other areas of human ontological emergence that are inaccessible to the simulation, the CPS will leave such questions in the hands of human actors, later integrating them once such questions have been resolved.

Recent advances in the field of hybrid modeling have increased the usefulness of CPS (see Tolk et al. (2018b) for a comprehensive view of the current Literature on the field). Their abilities to obtain and process information while the systems are in operation have made them increasingly autonomous. The recent autonomous landing of NASA's latest rover on Mars is but one example. But the autonomy of CPSs is also a reason for caution. Worries can range from tragedies involving military drones, to robot overlords, benevolent or not. The concerns are legitimate.

Just as with Big Data, the problem with these types of hybrid simulations lies in the integration of models with vastly different types of data. However, as Tolk et al. (2018a) point out, many of these interoperability issues have now been resolved. In particular, the authors present and promote the Hybrid Flow System Specification (HyFlow) formalism, which provides a description of modular hybrid models of all kinds. "Hyflow provides a new approach to the representation of models described in multiple paradigms. Instead of treating models as heterogeneous, HyFlow promotes a unifying view where all models are regarded as a realization of the basic HyFlow model." Tolk et al. (2018a) point out that "since the co-simulation of HyFlow models is guaranteed, the interoperability of new families of models can be achieved by expressing these new models in the formalism. HyFlow also provides the ability to dynamically change model composition. Dynamic topologies make it easier to express adaptable CPSs that modify the interaction between components, or even that change dynamically their set of components."

The interoperability of HyFlow is guaranteed by construction, since models share the same underlying description (see Barros (2002) for a formal description of these types of continuous flow models). Furthermore, the wave of Big Data is indeed fed by vast amounts of data, as our current continuous interaction with computers and the internet generate vast amounts of information on our behavior, providing researchers with data sets larger than anything seen before. Big Data allows us to very precisely measure human interactions and preferences. Taken together, these huge amounts of data coupled with an established way in which to integrate it into holistic models force us to reconceptualize the possibilities of the scientific endeavor.

Traditional social sciences were not developed with these types of capabilities for experimentation in mind. Experimenting with humans was simply not a possibility in macroeconomics or sociology. In contrast, we now face a reality in which a company such as the ride-sharing Uber can let a computer algorithm loose on the task of making its non-centralized drivers most efficient. As Gelfert (2016) points out, models are not only neutral abstractions of our world, they are the guiding light with which we conceive, mediate, contribute, and enable our scientific knowledge. Just as language shapes the world we live in, scientific models both constrain and contextualize our perception of our physical reality. In short, the existing CPS that are now available to scientists force a reconceptualization of the type of modeling and theorizing that is possible.

This work and others in this book fit within the paradigm of complex adaptive systems, in which the parts of the system and the system itself can adapt in multiple and novel ways. The emergent phenomena resulting from the interactions of the components is an essential feature of physical, natural, and social systems in our world, and the development of computational capabilities has allowed for novel ways to model these complex phenomena. The message of this article is that CPS let us peek into a realm hitherto unavailable to science. Since the final decades of the twentieth century, the reigning scientific paradigms had a tremendous boost from the insights of Complex Adaptive Systems and the added capabilities of computers. True ontologically emergent phenomena, however, lie outside of the field of exploration available, as computer simulations represent computable functions that can only transform the data fed to them. CPS can now bring together the physical world and its ontological nature with the power of computer algorithms attempting to transform it.

13.2   The Emergence of Cyber Physical Systems

One of the most important concepts in complexity studies is that of *emergence*, which is often characterized to mean that the aggregate is more than the sum of its parts, and as such we can understand a human mind as more than a collection of neurons. If we accept the concept of emergence, then we must conceive of a world that exists in multiple realities simultaneously. Pessa (2002) and Tolk (2019) demystify the concept of emergence and separate it into two categories: *epistemological* and *ontological*. Epistemological emergence refers to situations where the system follows generalizable rules, but it is irreducible to one single level of description because of conceptual reasons. A system may therefore be epistemologically emergent if it follows rules that in principle are knowable; the unpredictability is a product of humans not being able to fully grasp these laws in their entirety. Pessa (2002) points out that emergence can only be defined relatively to a given observer. This means that what may be emergent to someone may be an obvious and direct result of the interacting parts of the system to another. Tolk (2019) stresses the fact that computer simulations of these systems can only produce this type of epistemological emergence, as anything that is coded in the computer can be translated into knowable and reproducible mathematical functions. Most recently, Mittal (2019) summarizes the research on the applicability of M&S for emergence research, focusing on the synthetic creation of emergence by computational means.

Crutchfield (1994), on the other hand, defines intrinsic emergence as that which is independent of the external observer. The emergent properties are truly novel and cannot be explained solely by the components and their interactions. Take biological life, for example: even if we understand all the physical and chemical laws that underpin it, we cannot say that we truly understand it, much less that we can recreate it. Following Tolk (2019), we refer to this type of true emergence as ontological. The main point of this work is to establish that although ontological emergence is beyond the capabilities of a computer simulation, it is now potentially within the scope of a CPS that involves humans.

Just as CPS brings together human and artificial systems, this new area of research offers a unique opportunity to bridge the gap that has always existed between the material and social sciences. While computer scientists may not know what to do with the vast computing power they bring to the battle, social scientists can benefit from a greater empirical foundation for their work. CPS offers us the possibility to marry the strengths of both to create a truly holistic vision of reality, a vision backed by the data and models of computer scientists and supported by the bolder, more comprehensive theories offered by social scientists. Once we have accepted a reality full of emergent behaviors and phenomena, we must begin to reconsider our understanding of social actors and their multileveled nature. This does not only mean that a person must wear many hats within a day, but rather that a person is many things at the same time. If I am hungry and I am on a diet, I may only eat half the cake. This is a decision that is the result of the interacting levels of agency in which I exist. Reality thus is multidimensional, and our models need to find way to capture this.

It is in the context of these definitions of emergence, we propose the scientific study of CPS as a way to better understand and control ontological emergence. CPS expose the characteristics of intelligent, adaptive, and autonomous systems, operating in complex environments, often in collaboration with other systems and humans. CPS represent a new frontier in the study of emergence, and while this is an entirely new field, social and computer sciences can help us better understand the challenges and opportunities that lie ahead. These new pan-disciplinary frontiers of science present particular needs for novel interdisciplinary methodologies. The authors discuss in this work novel languages in which such concepts can be explained.

Reality has aspects that are mostly linear and some others that are nonlinear and thus irreducible. Some aspects of reality may not be linear nor emergent, but rather not computable at all (see Wolfram (1984) for a discussion on non-computability). Models and computer simulations, on the other hand, are usually exclusively linear or nonlinear, and, as argued by Tolk (2019) they can only produce epistemological emergence. Therefore, when working with models and computer simulations, it is of utmost importance that the researcher understand the implications of her model, as well as its aspects that are linear and nonlinear. The goal of the simulation must be to capture the aspects of reality that are computable, whether these are linear or emergent, and to minimize aspects of the model which are not compatible with reality. Of course, no model can ever be perfect, and thus it rests on the researcher to be upfront and straightforward about a model's limitations and applicability.

While research in many scientific areas is incorporating some of these nonlinear ideas, it is the growing paradigm of Modeling and Simulation (M&S) that is developing the methodologies for capturing the complexities of our natural and social worlds. We dedicate some of this work to discuss some salient issues for the M&S paradigm and the dangers of over-utilizing and over-interpreting the results of a computer simulation. We also discuss advances in social science and complexity studies to provide a context for modeling and simulation methodologies and CPS hybrid interactions. We hope that providing these overlapping contexts will help social and computer scientists, as well as policy makers, to recognize how cyber physical systems can be integrated and welcomed as assets, even when they are increasingly autonomous. Understanding the world is a transdisciplinary project, and this paper exemplifies the approach. We will borrow concepts and principles from social science, complexity studies, and the M&S domain used for cyber physical systems as we attempt to understand emergent social systems, and how to optimize them.

For this work, we take an ample view of what can be considered a CPS, as we claim that many modern human institutions already qualify to be a CPS. It can be argued that humans have been experimenting with ways to build ever more complex social structures. Using a cyber-dimension is a new aspect of our human reality and the history of civilization. A theoretical orientation drawn from complexity studies will help provide that context; one which situates CPS within a history increasing connectivity and efficiency in information processing. But that history also contains many cautionary tales of unhealthy norms, of strangling elites, and technological lock-ins that create suboptimal social outcomes in the long run. To grow optimally, one must learn the lessons in balance that studies in complexity seem to teach repeatedly: there are times for exploration and exploitation; moments when integration is appropriate, and others when segregation is; points where you need to constrain, and others where you should liberate. Giving us increasing precision for identifying these moments is the science of complexity. It studies the self-organization of energy and information into systems, including life, consciousness, and language. Furthermore, social science has developed a way of thinking about complex systems that explains the emergence and evolution of social structures such as institutions, organizations, and markets. Drawing from that work here, we offer a perspective that should help hybrid models to be welcomed into any of these domains. At the same time, a benefit of this perspective is the creation of a reserved and privileged position of human judgement, as increasingly autonomous systems present ever-more informed suggestions for action.

The world of nonlinearity is vast. It is by definition almost too vast. This stands in contrast with linear methodologies, where centuries of scientific endeavor have coalesced them to simple principles for evaluating models and theories. In other words, the linear paradigm has found an Occam's Razor charge to look for the simplest possible model and a desire to find theories that make clear, understandable predictions. This approach would have satisfied Karl Popper's falsificationist methodology. Controversies between alternative linear descriptions of the world are often settled when one model can be shown to be "nested" in another, and the search for simplicity makes the simplest possible explanation the most appropriate one. But what happens in the world of M&S? When trying to capture the complexities of real-world behavior, the goal is no longer to find the simplest possible description of the phenomenon, for this approach would miss key emergent aspects of the structure under consideration.

Nonlinear science is thus less tractable, as it aims to understand systems and behavior that is sensitive to initial conditions, and therefore not fully predictable; epistemologically emergent, and as such not describable with simple linear relationships; and ontologically emergent, including aspects that are beyond our current models and methodologies. The nonlinear approach to understanding reality is still in flux. In essence, there are no established ways to compare competing nonlinear descriptions, only between a nonlinear model and the linear benchmark (see Axtel et al. (1996) for a notable counterexample). The moral of the story here is that the world is multifaceted. Some aspects of it are easily describable by linear methodologies such as a linear regression. Others can be captured by a Multi-Agent Simulation (MAS). But true, ontological emergence is arguably beyond the scope of a pure computer simulation. CPS allows us to explore a new frontier in science. That is, the possibility to model and experiment with ontological emergence.

In the British comedy *the hitchhiker's guide to the galaxy*, an advanced extraterrestrial civilization creates a computer that can calculate the ultimate question of life the universe and everything, for which the answer turns out to be 42. When the civilization that originally asked such a broad question about the meaning of life is befuddled by the concise answer, the computer that gave the answer explains to them that they did not fully understand what they were asking. But this computer that found the answer also conceived of a bigger, more complex computer that would be able to decipher the question that was actually answered. It turns out that this computer is earth, the quintessential Cyber Physical System.

13.3 Distributed Agency: A Language to Describe Multileveled Structures and Agencies

There is a common reductionist past for disciplines ranging from evolutionary biology to traditional social sciences such as economics. We use working definitions of linear and nonlinear in this work, where the concept of *linear* represents systems where the aggregate is exactly equal to the sum of its parts, meaning that the way that the components are organized is irrelevant. Contrary to this conception, in *nonlinear* systems the aggregate is more than the sum of its parts, because the way that the components interact creates relevant structures that emerge when coupled together. Therefore, in the nonlinear world of systems and complexity, the interaction and adaptability of the components is usually a core element that cannot be ignored. The reason for this pervasive linear paradigm is that the mathematics necessary for understanding the complexities of a myriad of interactions is intractable. Such emergence implicitly and explicitly separates our world into multiple levels, each with its own rules, granularities, and units.

Naturally, some aspects of traditional economics do indeed consider some interactions between agents, such as in the subfield of Game Theory, where strategic decisions between parties are considered. But these tactical interactions are often only studied at the *micro level*, essentially assuming that they will cancel out in the aggregate. The nonlinear paradigm has taken this baton and developed it in a growing body of literature known as Multi-Agent Systems (MAS), where agents interact and give rise to emergent aggregate behavior. Even this expansion into the nonlinear world bears many of the same traits of its linear past, as the agents considered continue to have a linear DNA – they are most often exogenous, irreducible, and defined in isolation. The idea of emergence reflects the fact that different and irreducible levels of interaction will naturally arise in complex systems such as the ones studied by social disciplines, and thus the agent, as we define it in this work, is a combination of levels of interaction.

It is through this lens that we would like to consider humans, who are partly independent creatures, possessors of free will, but who are also defined by an array of upper-level emergent structures that "suggest" agreeable ways in which to behave. This conception stems directly from the concept of complexity, in which wholes are more than the sum of their parts, and can only be described in multileveled models. The language used in this work redefines agents in two ways. First, there are no obvious atomic agents, for all actors represent the emerging force resulting from the organization of relatively independent subsets. Second, agents are not created in a vacuum, but are rather the result of what an upper level spawns. As such, agency is granted in a quantitative, rather than the traditional qualitative way. The relevant agents in the proposed system are intermediate or *holonic*, in the sense that they are both influenced by an upper level with its own degree of agency, while at the same time they are determined by relatively independent subcomponents that must be "subdued" into acceptable behavior. Any observed action is considered to be the result of the interplay of multiple distinguishable actors. We refer to this methodology for modeling as Distributed Agency (Suarez and Castanon-Puga 2013).

Consider for example a deconstructed agent that is formed by multiple and relatively independent components. Part of the resulting agent's task is to present alternatives, or "fields of action" to its components. The lower-level agent is itself constrained by a field of action that the superstructure to which it belongs presents. The lower-level agent may act in a finite number of ways, and each potential action is answered by the upper level. The lower-level agent can order the answers of the upper level in terms of desirability, and we refer to this field as a *reward function*. We drop here customary assumptions made in traditional social disciplines and MAS about what is considered a decision-making unit (see Bankes (2002) for a review of MAS). To arrive at this, we redefine what a unit of decision is by unscrambling behavioral influences to the point of not being able to clearly delineate what the individual is, who is part of a group and who is not, or where a realm of influence ends; the boundary between an individual self and its social coordinates is dissolved. A multileveled model may be able to capture such levels of agency, but it may not be able to describe all aspects of a complex human reality through the lens of one single level of atomic agents.

This type of deconstructed conceptualization of agency allows researchers to explore, codify, and communicate aspects of socio-spatial realities that were until recently only describable through written language, and therefore mostly only studied by a discipline such as sociology. However, disciplines such as economics and business rely on an appropriate and complete understanding of the ways in which cultures and other aggregate structures codify interactions to produce desirable emergent social and entrepreneurial outcomes. The possibility of modeling such *meso* structures allows CPS a language in which it can model and operationalize human interactions in a hybrid, physical, interconnected, and computational way that did not exist until very recently. The vision for a CPS future described in this book proposes that computational capabilities, once we learn how to apply them properly, will allow for the reconstruction of our families, enterprises, nations, and emergent global institutions.

Figure 13.1 defines the concept of a distributed agent in more detail. At the extreme of the *x*-axis we find a liver, which represents a stereotypical object without relevant

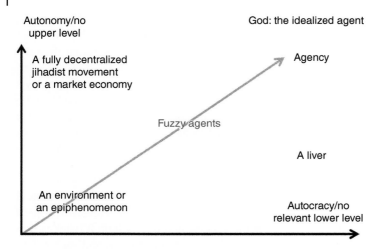

Figure 13.1 Fuzzy Agents in a complex world.

subcomponents, but also lacking independence. Unless it is not functioning properly, the "objective function" of the liver is in complete coordination with its upper-level agent, the body. To the extent that an upper-level agent, such as the body, coordinates its subcomponents, its lower levels seemingly disappear. In such cases, the lower-level components enjoy no agency, just as the soldiers of an army would if they were perfectly trained to follow all orders. On the other extreme, at the end of the *y*-axis, we find a group full of unconstrained agents, such as the one conceived of in traditional economic science, where there is no earth to take care of, but only selfish individuals maximizing their own utility functions. In traditional economics, society does not exist. Unlike this case, the types of fuzzy agents we consider are normally beset within the limits imposed by the upper level. For example, we can conceive of a company (formed by groups of investors, managers, and employees) as set within the boundaries of an industry, and unable to define itself in ways that go against established social norms and applicable laws. Upper levels can therefore represent coordination, identification with others, identities, institutions, implicit laws, religion, a credit bureau, or any human relationship. An environment, however, is not by itself an upper-level agent, for the environment may be independent of the creatures that inhabit it. The idealized agent appears in the upper right hand corner of the quadrant, since it possesses an objective function completely independent of its environment. Aside from these extreme cases, all other agent-like entities in the proposed language of Distributed Agency (DA) enjoy only limited agency.

In this language, we begin with a descriptive benchmark position in which all behavior is the result of optimizing fuzzy agents, intertwined and potentially defined in multiple dimensions. In this sense, when we arbitrarily zoom-in and analyze a traditionally well-defined agent (such as a human), we may classify its behavior as suboptimal, but only because we would be artificially studying it in isolation, or without regard for the struggles of its internal nature. The upper level may force lower-level members into behaviors that are only considered optimal for the former, and it is in this sense that we can understand the behavior of a kamikaze pilot. All possible definitions of optimality in a hierarchically disassembled world are relative. The relativity is a direct result of the hierarchical nature of the system, where each agent binds the subagents that compose it, and is bounded by the super-agent to which it belongs.

In a multileveled world of emergence, there is no clear distinction between agents and objects, as the same actor can be considered an agent in one level and an object in another. Therefore, agency is conceived as a quantitative, rather than a qualitative characteristic.

Traditionally, we began with a clearly defined agent and tried to understand its actions as a maximization of objectives given constraints. In a distributed agency representation we assume maximization occurs, and then work towards the delineation of the benefited entity, or agent involved. As such, this proposition is not a theory or hypothesis; it is a methodology or language in which different models can be expressed. As is expressed in Figure 13.2, **4** represents the actual

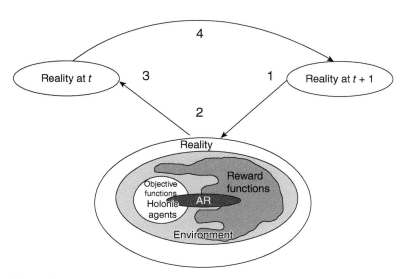

Figure 13.2 The backwards induction methodology that Distributed Agency proposes.

dynamics of the social phenomena analyzed, while **3, 2, 1** represent the modeling methodology proposed, which is in reverse order to more traditional approaches (see Miller and Page (2007) for a fuller description of the traditional approach). Whereas **1** – the moment in which we compare what our model produces with reality – traditionally represents the validation, the distributed agency model is validated when the conceived agents it produces are found in reality, here represented as **3**.

In this figure, an Agent Representation (AR) of reality combines levels of agency with their corresponding environments, reward functions, and objective functions. The resulting representation of intermediate, or *holonic* agents (where a *holon* is both a part and a whole), describes reality by definition (Schillo and Fischer 2002).

The proposition implies that the researcher observe behavior, and then use backwards induction to portray the forces at play that could have given rise to the decisions taken, as well as patterns and structures that emerged. As such, this proposition is not a theory or hypothesis, but rather a language in which different models can be expressed. The complexities of the proposed architecture can be endless. This notwithstanding, the paradigm for a new pandemic and interdisciplinary science built in a distributed agency architecture would accept the intercommunion by means of a model that is broad enough to accept the nature of realistic agents, but at the same time tractable enough for the capabilities of an appropriate simulation, expressed at a minimum desired level of realism. The model therefore intends to advance the development of a common language in which novel ideas can be transmitted across disciplines. Once we find ways to represent reality in an accurate and comprehensive computational fashion, the modeling side of CPS can come in and appropriately communicate with its physical side, creating positive feedback loops that make a more dramatic use of the optimization capabilities of numerical and systemic analysis.

There are multiple and growing ways in which CPS already control our modern lives. Since long ago we use digital clocks to tell us time, TVs and radios to communicate, and thermostats to control the temperature. Furthermore, computers have already invaded our lives in ways that affect us directly as individuals and the way we interact with the immediate neighbors of our networks. A computer algorithm may soon tell us the probability of success in our date tonight. Computer algorithms will answer the phone when we make a reservation. Computers will be our bosses. Computers will be our therapists. And this is not a sci-fi scenario of the next century – computers are already telling us what to wear or where to shop. A computer algorithm may have suggested that you befriend someone on Facebook today. Modern multinational companies are currently using computer algorithms to optimize their pricing strategies. Because of these ubiquitous cyber interactions, it is only a matter of time until we call ourselves the products of a CPS.

But the twenty-first century promises to be one in which our societies find ways to transform themselves with the use of computer-aided interactions and networking, not only affecting our individual experiences, but also the way in which these experiences are organized socially. For example, the Arab Spring would not have occurred before the advent of a computer platform such as Facebook. The field of CPS is now at a stage of development where it can answer social, rather than individualistic questions. With global problems such as climate change and nuclear proliferation, the need for novel ways to organize our world is dire. The types of social constructions that rule our lives were created long ago and are therefore accidental suboptimal structures than need revision, in particular through the modern capabilities that CPS now afford us. Better and unimaginable organizational structures that currently are so-called "black swans" (Taleb 2007), may now be within the reach of CPS that model and iteratively interact with our globalized reality.

Distributed Agency is a language that can help us reference entities and other aspects of agency and interaction that the computer can capture, but that a researcher may not be able to reference. In this same way, a neural network can record nonlinear aspects of interactions between variables that a linear regression cannot (Aminian et al. 2006). And yet, the added understanding affordable by the neural network may not be comprehensible for a human; certainly not in the way in which we can understand the coefficient of a linear regression as describing a positive or negative relationship between the independent and the dependent variable being modeled. The topological maps of agency and interaction of a complex CPS may not even be currently understandable or conceivable to humans, for they are "too emergent" to be seen by our naked eyes. A CPS cannot only express some aspects of reality to an arbitrary degree of accuracy, but can also implement the proposed models and learn from the realized outcome. If these general interactions can be faithfully codified, the computational capabilities of CPS can begin to propose visions and hypothesis that were previously unattainable. CPS have thus the promise of making social phenomena that is now "spooky emergent" only strong or even weakly emergent.

To allow for the most effective use and operationalization of CPS to build more resilient, less wasteful, and safer societies, we must reconsider the capabilities of CPS to redesign the way we build our world. And we do not have to hold CPS improvements in our lives to be solely confined to more efficient ways of organizing ourselves, for CPS can be just as well directed to explore ways in which we can organize ourselves more equitably. Any value judgement can be added to the task that a CPS is trying to accomplish. It is therefore time we reconsider the capabilities of CPS to change the way we live.

13.4 Social Adaptation: A Natural Extension of Human Adaptation to and Manipulation of Its Environment

Could it be that nothing in this universe is ultimately ontologically emergent? Is the universe full of computable properties that if only we knew, we would be able to predict? Could our human world simply be a computer simulation, as proposed in "The Hitchhiker's Guide to the Galaxy?" To conceive of our world as a computer simulation is a perspective that has been adopted by respectable researchers, such as Nick Bostrom (2003), who argues that it is more likely that not that we are living in a simulation, as was portrayed in the 1999 sci-fi movie "The Matrix." Why is anything that humans do considered "artificial" and therefore completely different from natural things? Is a popular song reflective of epistemological or ontological emergence? Can a computer create a song that rises to the Billboard's top 50? Would that be proof that all art is not ontologically emergent?

We humans consider ourselves the ultimate standard bearers of ontological emergence; a type of free wills that is not explainable by current systems. Like entrepreneurs who look at the world to attempt to "solve problems," we seek to create solutions (or works of art, for that matter) that are previously unavailable. As social builders, we attempt to create social structures that have agency of their own and can change the world around us, including that of the creators of the social structure itself. Is the novelty brought to the world by Shakespeare, Lenin, Beethoven, David Bowie, or Steve Jobs not fully ontologically emergent? In other words, if you somehow could model our world and feed your fantastically large simulation with all the necessary variables and human thoughts, are you bound to find Einstein's theory of general relativity? Did Einstein create his theory or simply discover it?

These types of questions belong in philosophical discussions more than in this work, but we find it necessary to stress the point that what we consider emergent, epistemologically emergent, and ontologically emergent, is dependent on the observer and may change over time, particularly as our understanding of complex adaptive systems increases. Furthermore, human civilization is in many ways dedicated and built on an ever increasing level of complexity. In other words, we humans thrive on creating emergent institutions to better exploit our environment. Social structures appear faster than predicted by Darwinian evolution, where a simple "survival of the fittest" conceptualization is insufficient for describing the ways in which humans organize and allow themselves to be organized by their social creations. Agency comes about in higher levels, with married couples that decide not to get divorced and nations that opt to go to war. The so-called Baldwin Effect is a good representation of a natural way in which systems can self-organize and adapt to a complex environment. Daniel Dennett (2003) summarizes the effect in the following way:

> Thanks to the Baldwin effect, species can be said to pretest the efficacy of particular different designs by phenotypic (individual) exploration of the space of nearby possibilities. If a particularly winning setting is thereby discovered, this discovery will *create* a new selection pressure: organisms that are closer in the adaptive landscape to that discovery will have a clear advantage over those more distant.

The main point of this work is that CPS and Modeling and Simulation (M&S) can help us speed these processes of structure exploration to create better social structures. Situating CPS in the context of the general theory of evolution and the history of societal change is a good place to start. Doing so can help us see hybrid modeling as part of a long evolution of information processing that is inherent to the evolution of humanity and society. Moreover, it can help us see where short-sighted models without sufficient feedback and information can cause crises. Helping us with such conceptualization is the framework offered by complexity studies, where systems are studied for how they "live" and "evolve" at "the edge of chaos," which are "points of criticality" between order and disorder. Systems that show both resilience and agency – responding to the environment on the one hand, but also preserving their forms and information on the other – are, in many ways, life-like.

13.5 Complexity and Society: Where CPS Fit in the Social Sciences

As stated in previous sections, Andreas Tolk argues in a chapter on the upcoming book on simulation (2019) that computer models cannot produce "ontological emergence," that is, computers can only transform the data that they are given access to and cannot create completely new information. But they can produce "epistemological emergence," in the sense of creating new rules for exchanging and ordering the information they are given. Tolk's point reflects the fact that the limiting factor in computer simulations ability to mimic life forms, is not computational power, where Moore's Law means that any system's carrying capacity may be virtually infinite. Instead, it is Metcalfe's Law, which states that the capacity to understand new information is limited by one's existing information. Indeed, one could describe sociology as the attempt to model society as an emergent phenomenon, and that many sociologists would describe society as ontologically emergent.

Symbolic interactionist sociologists in particular would largely agree that culture and society emerge as the result of the pragmatic collaborations among people, each of whom have their own unique perspectives. Metcalfe's Law applies just as much to the limitations of human beings in understanding each other, as it does to cyber systems. And yet humans do, well enough. The meanings of situations are constantly being negotiated in the course of interaction (Mead 1938; Blumer 1969; Goffman 1974) and epiphenomenal social constructions (having a "party," "class," "working out,") are performed in ways that reinforce those definitions (Berger and Luckmann 1966). Moreover, because situations change, negotiations are always ongoing and culture and its institutions are continually evolving (e.g. Goffman 1981; Fligstein and McAdam 2012). As new situations arise, so do opportunities to redefine situations and identities, to make them more fitting, powerful, or liberating for whatever agents are in negotiations. Culture is shared meaning, including shared understandings of reality. No culture can reflect that reality perfectly, but it can well enough to provide us a common orientation through which the agency of the larger group is distributed among us. As we discover better ways of understanding reality, our culture changes.

It would appear the same thing happens with our own individual models of reality. As individual cognitive agents, we too, are limited in our capacities to understand the world, but we muddle along, well enough. The models of the world we develop in the course of observing the results of our actions are akin to the definitions of the situation we negotiate with each other. Watching our culture change would be akin to watching a person think. In each case, we would be watching information being processed. Culture provides us with models of the world that we use to understand it and how we should engage with it. Likewise, our thinking gives ourselves models of the world, of situations and how we should behave in them. Friston and others (2010, 2006) and Pezzulo et al. (2018) describe this as the active inference that characterizes an embodied Bayesian brain which seeks to reduce uncertainty, and minimize surprise and the loss of free energy. In that conception we use cognition to create models of the world that reduce our surprise when we act, and we use the results of our actions to then revise our models. If that sounds like hybrid-modeling in CPS, it should. The computational side of CPS captures reality, transforms it through a model, and proposes actions implemented in the physical realm, creating potentially powerful feedback loops that only a systemic approach can encompass.

To sociological social psychologists, it should also sound like the interplay of the "models" of identity we develop in accordance with the "realities" of social interaction. Very much in line with Friston's model of cognition, symbolic interactionism has long conceptualized identity as "the looking glass self" (Cooley 1902), for the way we gain a sense of our own social identities by observing the reactions of others to us. Here too, a perfect match with reality is unattainable, but we make do, and act pragmatically. Although we can never know exactly what people think of us, we estimate it, based on their reactions to our actions, and refine our estimate, act again, observe the reactions, etc. We are cognitively modeling, testing the model in reality, and refining the model again, "back at the shop" in our own cognition. In that same sense, while CPS can never know what they do not know, and how their computational reality differs from our own, they can arrive at ever-closer approximations. As CPS technologies advance, the feedback loops that they create will represent our world with increased accuracy. The perspectives of CPS can become ever-closer to our own. Even if a particular CPS framework may not be able to produce ontological emergence in quite the same way we do, it could become close enough to appear lifelike. To the extent that we model our everyday interactions by the solutions proposed by CPS, the more they will rule our lives, and potentially the more they will *understand* our lives.

So too do other inventions of ours, which we have also developed as physically separate information processing systems: institutions, organizations, countries, companies, religions, even academic disciplines and fields of science, have all been "invented" by us to process information of one kind or another. Sociologists following the systems theory lineage of Parsons (1951) and Luhmann (1982) should be well acquainted with idea that social systems emerge with properties that are not predictable from the agents of which they are constituted. Emile Durkheim was one of the early progenitors of the view that saw society as emergent (Sawyer 2002). Emergence is the way groups of interacting agents can result in relatively stable properties that characterize that group (Holland 1998). Durkheim (1915) developed the idea that such properties can be collectively represented and modeled in the course of religious rituals. In general, sociologists using a Durkheimian framework (e.g. Vandenberghe 2007), see collectivities such as communities or large societies as having emergent properties in the form of representations that feed back to constrain the individuals that constitute those collectivities. As such, social structures and cultures are seen to exist apart from us, but change and evolve in their own right, though each does so according to the same information processing principles we will be outlining here.

One such principle is the compression of information for speed and efficiency. Modeling the world and the agent's relationship to it is cognition. When the agent is an organization or society, it can be considered collective, or distributed cognition (Tollefsen 2006; Theiner and Sutton 2014). Lenartowicz et al. (2016) have provided empirical evidence of cognition at an organizational level. In analyzing the communications of NASA they have shown how an organization can undergo a process of individuation that is a hallmark of Simondon's (1992) model of the individual as not rigidly fixed with given properties, but plastic, and always in the process of evolving. Lenartowicz et al. analysis shows how out of that process an organization can emerge as an autonomous cognitive agent that has a distinct identity, upon which it can self-reflect. It is difficult to read their account and not feel as if you are seeing an organization think.

The process is certainly consistent with Friston's model of cognition, but instead of a brain of distributed neurons, it is an organization of distributed brains, and working as a distributed information processing system, all the same. It models its environment, takes action on the basis of that model, observes the results of its actions, and refines its model accordingly. But the similarity between cognition and the actions of organizations shouldn't surprise us. The self-organization of order appears to be a constant in our universe. Cosmologist Eric Chaisson (2011) has shown how access to free energy is steadily increasing in the universe. Free energy is the energy available to do work, which is remarkably similar to Bateson's classic definition of information as "difference that makes a difference" (1972, p. 453). Indeed, many working in complexity studies see the evolution of complex systems as the processing of information into increasingly powerful forms. HyFlow models in CPS contribute to that process by giving us increasingly precise and relevant information, as they allow for the synthesis of arbitrarily large numbers of types of data. As such, CPS can be understood in the fullest context of "the world as evolving information" (Gershenson 2012), where our access to free energy and information increases in emergent ways.

Because humans are different kinds of information processing systems from the cyber ones we have developed in the ways we sense and sort information, we can understand some aspects of reality which CPS cannot. Many have noted that while computers are better than humans at computing, we humans are likely better at synthesizing, in part because of the feeling of expertise that arises from the holistic mastery of knowledge (e.g. Dreyfus 1992). Only after mastery do certain kinds of information emerge. This is reflected in our fast and slow cognition processes documented by Kahneman and Egan (2011). There is a fulfilling satisfaction involved in mastery that involves a balance between challenge and skill (Csikszentmihalyi 1990, 2000). That is part of our aesthetic response to putting information in order which is the basis for culture itself (Demerath 1993, 2002, 2012).

13.6 CPS Structures: Applications to the Human Realm

One of the ways in which we believe that CPS will change our world is by helping us build better social structures. Soon enough, researchers and computers in CPS will learn to learn what aspects of reality can indeed be modeled and what areas represent ontological emergence that need to be explored by humans, that is, on the physical side of CPS. Coupled with the ideas of Distributed Agency described in Section 13.3, the claim is that CPS will allow us create levels of agency that will help institutions to become better organized.

Social structures evolve over time. Niches are occupied to help process information and minimize the friction of intersecting agents with different trajectories. Positions emerge out of the support they receive for smoothing out those intersections. As they layer, the system becomes ever-more interconnected, and an efficient means of processing information. The iterative interactions between the Cyber and Physical worlds of CPS allow for the deeper exploration of the multidimensional spaces in which these structures exist. Below is a diagram illustrating this for schools, particularly how some positions become institutionalized, such as guidance counselors, and others forms, such as Future Farmers of American club (FFA), may fade away, while robotics clubs may be on the rise.

But change is not always for the best. Leaders and elites can exploit their advantages over other agents to the detriment of that system. Those other agents can become so constrained that they are unable to use their information and autonomy to the benefit of the system. This is illustrated in the next diagram, where administrators are overly constraining councilors (though do so to teachers might be more common), causing them to adhere too closely to the administrator's agenda and "informational vector." This means the councilors don't offer much help in reducing the difference between parents and the school, who are trying to use their more experienced position to reduce the difference for their children who are students. It is here that CPS could be introduced to propose previously untried organizational schemes, record the produced outcome, and iteratively arrive at increasingly improved social outcomes.

In Figure 13.3, we see the original, unrefined interaction between students and teachers represented, as well as the more nuanced levels of social institutions,

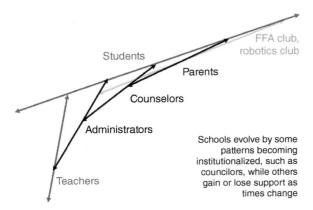

Figure 13.3 Agents use their unique information to position themselves in mutually advantageous ways.

norms, and agency that the natural repeated interactions between these actors create. Originally, the interactions and incentives between these sets of actors were not aligned (and this is represented in the figure as an geometrical angle), but the system their interactions create eventually proposes new structures to smooth out the interaction and propose increasingly better ways to arrive at a desired mutually agreed outcome.

Such problems can reduce the agency of the system, as the school has fewer parents and students enthusiastically engaging with teachers and administrators. It can cause students, teachers, and councilors to leave for better functioning schools. Or, the school can respond to the problems it hears from parents and counselors. An appropriately tuned CPS could not only capture the nuanced reality of these potential side effects of a proposed policy, and accordingly incorporate all of the ultimate repercussions of restructuring an organization such as this school system.

But to listen means slowing down. And that requires a commitment to the system overall, and a humility in one's ability to perceive it. Prioritizing a system-wide view over one's view means the controlling agents must cede the floor to the less powerful ones, and take their perspectives seriously. The recent education research of McQuillan and Kershner (2018) uses complexity theory in interpreting their data; they show the importance of decentralization to the emergence of an authentic school culture that make a school more adaptive. Over-centralization, on the other hand, would make a school less responsive to the diverse conditions and interests of its constituents (Figure 13.4).

Decentralizing mechanisms can be forums for free and open dialogue, the more informal and public the better. These can be places where we are forced to slow

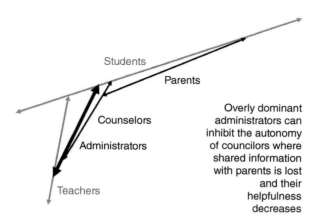

Figure 13.4 Overly dominant agents can suboptimize the system by constraining the freedom of other agents.

down our "informational vectors," and increase our humility and openness. Reduced speed increases interaction time, as much for us as it does for objects in space. Asteroids adrift are more likely to end up an orbital system if their velocities are slower. Slowing down to hear what others have to say increases the likelihood of finding shared information, and thus, the chances of finding new information. Forums for free and open discourse are critical to a healthy society (Habermas 1998). The oppression and negligence that arises from indulging ourselves in flow. When we believe our vector is headed in the right direction, our ride is smooth, fast, efficient, and comforting.

Inefficiencies are inevitable, particularly since our environments are always changing. Having space and time for taking in more diverse kinds of information, for slower, less predictable information processing, is part of any healthy system. Designers and engineers need to listen and watch for signs that their systems are suboptimized. For indications that lower-level agents are valuing something that wasn't anticipated in the design. Designer Tom Hulme (2016) shows how effective design can be achieved by paving "desire paths" (trails created by the erosion caused by informal human or animal traffic). Indeed, many roads that wind through country sides were originally game trails. Sociologist Laura Nichols (2014) develops the concept of "social desire paths," as patterns of unanticipated behavior that both indicate shortcomings of formal arrangements, and the values of the people creating the pattern. One of the cases of social desire paths that Nichols describes is the pattern of homeless persons riding buses throughout cold nights to stay warm, and to avoid shelters they perceived as overly constraining or dangerous. CPS can help detect the patterns that constitute social desire paths. Those patterns can then be studied to understand how to improve whatever set of institutions, or social structures are at issue.

To detect socially desirable paths, CPS follow us, picking up the data we leave behind, compiling patterns, and reporting back to us what patterns they've found, perhaps with suggestions on what we might next do. When the suggestions are about products we might buy, it seems a helpful suggestion, a shortcut in the time and effort of shopping. But when the suggestions are about something more serious or more complex, like who to vote for, or who to date, CPS could push us into homogeneous search filter bubbles that suboptimize the information in our environment. The preference for interacting with similar others is a robust finding in sociology (McPherson et al. 2001). Such homogeneity can be harmful in isolating us from each other. Just as an example, it can lead us into adopting social media as the means of sorting information, and that can get us into trouble.

But the temptation for short cuts is a fact of physics, let alone the sensation of ease. One way of thinking about emergence is as a phase transition in how information is stored or represented that increases the "certainty" of that information across time and space. Publishing a paper, for example, is a way of preserving that information. Published papers can be referenced with a greater certainty of their reliability and validity. A peer review process using multiple reviewers is one example of distributed information processing. That kind of peer review leading to publication takes place in oral cultures as well, with wisdom and sayings passing test of time and living on through their use, just as can scholarly citations. "Don't throw out the baby with the bath water," is a pithy summary of a philosophical truth. When black women, for example, have been excluded from formal education and the associated means of their knowledge validated, they did not demure from the act of trying to understand the world (Hill Collins 2002). Like any of us, would have, they continued to offer their own insights on the world to others in the course of interact, collaborating on how they might be put into forms that would last. As did the men. We don't know who wrote the line, "Oh the old sheep, they know the road. Young lambs gotta find the way." But it doesn't matter. It finds our way to us now through the distributed processing of people using it, and that peer review shows that it contains solid information.

And we need it, as we are all young lambs when it comes to making decisions on what paths to pave over, or what politicians to vote for. Here is where space and time for slower, processing of more diverse information is critical. And this why we risk a ride that is too easy if we let CPS do our dirty work for us. Walter Benjamin (2008) feared that our systems of consumption and entertainment would become so effective, richly varied due to their profitability that people would generally lose interest in governance, and civic engagement would decline. The tendency would be towards simplistic thinking, a lack of public concern, and superficial populism, which would increase the odds of incompetent leadership and the abuse of power. Ironically, the solution may lie in the kinds of artificial intelligence that can provoke interest most effectively. We can use help in preserving our space and time for slowing down and humbly considering what we don't know, or even reconsidering what we believe to be wrong.

Those moments when we need to slow down are inflection points; angles of intersecting vectors where energy is lost, and where we should be looking for shortcuts. Like an autonomous vehicle that might gently wake us up for an approaching intersection, artificial intelligence can invite us back into a process of decision-making at any point, but particularly at those moments where a fuzzy logic approach is most useful (e.g. Suarez et al. 2008). Such an approach is appropriate when the complexity of intersections mean outcomes are unpredictable, and likelihoods have to be inferred from broad range of possibilities. Metcalfe's Law means that for agents to process less familiar information, they have to go in with fewer assumptions, or less velocity on their informational vector, and less worried about friction. Having multiple available categories will slow one down, and in accordance with Ashby's Principle of Requisite Variety, the more categories one has, the more likely it is one will be able to categorize and process new information. The same principle operates at the periphery of any system, each of us included. We all host a community of various species of microbes, without which, we would not be able to process the new information coming at us in the form of viruses and bacteria, some of which are unfriendly. The most diverse one's microbiome, the more likely it is one will be able to come to terms with those hostiles, and the stronger is one's autoimmune system.

AI might strategically diversify our information. Rather than narrowing it down and indulging us in ease, it could challenge us with new information, but just enough to make it pleasurable and in a state of flow. For example, AI might help us counter the phenomenon that its own use has created, where search algorithms create filter bubbles that lead us into like-minded forums political discourse, and then into increasingly "ideal" versions of that discourse, which can end up being extremist and polarizing (Sunstein 2018). To counter that, one could fight fire with fire, albeit more sophisticatedly. AI could lead us into forums of interaction where differences are both affirmed and assimilated, and new connections are facilitated. The simple technique of "FlipFeed," produced by MIT Media Lab, is to offer views of others' Twitter feeds. A more nuanced approach would be that of the app "Woebot." It responds to people seeking counseling on personal issues much as a counselor would, and based on the expertise of a professional counselor who designed the system. Here is how CPS can help us answer one of any person's most pressing questions: how to deal with life's stressors?

At the collective level are questions just as pressing and fundamental that CPS can help us answer; when are we allowed to exclude others who don't share information with us; when are we allowed to self-segregate? Since it is so often an indulgence of ease to do that, the question might be better phrased as when we should allow others to do that. AI could help educate us to that end. We might be coaxed into seeing how homogeneous "safe spaces" for interaction have their place, especially for those who aren't often safe. We might be educated into understanding how minorities are inhibited by the stereotypes and assumptions of the dominant group. How hegemonic ontologies (e.g. Connell and Messerschmidt 2005; McIntosh 2012) can be oppressive and overly constrain the free exchange of information for those who are "typed," "objectified," "tokenized." Even questions of political correctness and cultural appropriation, usually mystifying to those of us in the majority, can become clarified with the right AI tool. A simple example is website that calculates your privilege, gives you a bill for what you owe, then allows you to reduce your debt by having others calculate their own privilege (as checkmyprivilege.com has done). But one can imagine the even greater effectiveness of such tools if they were integrated into real life circumstances, using hybrid-models that run simulations to anticipate real time outcomes, and can then make suggestions in real time. Imagine a school principal using a model of her school to see how allowing one kind of club would be healthy, and another would be a disaster.

In the end, we want systems that can help us spot the short cuts people are beginning to make, or those they might want to make; or to entice them into going the long way; to walking rather than driving, or eating a salad instead of a burger and fries. Systems that collect information that we cannot are fuzzier, in extending lines of information from a previously osscified, bureaucratic administrative set of policies, that make for a vector going too fast and heavy to slow down, such as depicted below.

Distributed agency is maximized through the fuzziness at the edges. Different competing fields of information can be processed and linked to the larger one through smaller agents first making contact. That first contact is in the direction of the conflicting agent at first. It absorbs the contradictory information and slows it down, as one does when catching a hard-thrown ball, or just something very heavy, or delicate. In any of those cases there is much informational momentum that has to be assimilated into one's own system. Fragile systems have less capacity to absorb contradictory information. One could say they are less adaptive than those systems that can absorb different information by being able to move in the other agent's vector. Like catching an egg, one has to give in its direction. In effect, one has to empathize when engaging with different agents that are fragile, and vulnerable, and cannot slow themselves down. It is even akin to Jesus' recommendation that we should love our enemies. Figures 13.5 and 13.6 below express this.

The above diagram illustrates two relatively adaptive systems, in that they are associated with relatively independent peripheral subsystems which help them integrate each other's information.

The following Figure 13.7, however, shows the greater necessity of such peripheral subsystems when attempting to engage with a more fragile, vulnerable, less adaptive system.

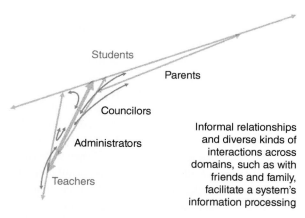

Figure 13.5 Suboptimizing agents can reduce conflict using information from smaller, more nimble agents.

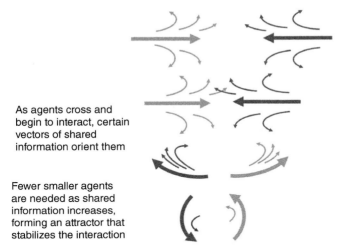

As agents cross and begin to interact, certain vectors of shared information orient them

Fewer smaller agents are needed as shared information increases, forming an attractor that stabilizes the interaction

Figure 13.6 Assimilating different information requires dispersing smaller agents for freedom and range.

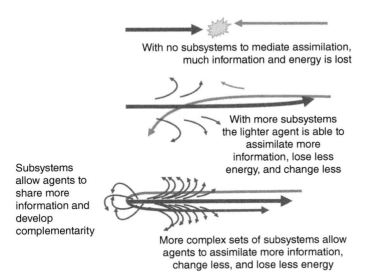

With no subsystems to mediate assimilation, much information and energy is lost

With more subsystems the lighter agent is able to assimilate more information, lose less energy, and change less

Subsystems allow agents to share more information and develop complementarity

More complex sets of subsystems allow agents to assimilate more information, change less, and lose less energy

Figure 13.7 More range in subsystems increases a system's capacity to integrate the information of other systems.

The above is meant to illustrate the basic truth inherent in Ashby's principle of requisite variety (1958), and Bar Yam's Theory of Multi-Scale Variety (2004).

In the final diagram below, we illustrate how a system can become more optimized to process the information in its environment, as subsystems become more complementary (Figure 13.8).

(a)   (b)   (c)

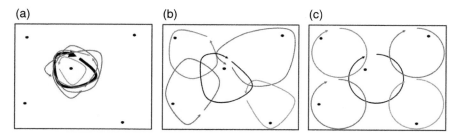

Figure 13.8 From suboptimization to distributed agency. (a) Depicts an overly dominant agent causing system suboptimization; (b) and (c) shows increasingly optimized systems processing more distant data points with greater coherence, less redundancy, and less wasted energy for more distributed agency.

CPS can be those soft hands for us, and our higher level agents as well. But the fragility of the systems with which they interact cannot be underestimated. For example, political scientists know that voters can differ in when they choose to vote, or how likely they are to vote. Democrats, for example, are a bit more likely to stay home because of inclement weather. CPS can overlook those differences to the detriment of properly sampling an electorate. We see this when potential voters stay home after seeing that a race has been projected as decided before the polls have closed. Caution is the better part of valor. Indeed, a social science and complexity perspective should encourage us follow the precautionary approach that has been adopted in laws in the European Union, as well as the Kyoto Accord in how countries treat the environment. The Precautionary Principle, as summarized at the Wingspread Conference (1998), might be adapted for CPS as follows (with the suggested changes in brackets): "When an activity [of a CPS] raises threats of harm to human health, [existing cultures and institutions,] or the environment, precautionary measures should be taken even if some cause and effect relationships are not fully established scientifically." We know that any system is limited in its capacity to understand another, different system, so we should err on the side of caution.

All of this means that CPS need to be connected to the "fuzziest" possible systems. Those that include humans, of course, but also smaller, more sensitive organisms, like a canary in a coal mine, for taking in information of the most different and uncommunicative systems. But like gathering intelligence on an adversary from agents in the field, they need maximum flexibility and independence to be able maintain their relationships in that different system. We can be the soft hands for catching a fragile object for CPS.

We should take heart in the fact that we are getting better at understanding systems, at understanding each other, identifying our mistakes, and making progress (e.g. Pinker 2011). The flow of smoothing things out is inherently fulfilling for us. We will always try to create better orders, and CPS can help us. They may even help us see what we are doing wrong, as much as what we can do right.

13.7 Conclusions

Many scientific theories and disciplines are based on the idea of linearity, which basically means that we treat the atoms of our understanding – at whichever level we are focusing our analysis – as completely independent, autarkic or isolated entities. In stark contrast, the nascent nonlinear paradigm is based on the idea of interdependence of the parts, such that wholes cannot be analyzed as a collection of independent atoms; there is an element of structure or design that is lost in a reductionist analysis, an aspect which emerges in the whole.

One of the main reasons why the complexity paradigm waited until the final decades of the twentieth century to surge is that our technical and mathematical abilities were simply too crude to be able to deal with the vast amount of computations that are necessary to keep track of the dynamics of interaction in a nonlinear model, rendering mathematical approaches of this kind impossible. It is for this reason that sociology had drifted away from traditional economic theory, for the former intended to describe complex reality while the latter opted to dwell on models which could be handled under the linear umbrella, even if this meant creating oversimplified agents that lost track of the actual intricacies of social phenomena. Enter the age of the computer. The development of computational capabilities has redefined the way in which researchers see the world. While the linear paradigm is based on an assumption of independence of whatever atoms or agents are used at the core of the model, the nonlinear view of the world is based on a perception of ingrained interdependence. In such a world, the independence of events that is assumed in classical statistics theory no longer holds true – a holistic approach is necessary to describe complex phenomena. As computational capabilities have grown exponentially in recent years, the possibility for exploring much more realistic descriptions of reality have begun to come into focus. This sort of computer simulation holds the key to a new and exciting branch of social science that will dominate the landscape in this new millennium.

Much of the development in this field has centered around Multi-Agent Systems, where agents are explicitly modeled as independent computer programs that interact with each other to produce sometimes surprising aggregate behavior. These approaches claim that the emergent upper level phenomena can be understood as a direct reflection of the interactions of the lower-level agents. While this approach has made significant advances in our understanding of biological and social systems, we must remain aware that this approach may suffer from some of the same issues as the paradigm it claims to replace. The understanding of a real world phenomenon, such as an economy or a stock market crash, implies considering agents that are not as clearly defined as those represented in a traditional multi-agent system, particularly when we accept the fact that the world is naturally epistemologically emergent, and thus any description of reality will not be able to mimic all of the many levels that exist. Any representation of the world must focus on one or several aspects of it, not on the whole. A model of all aspects of the world not only would quickly become intractable, but it would become a completely abstract study of the model itself and not of reality (Axtell et al. 1996).

Most importantly, this paper reviewed the idea that ontological emergence lies beyond the capabilities of the computer simulation. It is here that CPS can create a completely new area of scientific inquiry, by exploring ontological emergence with truly ontologically emergent physical and human elements of the system. It is in this way that we can conceive of a company such as Uber relying on computer algorithms to better manage their drivers. But this is only the beginning, as we can imagine more involved social experiments that make use of the growing capabilities of CPS and M&S. In this way, we can envision the restructuring of the democratic institutions of a nation such as the United States. A CPS could propose novel ways in which to organize elections, redraw congressional district maps, and augment the number of congressmen in the House of Representatives. These cyber-social experiments could be first implemented in a few states, to see the ontologically emergent results and then feed them back to the computer for reassessment, and further implementation in subsequent states.

We have argued that emergence is directly linked to the interaction between the agents in the system. Although social interaction may be dwindling in some of our communities, our cyber interactions are exponentially increasing, both in cyber-social networks as well as between us and our "smartphones." The vast number of cyber-captured interactions will allow for a much more in-depth understanding of the way that humans make decisions in context, and of the way that we integrate with social constructs. We are about to let loose a series of revolutions in tandem, where computers will be even more pervasive in our everyday lives. CPS are now poised to dominate not only our individual lives, but also the societies that we inhabit.

References

Akoka, J., Comyn-Wattiau, I., & Laoufi, N. (2017). Research on Big Data–A systematic mapping study. *Computer Standards and Interfaces*, *54*, 105–115.
Aminian, F., Suarez, E. D., Aminian, M., & Walz, D. T. (2006). Forecasting economic data with neural networks. *Computational Economics*, *28*(1), 71–88.
Ashby, W. R. (1958). Requisite variety and its implications for the control of complex systems. *Cybernetica*, *1*(2), 83–99.

Axtell, R., Axelrod, R., Epstein, J. M., & Cohen, M. D. (1996). Aligning simulation models: A case study and results. *Computational and Mathematical Organization Theory*, *1*(2), 123–141.

Baheti, R., & Gill, H. (2011). Cyber-physical systems. *The Impact of Control Technology*, *12*(1), 161–166.

Bankes, S. C. (2002). Agent-based modeling: A revolution? *Proceedings of the National Academy of Sciences*, *99*(Suppl 3), 7199–7200.

Bar Yam, Y. (2004). A mathematical theory of strong emergence using multiscale variety. *Complexity*, *9*(6), 15–24.

Barros, F. J. (2002). Towards a theory of continuous flow models. *International Journal of General Systems*, *31*(1), 29–40.

Bateson, G. (1972). *Steps to an Ecology of Mind*. New York: Ballantine.

Benjamin, W. (2008). *The Work of Art in the Age of Mechanical Reproduction*. London: Penguin.

Berger, P. L., & Luckmann, T. (1966). *The Social Construction of Reality: A Treatise in the Sociology of Knowledge*. Garden City, NY: Anchor Books.

Blumer, H. (1969). *Symbolic Interactionism: Perspective and Method*. Englewood Cliffs, NJ: Prentice Hall.

Bostrom, N. (2003). Are we living in a computer simulation? *The Philosophical Quarterly*, *53*(211), 243–255.

Chaisson, E. J. (2011). Energy rate density as a complexity metric and evolutionary driver. *Complexity*, *16*(3), 27–40.

Collins, P. H. (2002). *Black Feminist Thought: Knowledge, Consciousness, and the Politics of Empowerment*. Routledge.

Connell, R. W., & Messerschmidt, J. W. (2005). Hegemonic masculinity: Rethinking the concept. *Gender and Society*, *19*(6), 829–859.

Cooley, C. H. (1902). *Human Nature and the Social Order* (pp. 152). New York: Scribner's.

Crutchfield, J. P. (1994). The calculi of emergence. *Physica D*, *75*(1–3), 11–54.

Csikszentmihalyi, M. (1990). *Flow: The Psychology of Optimal Experience*. New York, NY: Harper & Row.

Csikszentmihalyi, M. (2000). *Beyond Boredom and Anxiety: Experiencing Flow in Work and Play*. San Francisco, CA: Jossey-Bass.

Demerath, L. (1993). Knowledge-based affect: Cognitive origins of good and bad. *Social Psychology Quarterly*, *56*, 136–147.

Demerath, L. (2002). Epistemological culture theory: a micro account of the origin and maintenance of culture. *Sociological Theory*, *20*(2), 208–226.

Demerath, L. (2012). *Explaining Culture: The Social Pursuit of Subjective Order*. Lanham, NJ: Lexington.

Dennett, D. (2003). The Baldwin Effect, a Crane, not a Skyhook. In B. H. Weber & D. J. Depew (Eds.), *Evolution and Learning: The Baldwin Effect Reconsidered* (pp. 69–106). Cambridge, MA: MIT Press. ISBN: 0-262-23229-4

Dreyfus, H. L. (1992). *What Computers Still Can't Do: A Critique of Artificial Reason.* Boston, MA: MIT Press.

Durkheim, E. (1915). *The Elementary Forms of the Religious Life.* New York: Free Press.

Fligstein, N., & McAdam, D. (2012). *A Theory of Fields.* Oxford: Oxford University Press.

Friston, K. (2010). The free-energy principle: A unified brain theory? *Nature Reviews Neuroscience, 11*(2), 127.

Friston, K., Kilner, J., & Harrison, L. (2006). A free energy principle for the brain. *Journal of Physiology-Paris, 100*(1–3), 70–87.

Gelfert, A. (2016). *How to Do Science with Models: A Philosophical Primer.* Cham, Switzerland: Springer.

Gershenson, C. (2012). The World as Evolving Information. In A. Minai, D. Braha, & Y. Bar-Yam (Eds.), *Unifying Themes in Complex Systems VII* (pp. 100–115). Berlin, Heidelberg: Springer.

Goffman, E. (1974). *Frame Analysis: An Essay on the Organization of Experience.* New York: Harper Colophon.

Goffman, E. (1981). A reply to Denzin and Keller. *Contemporary Sociology, 10*, 60–68.

Habermas, J. (1998). *Habermas on Law and Democracy: Critical Exchanges* (Vol. *5*). Berkeley: University of California Press.

Holland, J. H. (1998). *Emergence.* Reading MA: Addison-Wesley.

Hulme, T. (2016). What can we learn from shortcuts? TED. Available at: https://www.ted.com/talks/tom_hulme_what_can_we_learn_from_shortcuts. Viewed 2 December 2018.

Kahneman, D., & Egan, P. (2011). *Thinking, Fast and Slow* (Vol. *1*). New York: Farrar, Straus and Giroux.

Lenartowicz, M., Weinbaum, D. R., & Braathen, P. (2016). Social systems: Complex adaptive loci of cognition. *Emergence: Complexity and Organization, 18*(2), 1–19.

Luhmann, N. (1982). The World Society as a Social System. *International Journal of General Systems, 8*(3), 131–138.

McIntosh, P. (2012). Reflections and future directions for privilege studies. *Journal of Social Issues, 68*(1), 194–206.

McPherson, M., Smith-Lovin, L., & Cook, J. M. (2001). Birds of a feather: Homophily in social networks. *Annual Review of Sociology, 27*(1), 415–444.

McQuillan, P., & Kershner, B. (2018). Urban School Leadership and Adaptive Change: The "Rabbit Hole" of Continuous Emergence. In *International Conference on Complex Systems* (pp. 386–397). Cham: Springer.

Mead, G. H. (1938). *The Philosophy of the Act.* Chicago, IL: University of Chicago Press.

Miller, J., & Page, S. E. (2007). *Complex Adaptive Systems: An Introduction to Computational Models of Social Life.* Princeton Studies in Complexity. Princeton, NJ: Princeton University Press.

Mittal, S. (2019). New Frontiers in Modeling and Simulation in Complex Systems Engineering: The Case of Synthetic Emergence. In J. Sokolowski, et al. (Eds.), *Summer of Simulation: 50 Years of Seminal Computing Research* (pp. 173–194). Cham, Switzerland: Springer International Publishing AG.

Nichols, L. (2014). Social desire paths: a new theoretical concept to increase the usability of social science research in society. *Theory and Society*, *43*(6), 647–665.

Parsons, T. (1951). *The Social System*. Glencoe, IL: Free Press.

Pessa, E. (2002). What Is Emergence? In G. Minati & E. Pessa (Eds.), *Emergence in Complex, Cognitive, Social, and Biological Systems* (pp. 379–382). Boston, MA: Springer.

Pezzulo, G., Rigoli, F., & Friston, K. J. (2018). Hierarchical active inference: A theory of motivated control. *Trends in Cognitive Sciences*, *22*(4), 294–306.

Pinker, S. (2011). *Our Better Angels: Why Violence Has Declined*. New York: Viking.

Sawyer, R. K. R. (2002). Emergence in sociology: contemporary philosophy of mind and some implications for sociological theory. *American Journal of Sociology*, *107*(3), 551–585.

Schillo, M., & Fischer, K. (2002). Holonic multiagent systems. *Manufacturing Systems*, *8*(13), 538–550.

Simondon, G. (1992). The Genesis of the Individual. In J. Crary & S. Kwinter (Eds.), *Zone: Incorporations 6*. New York: Zone. ISBN: 0942299299

Suarez, E. D., & Castañón-Puga, M. (2013). Distributed agency. *International Journal of Agent Technologies and Systems (IJATS)*, *5*(1), 32–52.

Suarez, E. D., Rodríguez-Díaz, A., & Castañón-Puga, M. (2008). Fuzzy Agents. In O. Castillo, P. Melin, & W. Pedrycz (Eds.), *Soft Computing for Hybrid Intelligent Systems* (pp. 269–293). Berlin, Heidelberg: Springer.

Sunstein, C. R. (2018). *# Republic: Divided Democracy in the Age of Social Media*. Princeton: Princeton University Press.

Taleb, N. N. (2007). *The Black Swan: The Impact of the Highly Improbable* (Vol. *2*). New York: Random House.

Theiner, G., & Sutton, J. (2014). The collaborative emergence of group cognition. *Behavioral and Brain Sciences*, *37*(3), 277–278.

Tolk, A. (2019). Limitations and Usefulness of Computer Simulations for Complex Adaptive Systems Research. In J. Sokolowski, U. Durak, N. Mustafee, & A. Tolk (Eds.), *Summer of Simulation: 50 Years of Seminal Computing Research* (pp. 77–96). Cham, Switzerland: Springer International Publishing AG.

Tolk, A., Barros, F., D'Ambrogio, A., & Rajhans, A. (2018a, April). Hybrid simulation for cyber physical systems: a panel on where are we going regarding complexity, intelligence, and adaptability of CPS using simulation. In *Proceedings of the Symposium on Modeling and Simulation of Complexity in Intelligent, Adaptive and Autonomous Systems*. Society for Computer Simulation International.

Tolk, A., Page, E., & Mittal, S. (2018b). Hybrid simulation for cyber physical system: state of the art and literature review. In *Spring Simulation Multi-Conference*, Baltimore, MD.

Tolk, A., Wildman, W., Shults, F., & Diallo, S. (2018c). Human simulation as the Lingua Franca for computational social sciences and humanities: Potentials and pitfalls. *Journal of Cognition and Culture*, *18*(5), 462–482.

Tollefsen, D. P. (2006). From extended mind to collective mind. *Cognitive Systems Research*, *7*(2–3), 140–150.

Vandenberghe, F. (2007). Avatars of the collective: A realist theory of collective subjectivities. *Sociological Theory*, *25*(4), 295–324.

Wingspread Statement. (1998). The precautionary principle. *Rachel's Environment and Health Weekly*, 586.

Wolfram, S. (1984). Computation theory of cellular automata. *Communications in Mathematical Physics*, *96*(1), 15–57.

Part V

Way Forward

14

A Research Agenda for Complexity in Application of Modeling and Simulation for Cyber Physical Systems Engineering

Andreas Tolk[1] and Saurabh Mittal[2]

[1] *The MITRE Corporation, Hampton, VA, USA*
[2] *The MITRE Corporation, Fairborn, OH, USA*

14.1 Introduction

This compendium is not the first book on Cyber Physical Systems (CPS) resulting in the need to align research better and create a common research agenda. The authors themselves already contributed similar recommendations for the related topics systems of systems (Tolk and Rainey 2014) and complex systems engineering (Diallo et al. 2018). However, while these research ideas are surely applicable, with this compendium we put the focus on cyber-physical systems, so that the research agenda must become more focused as well.

During one of the initial NSF Workshops on CPS, Edward Lee proposed a research agenda (Lee 2006). At this 2006 workshop, most CPS expert had a strong control theory and electrical engineering background, as it was typical for the robotics community. Accordingly, the recommendations were targeted at their research domains and focused on better support to control functions needed for robots, down to memory managements, alignment of pipeline structures for route management of control, and other hardware-near computational challenges. Interestingly, the recommendations have three common themes: (i) introduction of time to cope with the dynamic behavior of the CPS and its environment, (ii) predictability of control decisions in the environment, and (iii) better alignment of control decisions for parallel control channels.

Within systems engineering, modeling and simulation (M&S) methods have been increasingly integrated to cope exactly with these types of challenges:

Complexity Challenges in Cyber Physical Systems: Using Modeling and Simulation (M&S) to Support Intelligence, Adaptation and Autonomy, First Edition. Edited by Saurabh Mittal and Andreas Tolk.

1) better understanding the dynamic behavior of complex systems,
2) providing predictive analytic methods by simulating future states of the system, and
3) providing better means by conceptual alignments of different viewpoints by data mediation and multi-modeling methods.

The methods needed to be better understood are partly provided by M&S methods in great detail already, and more have been addressed in Diallo et al. (2018). Research prospects for CPS engineering need to be looked through a lens of Complexity Science and how the advances in M&S bring CPS engineering, Systems Engineering, and Complexity Science together.

We will also make a connection to the recommended methods systems engineers should be aware of in one of the following subsections of this chapter.

14.2 Research Challenges Identified Within the Compendium

The landscape of CPS research is too fragmented and dominated by individual viewpoints, simply because a CPS is a socio-technical system having economic implications. Within this compendium, the authors form various chapters, each of which providing a different perspective to the challenge of better coping with complexity for CPS engineering, also address new research needs. Within this summary, we try to aggregate and compile them into a format that is particularly suitable for scholars and researchers in the hope to initialize more exchange of research results and possible collaborations.

What we tried with this chapter is to identify some common themes and topics that are emerging in discussions, conferences, and our chapters. All chapter authors and many additional experts need to be given credit for this. The themes are enumerated below:

- Common formalism
- Complex environments
- Complexity toolbox
- Multiperspective challenges
- M&S Methods for better communication in complex projects
- Resilience
- Supporting human machine teams

14.2.1 Common Formalism

As discussed in other parts of this book as well, the idea to compile these contributions were conceived during an expert panel at the Spring Simulation Conference

2018 in Baltimore, Maryland (Tolk et al. 2018). Many of the participants of this panel extended their contribution and added latest research results. New authors were motivated by this panel to contribute their views, some of them realizing for the first time their close relation to simulation. The reason is that M&S experts usually express their findings in the language of the supported discipline or field. The authors of Chen and Crilly (2016) are giving an example applicable to our profession as well: In their study, Chen and Crilly evaluated commonality of issues between practitioners in the field of synthetic biology and swarm robotics and showed that these practitioners shared more complexity related issues between each other than they did with colleagues in their original domain. Nonetheless, the sharing of information and reuse of solutions was hindered by the different terms and concepts used to describe them within their home domain.

One way to overcome such challenges is the use of a common formalism that captures the results of research independently from the language of the supporting discipline. However, while the panel agreed on the general necessity to agree on a unifying formalism allowing for conceptual composability to support collaborative ensembles is a common theme, it also observed that there is no agreement yet on how to accomplish this. Furthermore it was observed that hybridization inherently makes it hard to agree on single formalisms, standards, and/or technologies. There may be something we can agree upon to better manage the complexity of CPS beside such an early agreement, but although several approaches were discussed, it was too early to recommend any concrete artifacts. Within this book, in particular the chapters of Barros and Traoré and Mittal et al. are providing possible contributions to such a future solution, extending their work presented at the conference, showing that such a desired unifying modular formalism would leverage our ability to co-simulate cyber-physical systems, and that such a unification is not only desirable but it is likely to be possible.

There are obviously many additional candidates for such a common formalism in the discussion. Nance (1981) denoted a class of formalism as discrete-event, which started to play a predominant role for digital simulation systems. In the panel discussion at the Spring Simulation Multi-Conference 2017, Zeigler focused on the use of the DEVS as introduced in (1976) and only recently updated in (2018). Due to the popularity of DEVS in the academic community, this formalism was used extensively to address the possible combination of different modeling paradigms, such as in Vangheluwe (2000) and Vangheluwe et al. (2002). These papers show that DEVS can express multiple paradigms, helping to make them at least comparable and hopefully ultimately composable into hybrid approaches. Although the DEVS formalism is not without critics, it has been established as the most used simulation formalism and it is worth to consider it in support of CPS support as well. Accordingly in Traoré et al. (2019) and Zeigler et al. (2018), the authors are exploring ways to extend DEVS in support of value-based healthcare. The formalism defined in this paper support the multi-perspective modeling and holistic simulation of healthcare systems and are applicable to other forms of CPS as well, as healthcare is used as an application example that doesnot limit the general applicability. Mittal and Martin discuss the significance of a robust co-simulation environment incorporating super-formalism such as DEVS in the modeling and simulation of complex adaptive systems in cyber domain (Mittal and Risco-Martín 2017; Mittal and Zeigler 2017).

All these are valid and valuable approaches, and a position paper prepared in support of the panel discussion lists more examples (Tolk et al. 2018). Neither this literature review nor the enumeration in this section is complete or exclusive. They are written to provide examples and define possible starting point for ongoing research. The disciplines of mathematics, theoretical computer science, and physics must provide a solid foundation to cope with these hybrid solutions, bringing computational and physical components consistently together under one unifying theory with applicable common methods and derived domain specific solutions. It remains one of the main challenges to work on such ideas to contribute to a common formalism that can bridge the gaps between the various disciplines – or even between different M&S approaches – and allow for the better management of complexity in the cyber physical realm addressed in this book.

14.2.2 Complex Environments

As discussed in most chapters, CPS operate in increasingly complex environments. Providing for these diverse and often rapidly changing requirements is another topic of concern.

14.2.2.1 Contextual Environments

CPS are domain-specific systems (Sehgal et al. 2014). For example, a CPS such as automated medicine delivery based on wearable sensors is a healthcare specific CPS, a CPS such as automated disaster diagnostic system using drone technology is an aviation specific CPS, etc. However, many physical devices they employ are not domain-specific but general, reusable components. For example, the same components can be used within aviation-specific as well as healthcare-specific CPS, or they can be used in support of a completely new domain, such as an automated disaster response management system. The case in point being, the CPS devices will be used in multiple contexts and modalities for their use in different scopes in different CPS.

When CPS communicate through commodity Internet (or World Wide Web) that increases their scale and socio-technical impact, they take the form of Internet of Things (Jazdi 2014). Making available these varied scopes in the applicable experimentation environment is a challenge. Given a CPS device, we need plug-gable virtual domains wherein the same physical devices can be challenged through a different cyber profile. This device-centric scope, as described the chapter about Taming Complexity and Risk in IoT Ecosystem using System Entity Structure Modeling in this book, provides a mechanism to develop a virtual environment with pluggable domain environments. We require technologies that allow various domain environment models (e.g. healthcare, aviation, manufacturing, demography, etc.) to be plugged into the same simulation environment. Co-simulation is an emerging technology that allows for such domain simulation to coexist for the same CPS device model.

14.2.2.2 Virtual Environments

The use of M&S methods to support testing and evaluation by providing a virtual environment that generates realistic stimuli for systems under test is a well established practice, some examples for related domains are given in Bergman et al. (2009) and Hahn et al. (2013). Many other cases exist – in particular in defense related domains – where test cases have to be provided by simulation and virtual environments because conducting the experiment with the real system is too dangerous, too expensive, or otherwise not practically feasible. Within the chapters of this book, two related topics emerged, namely the use of agent-based M&S methods to provide a proficient technical development and testing environment for CPS, and secondly providing a sufficient socio-technical environment (e.g. live, virtual, and constructive [LVC] environment) to address the multi-modality of CPS as well as evaluate social implications of new approaches.

The authors of Talcott (2008) and Sanislav and Miclea (2012) are providing a theoretic basis as well as practical recommendations to utilize M&S methods in support of CPS developing and testing. The case for the use of agent-based development and testbeds is also made in Tolk (2014), which also points out the taxonomic similarities between autonomous systems and intelligent agents. This similarity allows, e.g., to develop behavior rules for autonomous systems using the agents, using the virtual testbed to conduct a series of test, and than transfer the rules from the intelligent agents to the autonomous systems. Some commercial solutions – such as Compsim's KEEL® (Knowledge Enhanced Electronic Logic) Technology – are designed to be transferable from software environments to CPS environments without the need to rewrite the solution (Keeley 2007), as the same logic used in the virtual environment can be directly applied for the CPS as decision logic for the real operations.

The need for socio-technical environments is motivated by the observation that CPS are increasingly pervasive in urban contexts where the environment is largely defined by a dynamic society rather than a single user. New M&S methods and platforms should permit designing interactive CPS-social spaces, reaching the level of robustness and credibility that today is only achieved by long-standing specialized areas such as traffic simulation. Understanding CPS as social components actually has a long tradition, going back to attribute social behavior to intelligent agents (Jennings and Wooldridge 1996). Such an environment would also support the evaluation of human–machine cooperation in teams, as increasingly envisioned for scenarios that require human ingenuity but are too dangerous to be dealt with by a human team alone, such as envisioned in Zander et al. (2015) and others.

These are just some examples why more research is needed to make the vision of such technical and socio-technical virtual environments for developing, testing, and evaluating a CPS, a reality. They provide the option of experimenting with self-optimization and -organization of CPS in a safe environment, continuous optimization of decision logic while the systems already are deployed, and much more.

14.2.3 Complexity Toolbox

The International Council of Systems Engineering (INCOSE) is a professional society in the field of systems engineering with more than 17 000 members. It provides educational, networking, and career-advancement opportunities for systems engineers worldwide, following emerging trends and conducting leading-edge research in such domains. The field of complex systems engineering has been in their focus recently as well, and recommendations and findings will be beneficial for managing complexity in CPS. Norman and Kuras used their experience as professional systems engineers to capture the main differences and challenges of complex systems engineering in comparison with traditional engineering in Norman and Kuras (2006). Their summarized findings are enumerated in Table 14.1.

Obviously, this openness of the environment combined with multi-modality of the systems themselves results in additional requirements for systems engineering methods supporting to analyze, diagnose, model, and synthesize complex systems. INCOSE provides a complexity primer for systems engineers that identifies an orchestrated set of tools (Sheard et al. 2015), many of which are familiar to simulation engineers, as the ability to provide insights into the dynamic behavior of complex systems in a complex environment is pivotal to systems engineers, and it is exactly what M&S methods can provide. The supporting methods for complex systems engineering identified in this primer are the following. Of particular interest is the capability of simulation-based solutions to numerically provide data comparable to empirically observed data describing the real-world referent. These data can be evaluated with the same methods used for real-world evaluation in lieu of having access to such systems because such an experiment were too dangerous, too expensive, or for any other reasons not feasible, e.g. if the system is not implemented yet.

- Methods supporting the *analysis* of complex systems are data mining, splines, fuzzy logic, neural networks, classification and regression trees, kernel

Table 14.1 Observations in traditional and complex systems engineering.

Traditional systems engineering	Complex systems engineering
Products are reproducible	No two enterprises are alike
Products are realized to meet preconceived specifications	Enterprises continually evolve so as to increase their own complexity
Products have well-defined boundaries	Enterprises have ambiguous boundaries
Unwanted possibilities are removed during the realizations of products	New possibilities are constantly assessed for utility and feasibility in the evolution of an enterprise
External agents integrate products	Enterprises are self-integrating and re-integrating
Development always ends for each instance of product realization	Enterprise development never ends enterprises evolve
Product development ends when unwanted possibilities are removed and sources of internal friction (competition for resources, differing interpretations of the same inputs, etc.) are removed	Enterprises depend on both internal cooperation and internal competition to stimulate their evolution

machines, nonlinear time series, Markov chains, power law statistics, social network analysis, and agent based models.

- Methods supporting the *diagnosis* of complex systems are: algorithmic complexity, Monte Carlo methods, thermodynamic depth, fractal dimension, Information Theory, statistical complexity, Graph theory, functional information, and multi-scale complexity.
- Methods supporting the *modeling* of complex systems are uncertainty modeling, virtual immersive modeling, functional/behavioral models, feedback control models, dissipative systems, game theory, cellular automata, system dynamics, dynamical systems, network models, and multi-scale models.
- Methods supporting the *synthesis* for complex systems are design structure matrices, systems and enterprise architectural frameworks, simulated annealing, artificial immune systems, particle swarm optimization, genetic algorithms, multi agent systems, and adaptive networks.

The common denominator of these enumerated methods is the use of M&S methods and the application of computational intelligence. Obviously, this enumeration is not complete. It is more an open toolbox helping the systems engineers to fulfill their various tasks. A challenge with the approach is again the need for these tools to be orchestrated, which means that they need to inter-operate with each other.

There are plenty of research topics in this domain. The list provided by INCOSE is a good start, but there are certainly many more options. There are plenty of mathematical tools that can provide meaningful insights, such as multifractal analysis, such as described in Furuya and Yakubo (2011) for complex networks but applied for several other domains already as well. Many operational research methods, including traditional statistical means, can also contribute. Which methods to apply when, how to provide guidance for users of the tool box who are not necessarily engineers, and development of new methods belong to the set of possible topics of interest. This domain will require reaching out to many other disciplines, hopefully leading to an increase in interdisciplinary research and exchange of results.

14.2.4   Multiperspective Challenges

It is fairly acknowledged that multiple viewpoints or perspectives are needed to address the complexity of modern systems engineering. It is also broadly acknowledged that such perspectives need to be aligned in a holistic study of the same system, whether by manipulating independently perspective-specific models or by composing them. Therefore, we naturally assume an underlying consistency between those perspectives, since they depict the same reality, which the alignment just reveals and formally captures. However, this assumption that multiple perspectives are based on a common, single reality is not always justified, as our perspectives are constrained by computational as well as cognitive boundaries. Nonetheless, simulations are one of the best ways to capture our knowledge in executable and unambiguously communicable form.

> A simulation is a mathematical model that describes or creates computationally a system process. Simulations are our best cognitive representation of complex reality, that is, our deepest conception of what reality is. (Vallverdú 2014)

The epistemological challenge of each simulation representing a different perspective lies in the shared model that is used to build the simulation. In Tolk et al. (2013), we show that this model is a result of the task-driven purposeful simplification and abstraction of a perception or understanding of reality. Whatever is in this model, it becomes part of the reality of the simulation experience. Everything outside this model cannot be part of a valid interpretation of the simulation. This has direct implications for the assumption of the existence of a common, computable *Übermodel* representing a common truth.

Obviously, there are many ways to represent very different perspectives that differ in underlying theory or world view, but nonetheless can provide valuable alternative viewpoints. Looking at the variety of computational functions used to capture knowledge for CPS, it would be naive to assume that they all naturally align as CPS is a multi-disciplinary undertaking. They may also use different levels of abstraction, resulting in various degrees of resolution, how detailed the system and its sub-components are modeled, and fidelity, how accurate the behavior of the system and its sub-component are modeled. The resulting challenges of multi-resolution, multi-scope, multi-structure, multi-phase, and multi-fidelity compositions have been the topic of many M&S scholarly contributions, and the lessons learned need to be captured and applied for the CPS development and evaluation tasks as well.

Another aspect of interest are the general computational constraints. This group comprises challenges derived from the need to use numerical solutions and heuristics in case an exact representation or solution is not possible. Examples for this group are given in Oberkampf et al. (2002). It is well known in the M&S community that even slight variations in numerical solutions or heuristics can result in significant changes of the simulation outcome. This is even more challenging in highly non-linear systems with a high sensitivity on its initial conditions. The use of different numerical solutions or guiding heuristics can therefore lead to significantly different solutions, even if the same perspective is addressed.

Finally, early insights from pioneers like Kurt Gödel introducing the incompleteness theorem and Alan Turing demonstrating the undecidability of many problems by algorithms lead to another viewpoint on multi-perspective modeling. As shown in Tolk et al. (2013), it is possible to capture the various perspectives and viewpoints in a common reference model that allows for contradictions in the knowledge representation based on its conceptualizations. Such a reference model can lead to a set of conceptual models that are a consistent subset of the inconsistent reference model of everything. The consistent subset can be implemented as a simulation, and all resulting simulations can provide different perspectives of the overall problem under evaluation. Instead of providing one solution, such an approach leads to a set of solutions, all based on different conceptualizations. This approach adds an additional component to the stability and sensitivity analysis of the solution set. Like the well known weather maps show different paths and cone structures for an approaching hurricane using different models to produce the different forecast results the approach discussed in this section can do something similar for multi-perspective models: providing all information to the decision maker. Again, proper visualization of all these data will be key.

14.2.5 M&S Methods for Better Communication in Complex Projects

Current systems are becoming more complex every day. With the new era of IoT, Big Data, Deep Learning, etc., the definition of models is becoming essential to analyze the dimension of the problems we are supposed to solve or to tackle.

Heterogeneous teams must work together to develop/design smart cities – such as analyze contamination, sustainability of new parks, or public buildings –, complex rescue missions – with many unmanned vehicles, cameras, sensors, different targets –, supply chain and logistics – with sensors of all kind in the transportation system – and a long set of additional examples. These different communities of practice rarely share common processes or organizational structures, so that the simple communication of plans, coordination points, phases, can become a major challenge. The Model Driven Architecture (MDA) concepts, which were proposed over a decade ago by the Object Management Group (OMG) (Kleppe et al. 2003), when utilized on their own often failed when communicating ideas with engineers out of the core group, such as supporting software and system engineers, and even more so when the project must be explained to non-experts in the area (politicians, administrative staff, etc.).

Model-based and simulation-based engineering are two distinct methods applied to systems engineering life cycle. Model-based engineering (MBE) when applied to Systems engineering is labeled as Model-based Systems Engineering. When advanced model transformation technologies are applied to both MBE and MBSE, it takes the form as Model-Driven Systems Engineering (MDSE) (Mittal and Martin 2013). All these model-based flavors do not advocate simulation as a means to perform experimentation. However, the value they provide in developing a shared model (through MBSE) and bringing multi-domain models towards a shared computational model (through MDSE), is immense. Simulation-based engineering subsumes the notion of model, and is more focused towards experimentation and experiencing the model by a human actor. Connecting the two distinct methods is an intermediate third method: an executable modeling method – that steps into the model-experience category but is not quite there – to be incorporated in the simulation-based engineering method (Zeigler et al. 2018). Together these three methods provide different value propositions to address the complexity associated with modeling, simulation, experimentation, and project management.

M&S can help in the whole definition and development of such projects. However, there is no integrative M&S methodology to handle all these issues. We have been talking about integrative M&S for decades, but there is not a generally accepted integrative solution yet. We are seeing how big data analysis or deep learning are offering interesting models through training, validation, and tests, but they respond to small, focused questions. As stated by van Dam during his lecture at Stanford:

> If a picture is worth a thousand words, a moving picture is worth a thousand static ones, and a truly interactive, user-controlled dynamic picture is worth thousand ones that you watch passively. (van Dam 1999)

In diverse and multi-disciplinary communities, such as the community supporting CPS developing and evaluation, this is a powerful message that is today as valid as it has been in 1999. New immersive technologies even support the next level of interactive communication of such concepts between experts as well as users of the systems.

14.2.6 Cyber Physical Systems Resilience

Critical infrastructures, such as, power grid, oil and gas refineries, or water distribution, are characterized by complex technological networks. The cyber-physical interconnectivity exposes an attack surface that makes them vulnerable to cyber attacks. The potential for disruptions in these critical infrastructures can be attributed to both the interdependence and vulnerability of the networks interconnecting the physical plants and control centers. There is a need to develop cyber resilience metrics for these critical infrastructures to provide quantitative insights. These metrics must support a capability that provides adequate security controls ensuring operational resilience and development of cost-effective mitigation plan. It is highly unlikely for researchers to gain access to data from operational environments to generate resilience metrics. Instead, a validated hybrid simulation environment that characterizes both the cyber and physical environments would be applicable that provide useful cyber resilience metrics. This then becomes useful for decision support systems and eventually helps formulate an informed mitigation plan. Following are the challenges with developing hybrid simulations for CPS Resilience.

14.2.6.1 CPS Resilience Metrics

A recent literature review provides an overview of the state of the art of measuring resilience in CPS related systems, showing also several gaps that have to be addressed in future research (Gay and Sinha 2013). Prior to developing the hybrid simulation environment for cyber resilience metrics, there is a need to develop a CPS resilience simulation framework. Tierney and Bruneau proposed a R4 framework for disaster resilience across the Technological, Organizations, Societal, and Environmental (TOSE) dimensions (Tierney and Bruneau 2007). The R4 framework comprises of Robustness (Ability of systems to function under degraded performance), Redundancy (identification of substitute elements that satisfy functional requirements in the event of significant performance degradation), Resourcefulness (initiate solutions by identifying resources based on prioritization of problems), and Rapidity (ability to restore functionality in timely fashion). This framework does provide a starting point for development of cyber-physical resilience simulation framework. A hybrid simulation framework that will allow measurement of the R4 across the TOSE dimensions for cyber physical systems will address all the aspects that are required for CPS resilience. The ability to characterize the interplay between the TOSE dimensions will be crucial. For instance, merely computing the R4 metrics from the Technological dimension and organizational dimension independently will not provide effective insights into the lack of resilience. The R4 metrics in the Technological dimension would not be aware of the mission goals and organization constraints which will limit the effectiveness of the simulation and lead to silo effects. There are factors at the intersection of these four dimensions that need to be accounted for in a hybrid simulation framework. The hybrid simulation framework is only effective if it provides useful insights to not only the technology stakeholders, but also, decision makers, who would like to utilize the outputs from the simulations to develop informed decision support systems.

14.2.6.2   CPS Resilience Simulation and Actionable Intelligence

The purpose of developing the hybrid simulation framework for cyber physical system resilience is not only to generate quantifiable resilience metrics, but also, provide an action plan for mitigation to improve resilience. In order to improve resilience in CPS, the mitigation strategy could focus on the cyber or physical components. The hybrid simulation framework should be able implement both type of mitigation strategies and provide actionable intelligence on the effectiveness. In addition, the mitigation strategies should also be aligned with the TOSE dimensions to ensure that the proposed changes are not adversely impacting the interplay between the TOSE dimensions. Though there are several cyber focused mitigation strategies, they are typically dependent on the type of cyber component and are not amenable to generalized mitigation plans which are applicable to broader class of CPS. However, the physical systems do not undergo changes at the rapid pace at which cyber technologies evolve. Physical systems follow laws of physics and have inertia. We can leverage this property to find out if the physical systems can operate at an acceptable capacity even if a signal is tampered or lost. There is a considerable order of magnitude difference between physical and cyber speeds. The inertia property in CPS provides the latitude to operate even if some of the operational states are lost.

Within the hybrid simulation framework, it will be beneficial to observe to what extent cyber attack can be withstood, if the physical system can tolerate loss of signals. The simulation framework should be able to answer questions on detection (when should we drop), fault isolation (what should be drop), recovery (how soon can we recover to known good state). Preliminary ideas and concepts for such a framework are discussed in some detail in Couretas (2018). Additional research is needed in this domain, in particular on how to extend the framework concepts using a common formalism and a standardized way to generate digital twins for system evaluation, as envisioned in other sections of this chapter.

14.2.7 Supporting Human Machine Teams

CPS are applied in many domains that require human machine teams, such as emergency response and disaster relief (Zander et al. 2015; Bozkurt et al. 2014), military applications (Gunes et al. 2014), and many other fields (Kim and Kumar 2012). Such teams are providing a new set of requirements regarding the human machine interfaces as well as control of machines without unnecessary constrains on their capabilities.

The multi-modality of CPS is a recurring theme in this compendium, and in the case of the human machine interface discussion, it provides new opportunities. First of all, CPS can extend and enhance the sensory input of humans. Whatever the CPS detects can be displayed for the human, either using virtual reality or augmented reality methods. Rescue workers can "see" into dangerous regions by sending CPS entities into collapsed structures, soldiers can conduct scouting operations using CPS devices, and more. Similarly, CPS can become local actors for the human partners, e.g., surgeons can conduct operations close to the battlefield using CPS means to not only provide the visual, haptic, and other information needed to the surgeon, they also conduct the surgery based on information received from the surgeon. Other applications may require that the CPS system recognizes hand signs or gestures from team members as well as from others. All these diverse applications will require research to address the challenges, and a close collaboration with artificial intelligence researchers as well as with experts in robotics and unmanned vehicle sensor and control promises reusability of many solutions.

A recurring research topic in this context is the challenge of orchestrating the actions of CPS participating in an operation. In their chapter on challenges in the operation and design of intelligent CPS in this volume, the challenges of connected and collaborative operations are discussed in some detail. Their insight is well aligned with a command and control model developed by military operations analysis expert, the NATO Net Enabled Capability Command and Control Maturity Model (Alberts et al. 2010). This model also looks at various stages of orchestration between entities, going from conflicting behavior to using deconfliction by territorial limit of responsibility, to coordinated behavior towards a common goal, to collaboration under a common control, to coherent solution based on self-organization and optimization of local efficiency towards the common goal. However, validations of such approaches have been more anecdotal than based on systematic evaluations. Under which circumstances and constraints such diverse control paradigms are feasible and deliver the expected results requires more research. Research on system of systems captured in Tolk and Rainey (2014) already addressed this need. The various complexity challenges for CPS described in multiple chapters of this book reemphasize the importance of this work.

14.3 Summary and Discussion

Within this concluding chapter of the book, we tried to compile a set of research avenues that need to be addressed in support of better addressing the identified complexity challenges in CPS. The need to provide a better understanding of the dynamics of the systems in their complex environments is still a major challenge, and international expert societies, like INCOSE, identified M&S methods as a promising way to meet this challenge.

However, the humongous variety of computational function categories, their various perspectives, model abstraction levels, and differences in resolution in fidelity provide practical and computational obstacles that still have to be overcome. Coping with uncertainty, vagueness, incompleteness, and contradictions in data is adding to the problem. The issue of synthetic data generation for experimentation of the CPS is in nascent state, simply because the issue of multi-scope CPS is largely too complex to implement. Cyber security and resilience are often unanswered, and with the increasing dependence of our critical infrastructure on CPS, this becomes a growing concern.

The research agenda compiled here does not require radical new methods. Much more we should be looking for solutions that will enable us to move concepts from what is "possible in principle" to what is "possible in practice" through applied research. This can only be accomplished by moving our collaborative efforts from multi-disciplinary research via inter-disciplinary research towards truly trans-disciplinary research. This requires the establishment of a common body of knowledge, a collection of agreed to concepts, terms, and activities that make up the professional domain of CPS engineers. With this compendium we hope to contribute to the discussion and initialization of efforts leading to such a new discipline that hopefully will embrace many of the solutions offered by M&S and computational realms.

Acknowledgments

The work presented in this paper was partly supported by the MITRE Innovation Program. It also draws heavily from discussions and additional inputs from the authors of the chapters in this volume. The views, opinions, and/or findings contained in this paper are those of The MITRE Corporation and should not be construed as an official government position, policy, or decision, unless designated by other documentation. Approved for Public Release; Distribution Unlimited. Case Number 19-0710.

References

David S Alberts, Reiner K Huber, and James Moffat. *NATO net-enabled capability command and control maturity model (N2C2M2)*. Technical report, Command and Control Research and Technology Program, 2010.

Bergman, D. C., (Kevin) Jin, D., Nicol, D. M., & Yardley, T. (2009). The virtual power system testbed and inter-testbed integration. In *Proceedings of the Cyber Security Experimentation and Test (CSET)*, Montreal, Canada.

Bozkurt, A., Roberts, D. L., Sherman, B. L., Brugarolas, R., Mealin, S., Majikes, J., ... Loftin, R. (2014). Toward cyber-enhanced working dogs for search and rescue. *IEEE Intelligent Systems, 29*(6), 32–39.

Chen, C.-C., & Crilly, N. (2016). Describing complex design practices with a cross-domain framework: learning from synthetic biology and swarm robotics. *Research in Engineering Design, 27*(3), 291–305.

Couretas, J. M. (2018). *An Introduction to Cyber Modeling and Simulation* (Vol. *88*). Hoboken, NJ: Wiley.

Diallo, S., Mittal, S., & Tolk, A. (2018). Research agenda for the next-generation complex systems engineering. In *Emergent Behavior in Complex Systems Engineering: A Modeling and Simulation Approach* (pp. 370–389). Hoboken, NJ: Wiley.

Furuya, S., & Yakubo, K. (2011). Multifractality of complex networks. *Physical Review E, 84*(3), 036118.

Gay, L. F., & Sinha, S. K. (2013). Resilience of civil infrastructure systems: literature review for improved asset management. *International Journal of Critical Infrastructures, 9*(4), 330–350.

Gunes, V., Peter, S., Givargis, T., & Vahid, F. (2014). A survey on concepts, applications, and challenges in cyber-physical systems. *KSII Transactions on Internet and Information Systems, 8*(12), 4242–4268.

Hahn, A., Ashok, A., Sridhar, S., & Govindarasu, M. (2013). Cyber-physical security testbeds: architecture, application, and evaluation for smart grid. *IEEE Transactions on Smart Grid, 4*(2), 847–855.

Jazdi, N. (2014). Cyber physical systems in the context of industry 4.0. In *IEEE International Conferenceon Automation, Quality and Testing, Robotics*, Cluj-Napoca, Romania. IEEE.

Jennings, N., & Wooldridge, M. (1996). Software agents. *IEEE Review, 42*(1), 17–20.

Keeley, T. (2007). Giving devices the ability to exercise reason. In *Proceedings of the International Conference on Cybernetics and Information Technologies, Systems and Applications (CITSA 2007)* (vol. *1*, pp. 195–200). Orlando, FL.

Kim, K.-D., & Kumar, P. R. (2012). Cyber–physical systems: a perspective at the centennial. *Proceedings of the IEEE, 100*(Special Centennial Issue), 1287–1308.

Kleppe, A. G., Warmer, J., Warmer, J. B., & Bast, W. (2003). *MDA Explained: The Model Driven Architecture: Practice and Promise.* Boston, MA: Addison-Wesley Professional.

Lee, E. A. (2006, October 16–17). Cyber-physical systems-are computing foundations adequate. In *Position Paper for NSF Workshop on Cyber-Physical Systems: Research Motivation, Techniques and Roadmap* (vol. *2*, pp. 1–9). Austin, TX. Citeseer.

Mittal, S. & Risco Martin, J. L. (2013). Model-driven systems engineering in a netcentric environment with DEVS Unified Process. In *Proceedings of the Winter Simulation Conference*, Washington, DC.

Mittal, S., & Risco-Martín, J. L. (2017). Simulation-based complex adaptive systems. In S. Mittal, U. Durak, & T. Ören (Eds.), *Guide to Simulation-Based Disciplines* (pp. 127–150). Cham, Switzerland: Springer.

Mittal, S., & Zeigler, B. P. (2017). The practice of modeling and simulation in cyber environments. In A. Tolk & T. Ören (Eds.), *The Profession of Modeling and Simulation*. Hoboken, NJ: Wiley.

Nance, R. E. (1981). The time and state relationships in simulation modeling. *Communications of the ACM, 24*(4), 173–179.

Norman, D. O., & Kuras, M. L. (2006). Engineering complex systems. In D. Braha, A. A. Minai, & Y. Bar-Yam (Eds.), *Complex Engineered Systems* (pp. 206–245). Berlin, Heidelberg: Springer.

Oberkampf, W. L., DeLand, S. M., Rutherford, B. M., Diegert, K. V., & Alvin, K. F. (2002). Error and uncertainty in modeling and simulation. *Reliability Engineering & System Safety, 75*(3), 333–357.

Sanislav, T., & Miclea, L. (2012). Cyber-physical systems-concept, challenges and research areas. *Journal of Control Engineering and Applied Informatics, 14*(2), 28–33.

Sehgal, V. K., Patrick, A., & Rajpoot, L. (2014). A comparative study of cyber physical cloud, cloud of sensors and internet of things: their ideology, similarities and differences. In *IEEE International Advance Computing Conference*, Gurgaon, India. IEEE.

Sheard, S., Cook, S., Honour, E., Hybertson, D., Krupa, J., McEver, J., …, White, B. (2015). A complexity primer for systems engineers. In *INCOSE Complex Systems Working Group White Paper*.

Talcott, C. (2008). Cyber-physical systems and events. In *Software-Intensive Systems and New Computing Paradigms* (pp. 101–115). Heidelberg, Germany: Springer.

Tierney, K. K., & Bruneau, M. (2007). Conceptualizing and measuring resilience: a key to disaster loss reduction. *TR News, 250*, 14–17.

Tolk, A., Barros, F., D'Ambrogio, A., Rajhans, A., Mosterman, P. J., Shetty, S. S., … Yilmaz, L. (2018). Hybrid simulation for cyber physical systems: a panel on where are we going regarding complexity, intelligence, and adaptability of cps using simulation. In *Proceedings of the Symposium on Modeling and Simulation of Complexity in Intelligent, Adaptive and Autonomous Systems* (pp. 681–689). Baltimore, MD. Society for Computer Simulation International.

Tolk, A., Page, E. H., & Mittal, S. (2018). Hybrid simulation for cyber physical systems: state of the art and a literature review. In *Proceedings of the Annual Simulation Symposium* (pp. 122–133). Baltimore, MD. Society for Computer Simulation International.

Tolk, A. (2014). Merging two worlds: agent-based simulation methods for autonomous systems. In A. P. Williams & P. D. Scharre (Eds.), *Autonomous Systems: Issues for Defence Policymakers* (pp. 291–317). Norfolk, VA: HQ Sact.

Tolk, A., Diallo, S. Y., Padilla, J. J., & Herencia-Zapana, H. (2013). Reference modelling in support of M&S: foundations and applications. *Journal of Simulation*, 7(2), 69–82.

Tolk, A., & Rainey, L. B. (2014). Toward a research agenda for M&S support of system of systems engineering. In L. B. Rainey & A. Tolk (Eds.), *Modeling and Simulation Support for System of Systems Engineering Applications* (pp. 581–592). Hoboken, NJ: Wiley.

Traoré, M. K., Zacharewicz, G., Duboz, R., & Zeigler, B. (2019). Modeling and simulation framework for value-based healthcare systems. *Simulation*, 95(6), 481–497.

Vallverdú, J. (2014). What are simulations? An epistemological approach. *Procedia Technology*, 13, 6–15.

van Dam, A. (1999). Education: the unfinished revolution. *ACM Computing Surveys (CSUR)*, 31(4), 36.

Vangheluwe, H. (2000). DEVS as a common denominator for multi-formalism hybrid systems modelling. In *Proceedings of the International Symposium on Computer-Aided Control System Design* (pp. 129–134). Anchorage, AK. IEEE.

Vangheluwe, H., De Lara, J., & Mosterman, P. J. (2002). An introduction to multi-paradigm modelling and simulation. In *Proceedings of the AIS2002 conference (AI, Simulation and Planning in High Autonomy Systems)* (pp. 9–20). Lisboa, Portugal.

Zander, J., Mosterman, P. J., Padir, T., Wan, Y., & Fu, S. (2015). Cyber-physical systems can make emergency response smart. *Procedia Engineering*, 107, 312–318.

Zeigler, B. P., Mittal, S., & Traoré, M. K. (2018). Fundamental requirements and DEVS approach for modeling and simulation of complex adaptive system of systems: Healthcare reform. In *Symposium on M&S of Complex, Intelligent, Adaptive and Autonomous Systems (MSCIAAS'18), Spring Simulation Multi-Conference*, Baltimore, MD.

Zeigler, B. P. (1976). *Theory of Modelling and Simulation*. New York, NY: Wiley-Interscience.

Zeigler, B. P., Mittal, S., & Traoré, M. K. (2018). MBSE with/out simulation: state of the art and way forward. *Systems Open Acces*, 6(4), 40.

Zeigler, B. P., Muzy, A., & Kofman, E. (2018). *Theory of Modelling and Simulation: Discrete Event & Iterative System Computational Foundations*. London, UK: Academic Press.

Cyber Physical Systems

Modeling and Simulation to Balance Enthusiasm and Caution

Kris Rosfjord

The MITRE Corporation, McLean, VA, USA

We are in the midst of a rapid emergence in the use of and dependence on cyber physical systems. These systems are becoming further integrated into our critical infrastructure, health systems, defense capabilities, manufacturing processes, and personal property. The chapters in this book laid out how the thoughtful and creative application of modeling and simulation to cyber physical systems engineering is an avenue towards exploiting the CPS potential while guarding against the risks. Here I will briefly touch upon three of the introduced topics: timescale challenges, conceptual expectations, and workforce shortage.

In addition to balancing the inherent complexities of embedding computing into the physical world, there is a need to address both the legacy components and processes as well as the new systems. The need for cyber physical solutions to be reliable, safe, scalable, and adaptable exacerbates this challenge. The incorporation of legacy components creates a system comprised of parts with greatly differing lifetimes including

- Industrial components with an expected lifetime in the 10s of years (Givehchi et al. 2017)
- Network standards that have changed approximately once every 10 years
- Physical sensors with a lifetime on the order of year(s)
- Software with a lifetime on the order of months

The work presented in this paper was partly supported by the MITRE Innovation Program. The views, opinions, and/or findings contained in this paper are those of The MITRE Corporation and should not be construed as an official government position, policy, or decision, unless designated by other documentation.

The modeling and simulation approaches described in this text offer a strategy to design, predict, and analyze the effects of these multi-time-scale components inherent to cyber physical systems.

Further, while the techniques described in this text offer the ability to model the ecosystem as a whole as well as details at the component level one may consider the expectations and limitations of automated cyber-physical systems. For example, there is a strong motivation within the machine learning community to incorporate assurance (robustness, transparency, causality, etc.) and explainability. When applied to autonomous vehicles, this partially translates to a desire to understand why decisions were made that resulted in an accident. When we remove the autonomous component of the system and look at human-driven accidents, we realize that humans do not excel at identifying the causes of a vehicular accident when they are the driver of the accident (Loftus 2007). Similarly, it is challenging for humans to identify the intent of another system and yet we may ask our cyber physical systems their intentions. As cyber physical systems are more accepted and ingrained in various parts of everyday life, there is a need to distinguish what questions we are trying to answer with modeling and simulation that meet existing non-digitization capabilities, and which exceed our current capabilities.

Echoing Horowitz's words in the foreword to this text, one of the challenges associated with this field is the recognized shortages of people that can develop and/or employ these complex analysis techniques to cyber physical systems. If we broadly define the relevant technical backgrounds of scientists and engineers that will impact this field as math, computer science, physics, psychology and social sciences, and engineering we can track the trends of graduate degrees awarded in these fields. Figure E.1 displays this most recent data on PhD graduations over a

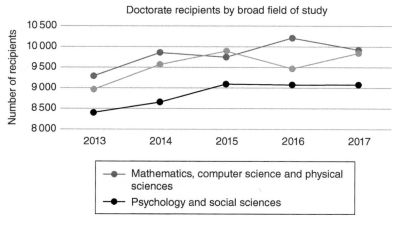

Figure E.1 Doctoral recipients by broad field of study. *Source:* Data from NSF Survey of Earned Doctorates (2013–2017).

period of five years, by broad field of study (NSF 2013–2017). This plot demonstrates a modest, 7%, increase in these targeted graduations over the five year time period. Thus, it is an unanswered question of whether the workforce shortage challenge will persist. This cross-disciplinary text brings together a variety of points of views from experts in the field of modeling and simulation for cyber physical systems with a diverse technical background, and offers the opportunity to develop the expertise of the needed developers and researchers as well as build upon the expertise of those with related skillsets.

This text touched upon many different ways that modeling and simulation will be utilized for the advancement of cyber physical systems. It is strenuous to define an upper-bound for the application of this field. As these systems continue to emerge and refine, it will be exciting to see the heights it will reach.

References

O. Givehchi, K. Landsdorf, P. Simoens, and A.W. Colombo "Interoperability for industrial cyber-physical systems: An approach for legacy systems", *IEEE Transactions on Industrial Informatics*, 13(6), 3370–3378, 2017.

Elizabeth F. Loftus *Eyewitness Testimony: Civil and Criminal*. Fourth edition. LexisNexis, 2007.

NSF Survey of Earned Doctorates 2013–2017.

Index

Complexity Challenges in Cyber Physical Systems: Using Modeling and Simulation (M&S) to Support Intelligence, Adaptation and Autonomy, First Edition. Edited by Saurabh Mittal and Andreas Tolk.
© 2020 John Wiley & Sons, Inc. Published 2020 by John Wiley & Sons, Inc.